MODERN ESCA
The Principles and Practice of X-Ray Photoelectron Spectroscopy

TERY L. BARR

CRC Press
Boca Raton   Ann Arbor   London   Tokyo

Library of Congress Cataloging-in-Publication Data

Barr, T. L. (Tery Lynn), 1939–
    Modern ESCA: the principles and practice of X-ray photoelectron
        spectroscopy / Tery L. Barr.
            p.   cm.
    Includes bibliographical references and index.
    ISBN 0-8493-8653-5
    1. Electron spectroscopy.   2. Surfaces (Technology)—Analysis.
    I. Title.
QD96.E44B37 1993
543′.0858—dc20   93-26854
                                                                                    CIP

© 1994 by CRC Press, Inc.

No claim to original U.S. Government works
International Standard Book Number 0-8493-8653-5
Library of Congress Card Number 93-26854
Printed in the United States of America 1 2 3 4 5 6 7 8 9 0
Printed on acid-free paper

AUTHOR

Tery Barr, Ph.D., received his B.S. in chemistry from the University of Virginia, M.S. from the University of South Carolina, and Ph.D. in physical chemistry from the University of Oregon. He is professor of Materials and Surface Studies at the University of Wisconsin-Milwaukee having recently returned from a year as Fulbright Visiting Faculty Scholar in the Department of Chemistry at the University of Cambridge.

Dr. Barr is a member of the American Vacuum Society (AVS), American Physical Society (APS), and Materials Research Society (MRS). He is co-editor of the international journal, *Vacuum*.

Over the last 20 years, Dr. Barr has authored more than 250 publications and presentations (worldwide) dealing with the principals and practice of ESCA. He has conducted these studies in both industry and academia and in conjunction with some of the world's leading chemists, physicists, and materials scientists.

PREFACE

The present book is intended as a text, and also as a reference source, for both the serious beginner and the experienced user of electron spectroscopy. There is a particular emphasis on demonstrating to these groups how the science has matured and grown in potential utility. The presentation concentrates on the branch of photoelectron spectroscopy generally labeled as either Electron Spectroscopy for Chemical Analysis (ESCA) or X-ray Photoelectron Spectroscopy (XPS). The first designation was coined by the originator of the field, Professor Kai Siegbahn, to try to accentuate the principal attributes of the spectroscopy. The second name was introduced at a later date as a means of labeling the spectroscopy both in terms of the particles employed in excitation and those realized during detection. Unfortunately, there also may be a somewhat disquieting reason for the preferential use of one, or the other, of these sometimes confusing designations for the same spectroscopy. The author is aware of the existence of certain prejudices that have led some in the surface physics community to the assumption that the designation "ESCA" connotes nonfundamental studies of poorly characterized materials, whereas to certain members of the surface chemistry community "XPS" signifies studies of pristine, model systems, with no connection to reality. Neither of these views is valid, however, and their existence only performs a disservice for the entire surface science community. Photoelectron spectroscopy has contributed substantially to the growth in importance of both basic and applied surface science. The key challenge of tomorrow will be for the scientists and engineers involved in this field to unite these two areas together and see to it that photoelectron spectroscopy continues to assume its unique role in the characterization of modern materials.

This book differs from most of the previous contributions dealing with this subject in that very little space is devoted to the subject of the instrumentation of ESCA. This is a purposeful omission based upon several factors. First, this has not been an area in which the author has been extensively involved, except for select applications areas (e.g., the design of a truly useful charge shift manipulation device). Any extensive discussion, therefore, would be derivative. Second, there already exists in other books several fine discussions of this subject. Thus, a repetition herein would defeat our purpose of dealing primarily with new concepts and results. Third, the principal intention of this book is, as the title specifies, "the principles and practice of modern ESCA". This means that we are assuming that the primary interests of our readers are in trying to learn how to *use* this form of spectroscopy in a way that will allow them to attack challenging, present day problems. Access to one, or more, spectrometers is assumed. In some cases these may need to be modified to suit the specific task and it is one of our intentions to provide the understanding needed to make that judgment. Instrumentation is an important area and the author suggests that readers who desire information in this area consult the sources listed later in this text.

Many readers will note that this text also does not devote extensive space to discussions of the generation of "conventional" core level spectra. In this sense the author is following the edict of his title and restricting his principal consideration to those features that might be described as "modern" ESCA, particularly methods that seem to be very useful in studies of the properties of the new "high tech" materials. In addition, it will be noted that the text emphasizes the use of core level and valence band binding energies, their shifts and line widths, at the expense of many other forms of ESCA analysis, particularly quantification. This was done because it is our intention to concentrate on areas of surface analysis in which ESCA is the preeminent form of analysis.

In the sections of the book dealing with the use of ESCA, the author had to choose between two alternative methods of presentation. Thus, an area such as catalysis could be covered either by presenting a cursory review of many applications (to try to demonstrate the *scope* of that application) or a more detailed discussion of a few cases. The latter approach was chosen because it was our wish to demonstrate how encompassing ESCA could be when it is employed as the major, or perhaps, the only, tool in a complicated analysis. Because of the constraints of copyright law this means that most of the problems considered were investigations of the author. The reader should be aware that many other ESCA labs could provide similar collections of results.

The author has also purposefully chosen to vary the style of presentation in several sections of the book. Thus, in those areas that describe aspects of practical application, a very informal approach was used, whereas, the more theoretical or technical areas are presented in a formal, more tightly referenced manner.

As special points of consideration the author recommends to the reader thoughtful consideration of the following generalized suggestions of practice:

1. ESCA always yields results.
2. Often these results will not be *directly* part of the desired materials surface.
3. Despite this, almost all results (even mistakes) are useful.
4. The key results often lie in the periphery of the principal part of the spectrum.
5. Avoid letting nonsurface scientists dictate to you about the measurements to be performed and the results to be expected. Also avoid having materials prepared and altered without your knowledge.
6. In view of (4), always try to run some spectra that explore, in detail, the less pronounced parts of the total spectrum.
7. Become adroit at reading the details of the spectroscopy.
8. Be as deliberate as possible in taking spectra and in reaching absolute judgements. ESCA is generally a nondestructive form of analysis and most spectrometers are stable for the time necessary for extended analyses.
9. Remember that almost every materials problem should provide several independent spectral features that will corroberate or deny any specific judgement.
10. Remember that the spectrometer is almost never wrong. On the other hand, our eyes and initial assumptions are often deceived with respect to the surface region.
11. To some degree, exposure to air alters every surface, but fortunately that alteration can often be detected and "seen" through by the ESCA.
12. If you have an idea — try it!

As with every text of this type, substantial debts are owed to many teachers, colleagues, and students for invaluable assistance. Tiche Novakov introduced the practice of ESCA to the author at Shell Development in Emeryville, California in 1969. David Shirley and his colleagues in the Field Free Lab at LBL proved its extreme versatility. John Flagg, Vice President of Research at UOP (retired) gave the author the opportunity to have his own HP ESCA and insured that he also had the opportunity to use it to explore his own ideas. Jerry Brinen (American Cyanamid) provided the inspiration and foundation for the use of ESCA in catalysis. Joe Greene (University of Illinois) and his colleagues were endless sources of novel materials and always showed faith in the importance of the chemical aspects of thin films. Professor Kai Siegbahn has been an inspiration, teacher, and colleague through many discussions and during my extensive stay in Uppsala. Ulrik Gelius (Uppsala) and the author have spent many hours exploring important fine points of the science. In these encounters the author learned to appreciate his genius for producing "the best". Alan Krauss (Argonne) and Orlando Auciello (MCNC) are the principal contributors to the ion beam deposition-high Tc superconductors described herein. Nicolo Giordano (U. Messina) provided numerous catalytic systems and supports to justify our arguments that high surface area materials are not "all surface". Dick Brundle and Paul Bagus (both of IBM) have provided invaluable expert connection between the applied and basic sides of this discipline. Jacek Klinowski

(Cambridge) has been an unprecedented source of knowledge and has become the chief inspiration on which the new NMR-ESCA silicate collaboration rests. Without M. Kelly, C. Bryson, and L. Fey (the team behind the HP ESCA) none of this would have occurred.

Crucial financial support for the construction of this book was provided by the Faculty Participation Program at Argonne National Laboratory and the Fulbright Commission for a Visiting Scholar Award, 1992–1993, to the Department of Chemistry of the University of Cambridge.

TABLE OF CONTENTS

Chapter 8

Chapter 1

INTRODUCTION

I. INTRODUCTORY STATEMENT

This book is intended as a bridge to the recent advances in the science of photoelectron spectroscopy, particularly the branch now commonly labeled as X-ray photoelectron spectroscopy (XPS) or electron spectroscopy for chemical analysis (ESCA). It is primarily aimed at the analytical user (or potential user) of the spectroscopy, and not the ESCA expert in either the instrumental or theoretical areas, although the latter should also find it a useful source of information. Although the presentation is primarily focused on descriptions of the more recent advances, it was felt that the proper perspective for the nonexpert would be aided by also reviewing the history, methods of practice, basic equipment, and fundamental theory involved. Thus, in the present, the latter topics are used to set the stage for the principal themes to follow. Some emphasis is placed on several of the problems inherent in the methodology, particularly as *raison d'être* for some of the recent developments. This discussion of problems should not be construed as criticisms of the spectroscopy. The nearly 1000 ESCA units in operation throughout the world are obvious testimony to the viable utility of the spectroscopy as the premier method for chemical analyses of surfaces.

II. HISTORICAL PERSPECTIVE

The photoelectron effect had its birth in controversy. Discovered by Heinrich Hertz[1] in 1887 as part of his monumental attempt to experimentally verify the implications of Maxwell's relations, the production of photoelectrons did the opposite, demonstrating that the (then) theory of radiation was incomplete. Numerous scientists tried diligently to mend this crack in the foundation of classical physics, but all of the *good* experiments merely reinforced Hertz's initial observation.[2] The theoretical explanation of the photoelectron effect awaited "the dawn of a new light" as Hoffmann[3] has described the year 1905, and the contributions therein by Albert Einstein. In one of Einstein's major achievements,[4] he brilliantly employed the new quantum of energy concept of Planck[5] to explain how radiation of low intensity, but high frequency, could eject electrons from metal foils, whereas the converse might fail to produce any electrons. As a result Einstein logically evolved a consistent, single relationship that (in slightly modified form) is still today the basis of photoelectron spectroscopy, i.e.,

$$h\nu = 1/2\, mv_e^2 + E_b + q\Phi \tag{1-1}$$

where ν is the frequency of the incoming quanta (later dubbed as "photons" by Lewis[6]), h Planck's constant, $1/2\, mv_e^2$ the kinetic energy of the outgoing "photoelectron" of mass m and velocity v_e, q a reference charge, Φ the work function of the emitting material, and E_b the so-called "binding energy", expressed here against the Fermi level of the material. It is the latter feature, of course, that is the backbone of the spectroscopy we describe in this book. Had Einstein left the argument at this point, his paper would have still been very famous, and perhaps not nearly so controversial. Unable, however, to accept the political view in place of the correct view (as he saw it), Einstein added a section to his

1

paper challenging the totality of the Maxwellian wave theory of radiation, and suggested the possibility of a wave-particle duality.[4] Even Planck hated these "alterations" of his original concepts, and the disputes over the latter half of Einstein's paper (that ushered in the Quantum Theory of Radiation) were not completely resolved until the advent of quantum mechanics in the mid 1920s.[2] In fact, although Einstein's 1921 Nobel Prize was granted primarily for this paper, it specifically mentions only the first half — the laws of the photoelectron effect. To the ever-honest Einstein and his growing band of followers, the dispute was illogical, as you "can't have one half of this paper without the other." In any case, all attempts (pro and con) to challenge the validity of Einstein's interpretation of the photoelectron effect succeeded instead in verifying it, culminating in the beautiful confirmation studies of Millikan.[7] Following these studies, scientists turned to attempts to use this new phenomenon, particularly as a form of spectroscopy. Preeminent in these early studies were the investigations of Robinson,[8] who easily verified that an elemental spectrum could be realized, but found that so many difficulties were inherent to these measurements that the studies did as much to discourage the evolution of this spectroscopy as to promote it. The lack of availability of a continuous, high vacuum capability, the inability to control stray interfering fields, and the unrealized need for integrated, stable electronics caused most scientists of the 1930s and 1940s to ignore the prospect of a "useful" photoelectron spectroscopy.

These problems were eventually rectified in the laboratories of Kai Siegbahn in Uppsala, Sweden.[9] Professor Siegbahn and colleagues were initially attempting to develop a spectroscopy to examine β decay, but they soon realized that a similar version could be employed to examine photoelectrons. Using giant Helmholtz coils to "deflect" all outside fields, they achieved their first spectrum in 1955. Soon thereafter they made a dramatic discovery. The binding energy, E_B, detected for the C(1s) lines in several different carbon containing systems was (reproducibly) found to exhibit different values. The same feature was found for the two chemically different sulfurs in a thiosulfate. Thus, accurately measured photoelectron spectroscopy was found to produce *chemical shifts*! Professor Siegbahn and his group were able to produce similar shifting effects for many other compounds, both in the gas state[10] and as solids.[11] They noted, as we now expect, that the photoelectrons from atoms in a more electropositive state are emitted with less kinetic energy (greater E_B) and vice versa for atoms that are made more negative. Thus, the use of photoemission to map changes in oxidation state and other chemical details seemed to be a viable possibility. Realizing the potential impact of these observations, Siegbahn dubbed his new technique electron spectroscopy for chemical analysis, or ESCA.* Some other scientists have preferred to identify the method with both the nature of the excitation source and the emitted particles, thus calling it various types of photoelectron spectroscopy (e.g., XPS, when X-rays are employed as the emitted photons, and ultraviolet photoelectron spectroscopy, or UPS, when UV radiation is used). The latter process was developed as a companion technique soon after the Swedish development in a number of labs, particularly those of Turner et al.[12] and Price and Turner.[13]

Obviously the choice of radiation frequency, ν, is a matter that depends upon the experimental parameters (as long as hv exceeds the so-called threshold energy, $E_b + q\Phi$). In point of fact, it is often advantageous to have available the ability to employ variable energies, such as provided by many synchrotron radiation centers. In the following, however, we will concentrate upon the area often reserved for the term ESCA (i.e., that employing X-ray emission, XPS). We do this for brevity, and to emphasize that area that has had, by far, the largest impact in materials analysis, particularly in the applied (industrial) setting. We will, however, where warranted, refer to interconnections with other radiation sources.

---

* The acronym "ESCA" was coined by Professor Siegbahn fairly early in the development of the spectroscopy. References 10 and 11 marked its general acceptance.

REFERENCES

1. **Hertz, H.,** Uber einen Einfluss des Utravioletten Lichtes auf die Electrische Entladung, *Wiedemannsche Ann. Phys.*, 31, 982, 1887.
2. **Jammer, M.,** *The Conceptual Development of Quantum Mechanics*, McGraw-Hill, New York, 1966, 28.
3. **Hoffmann, B.,** *Albert Einstein, Creator and Rebel*, Viking Press, New York, 1972, chap. 4.
4. **Einstein, A.,** Uber einen die Erzeugung and Verwandlung des Lichtes Betreffenden Heuristischen Gesichtspunkt, *Ann. Phys.*, 17, 132, 1905.
5. **Planck, M.,** Uber eine Verbesserung der Wienschen Spektralgleichung, *Verh. Dtsch. Phys. Ges.*, 2, 202, 1900.
6. **Lewis, G. N.,** The conservation of photons, *Nature*, 118, 874, 1926.
7. **Millikan, R. A.,** A direct photoelectric determination of Planck's 'h', *Phys. Rev.*, 7, 355, 1916.
8. **Robinson, H.,** *Philos. Mag.*, 50, 241, 1925.
9. **Siegbahn, K.,** Photoelectron spectroscopy: retrospects and prospects, *R. Soc. London*, 1985.
10. **Siegbahn, K., Nordling, C., Johansson, G., Hedman J., Heden, P. F., Hamrin, R., Gelius, U., Bergmark, T., Werme, L. O., Manne, R., and Baer, Y.,** *ESCA Applied to Free Molecules*, North-Holland, Amsterdam, 1969.
11. **Siegbahn, K., Nordling, C., Fahlman, A., Nordberg, R., Hamrin, K., Hedman, J., Johansson, G., Bergmark, T., Karlsson, S. E., Lindgren, I., and Lindberg, B.,** ESCA: atomic, molecular and solid structure studied by means of electron spectroscopy, *Nova Acta Regiae Soc. Sci., Ups.*, 4, 20, 1967.
12. **Turner, D. W., Baker, C., Baker, A. K., and Brundle, C. R.,** *Molecular Photoelectron Spectroscopy*, Wiley-Interscience, London, 1970.
13. **Price, W. C. and Turner, D. W.,** *Philos. Trans. Soc. London A*, 268, 1970.

Chapter 2

The ESCA Process

I. BASIC PRINCIPLES AND PROCEDURES

Although rather simple in appearance, it soon became apparent that certain difficulties prevented the ready, experimental implementation of Equation (1-1). One of the difficulties results from the fact that the threshold energy ($\frac{1}{2} mv^2_e = 0$) is defined with respect to a zero at the sample vacuum level. [The vacuum level may be defined as that point in energy space that, if exceeded by an electron, the latter will be completely (energetically) disassociated from its origin materials system.] This is a useful energy point for a gas (for which $\Phi = 0$), but for a solid ($\Phi \neq 0$) this reference point provides significant problems. The reason is that those features that define E_b are expressions of the atomic* *chemistry* of the system [i.e., $E_b = 0$ in Equation 1-1 defines the Fermi edge (or in the case of certain wide band gapped systems, the pseudo Fermi edge) of that part of the solid material]. The work function, Φ, on the other hand, may be totally independent of that original chemistry, but does depend upon certain influences of the surrounding material units. Thus, for example, different structural faces of a particular crystal may produce different values of Φ. A figurative rendition of these properties is presented in Figure 2-1. Unfortunately, although measurements of E_b *or* Φ are relatively straightforward, simultaneous measurements of both are not readily achieved. In order to circumvent this difficulty, Siegbahn and his group noted a very interesting feature of *many* solid systems. If the sample to be examined is a conductor, it can be placed in the ESCA spectrometer such that the sample is conductively coupled (in thermal equilibrium) with the conductive probe of the spectrometer. An energetic expression of this situation is displayed in Figure 2-1. In this arrangement the energy of the photoelectron detected by the β spectrometer may be measured against the vacuum level, EVL^{SP}, of the spectrometer. The work function of the latter, Φ_{SP}, is accurately predetermined by measuring the energy of photoelectrons emitted from a particular level in a reference sample with a well-established E_b (e.g., Au $4f_{7/2} = 83.96$ eV). Thus, this measurement determines the Fermi level of the spectrometer, E_F^{SP}. If the spectrometer system is reasonably stable, this value may be retained for a substantial period of time, and reevaluation, when needed, should be relatively simple. Thus, the binding energies of all conductive samples may be measured, because of the Fermi edge coupling relative to the spectrometer work function value, Φ_{SP} (see Figure 2-1).

Viewed in terms of Figure 2-1b, the aforementioned chemical shifts arise due to the small changes in E_b that *may* often occur (even for core level peaks) as a result of changes in the *chemical* (electronic state) environment of an atom, designated here as A, as registered by its jth state. Thus, the energy change, $\Delta E^m_{A(j)}$ ($1 \rightarrow 2$) from state $1 \rightarrow 2$, is given by

$$\Delta E^m_{A(j)}(1 \rightarrow 2) = E^m_{bA(J)}(2) - E^m_{bA(j)}(1) \qquad (2\text{-}1)$$

---

* We will refer herein to the basic unit site in the solid as *atomic*, although in reality this fundamental chemical component may be molecular or even a solid unit cell.

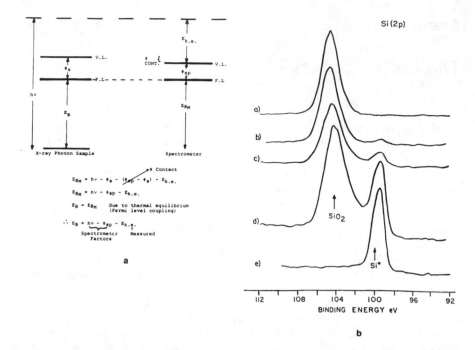

Figure 2-1 (a) ESCA binding energies and chemical shifts for conductive samples: energy level scheme (based on Siegbahn coupling of Fermi edges). (b) ESCA binding energies and chemical shifts for conductive samples: representative binding energies and chemical shifts.

II. PROBLEMS: RELAXATION AND SATELLITES

All of this would appear to be very straightforward, and indeed the measurement of these so-called chemical shifts, for many materials, is easily accomplished. Thus, we find ready production of chemical shift data, such as that reported in Table 2-1. The problem, however, is that any interpretation that $\Delta E^m_{A(j)}$ (1 → 2) is caused only by the chemical changes experienced by atom A in going from situation 1 to 2 may be incorrect.

There are several reasons for the latter qualifying statement. All of these reasons obviously detract from the straightforward, simple analysis of ESCA results that many casual users would appreciate. Fortunately, it often turns out that, even when such problems occur, they may be ignored. In addition, close scrutiny also reveals that an understanding of these problems may produce auxiliary benefits, and several of the new analytical methods, to be described later, are direct offshoots of these observations.

In order to understand these problems, it is important to note that the measured binding energy, E_b^m, is, in fact, determined as an energy difference between that for the N electron system in question immediately before measurements, $E_I(N)$, and that for the state of the N − 1 electron system at the point of photoelectron detection, $E_F(N - 1)$, i.e.,

$$E^m_{A(j)}(1) = E_F(N-1) - E_I(N) \qquad (2\text{-}2)$$

where I is our designation for the appropriate premeasurement, initial state, and F designates the corresponding post measurement, final state, for atom A, in chemical environment 1, as measured (m). The measurement is accomplished using the photoemission from the jth electronic state. Displayed in these terms, it is not difficult to see where

Table 2-1 **Representative chemical shift data**

Element	Probable oxidation state and compound	PMEP[a]	BE and selected line widths[b] for PMEP	Oxide BE O(1s)	Hydroxide BE O(1s)
Pd	Pd°	Pd(3d$_{5/2}$)	335.45(1.00)		
	PdO	Pd(3d$_{5/2}$)	336.9	530.1	
	PdO$_2$	Pd(3d$_{5/2}$)	?	?	
	Pd(OH)$_4$	Pd(3d$_{5/2}$)	338.55		?
	PdSO$_4$	Pd(3d$_{5/2}$)	338.7[c]		
	PdBr$_2$	Pd(3d$_{5/2}$)	337.7[c]		
Ni	Ni°	Ni(2p$_{3/2}$)	852.75		
	NiO	Ni(2p$_{3/2}$)	854.6	530.0	
	Ni$_2$O$_3$	Ni(2p$_{3/2}$)	855.7	530.0	531.7
	Ni(OH)$_2$	Ni(2p$_{3/2}$)	856.45		531.95
	NiSiO$_3$	Ni(2p$_{3/2}$)	856.9[c]		
	NiF$_2$	Ni(2p$_{3/2}$)	857.4[c]		
Cu	Cu°	Cu(2p$_{3/2}$)	932.47(1.01)		
	CuO	Cu(2p$_{3/2}$)	933.7	529.7	
	Cu$_2$O	Cu(2p$_{3/2}$)	932.5(1.2)	530.3	531.7
	Cu(OH)$_2$	Cu(2p$_{3/2}$)	934.75		
Si	Si°	Si(2p)	99.15		
	SiO$_2$	Si(2p)	102.5	531.7	
	Si(OH)$_4$	Si(2p)	103.0		531.9
Zr	Zr°	Zr(3d$_{5/2}$)	179.6		
	ZrO$_2$	Zr(3d$_{5/2}$)	182.2	529.9	
	Zr(OH)$_4$	Zr(3d$_{5/2}$)	183.6		531.2
Mo	Mo°	Mo(3d$_{5/2}$)	227.95		
	Mo$_2$O$_5$	Mo(3d$_{5/2}$)	231.6	?	
	MoO$_3$	Mo(3d$_{5/2}$)	232.3	530.5	531.9
	MoO$_3$XH$_2$O	Mo(3d$_{5/2}$)	232.6	530.4	
S	NiS	S(2p)	162.8[d]		
	Na$_2$SO$_3$	S(2p)	166.6[e]		
	CuSO$_4$	S(2p)	169.1[d]		

[a] Principal metal ESCA peak.

[b] Line widths are enclosed in parentheses.

[c] Wagner, C. D., Gale, L. H., and Raymond, R. H., *Anal. Chem.*, 51, 466, 1979.

[d] Wagner, C. D. and Taylor, J. A., *J. Electron Spectrosc., Relat. Phenom.*, 20, 83, 1980.

[e] Wagner, C. D. and Taylor, J. A., *J. Electron Spectrosc., Relat. Phenom.*, 28, 211, 1982.

Source: Except as footnoted, all values from: Barr, T. L., *J. Phys. Chem.*, 82, 1801, 1978. With permission.

the chemical argument arises. If we assume that we can represent our atom by typical "chemical" considerations [quantum mechanically, this means electrons in specific orbitals for models calculated (perhaps) to the Hartree-Fock limit], then, in order for this chemical (only) shift to be true, it is necessary that one is able to associate the energy difference in Equation 2-2 with differences in specific, premeasurement, orbital energies, ε, i.e.,

$$E^m_{A(j)}(1) = -\varepsilon_{A(j)}(1) \qquad (2\text{-}3)$$

where $\varepsilon_{A(j)}(1)$ refers to the jth (usually core level) orbital energy for atom A in chemical state 1. In this model the measured *chemical shift* defined in Equation 2-2 is given simply by the energy difference experienced by a particular orbital state for atom A in two different chemical environments, i.e.,

$$\Delta E^m_{A(j)}(1 \rightarrow 2) = \varepsilon_{A(j)}(2) - \varepsilon_{A(j)}(1) \qquad (2\text{-}4)$$

Thus, the two εs in Equation 2-4 may represent, for example, the C(1s) orbital energies for carbon in $FeCO_3$, (1), and FeC, (2), etc.

Those readers who are familiar with quantum mechanics will have already noted that this description is flawed or limited by our choice of the "one-electron" or orbital energy argument. In so doing, we have ignored the correlation energy,[1] but we have retained that feature labeled herein as the "chemical" environment. It turns out that this correlation deficiency is generally not very large, particularly when viewed from a qualitative perspective.[2]

Another feature is ignored in Equation 2-4 that is much more serious, at least from the standpoint of a fundamental understanding of ESCA. Close scrutiny of Equations 2-2 and 2-3 will show that inherent to these steps is the assumption that the N − 1 balance of electrons retained in the A system during, and following, photoelectron ejection are unaffected by that process, at least insofar as the time frame of the measurement process of the photoelectrons is concerned. In terms of the orbital, or one-electron, model this is often referred to as the *frozen orbital model* (i.e., in this approach the orbital states of the N − 1 electrons remain frozen in their initial states during the photoelectron ejection and detection).[3] When the orbital picture has been "perfected" to the point of the Hartree-Fock limit this feature comes under the general heading of Koopman's theorem (KT),[4] and therefore one will find the designation

$$E^m_{A(j)}(1) \cong -\varepsilon^{KT}_{A(j)}(1) \qquad (2\text{-}5)$$

often appearing in the literature.[2] Unfortunately, this model is fundamentally flawed, as all photoelectron measurements suffer from some degree of energy relaxation and/or secondary electron, satellite emissions as a result of the response of the N − 1 electron set to the photoemission of the electron in question. As a result of these (final state) effects, it should be obvious that the totally initial state, chemical (Koopman's theorem) argument expressed in Equation 2-5 is incomplete. This complicates, to some degree, the interpretations of ESCA. In addition, these relaxation shifts and satellite peaks play an integral role in some of the new auxiliary techniques described later in this book. Further discussion of the implications of these effects requires quantum mechanical considerations, which are presented in Chapter 4.

III. PROBLEMS: INTRODUCTION TO CHARGING AND FERMI EDGE COUPLING

Another difficulty in the ESCA measurement process results from the previous assertion of conductivity as a property needed for samples in order to apply the Fermi edge coupling-binding energy measurement process. This would seem to restrict ESCA to studies of reasonably good conductors. Fortunately this is not the case. One should note that the measurement process refers to the Fermi edge (thermal and electronic equilibrium) coupling with the conductive part of the probe at room temperature. It appears that all reasonable conductors (in most physical environments) and many narrow band-gapped semiconductors can generally be configured (as samples) to meet the latter criterion for

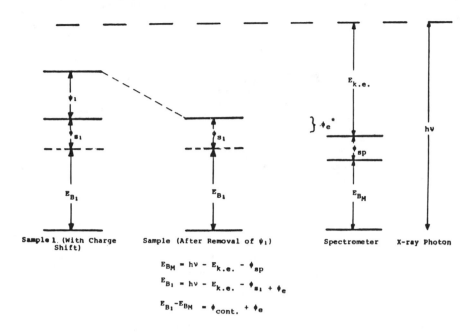

$$E_{B_M} = h\nu - E_{k.e.} - \phi_{sp}$$

$$E_{B_1} = h\nu - E_{k.e.} - \phi_{s_1} + \phi_e$$

$$E_{B_1} - E_{B_M} = \phi_{cont.} + \phi_e$$

Figure 2-2 Energy level scheme for nonconductive sample.

ESCA. This is true because sufficient intrinsic conductive and other "stray" electrons are present in these materials to provide that coupling. In the case of wide band-gapped semiconductors and insulators, however, the situation is different and energy measurement problems result. There is no obvious point where this differentiation of material-type occurs, except that the author has noted that he has been able to achieve Fermi edge coupling using a commercial ESCA system (Hewlett-Packard 5950A) for many configurations of elemental silicon (Si°), ($E_g \sim 1.1$ eV), whereas slight Fermi edge decouplings were found during similar ESCA investigations of thin films of InN ($E_g \sim 1.7$ eV) in the same system. It is readily apparent that a variety of micro- and macromorphological factors substantially influence these decoupling results. For example, a thick wafer of alumina, Al_2O_3 ($E_g \sim 9.0$ eV), behaves as an obvious insulator and produces substantial decoupling,[5] with its ensuing problems, whereas the space charge or Schottky Barrier flow of electrons from conductor to nonconductor prevents decoupling in the first few monolayers of Al_2O_3 grown on top of conductive Al°.[5] All of this is considered later in more detail.

The most obvious problem that occurs as a result of Fermi edge decoupling is *charging*[5,6] (see Figure 2-2). This is a multifaceted process, by which many nonconductive samples subjected to photoelectron (and related electron) emission build up macroscopic (surface- and near surface-oriented) clusters of positive charges that produce a Volta potential that results in a "drag" on the outgoing photoelectrons. As a result, a *charging shift* upfield (to higher binding energy) occurs in the measured binding energy scale. Numerous factors, including macromorphology, chemistry, uniformity of microstructure, etc., influence the size of this charging shift, which has been found to vary from a few tenths of an electronvolt, to more than 100 eV. When significant charging occurs, it is never entirely uniform (i.e., there is always some form of differential charging across the sample),[6] and substantial (broadening-type) distortions of the ESCA peaks generally result (see Figure 2-3). Thus, the ESCA peaks may not only shift to an unrealistic, large, binding energy, they may also distort, sometimes beyond recognition. It it quite obvious that when the latter occurs some form of rectification is necessary. In fact, since the

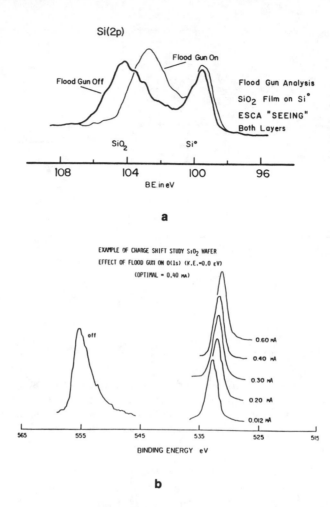

Figure 2-3 Representative charging case from ESCA: (A) SiO_2 film on $Si°$(flood gun analysis, with ESCA "seeing" both layers); (B)wafer of SiO_2 powder [effect of flood gun analysis on O(1s) (K.E. = 0.0 eV; optimal = 0.40 Å)].

charging effect has varied, often nonchemical causes, there may be advantages in always trying to remove it.[6,7] This often may be (substantially) achieved using some form of low energy, electron source (e.g., an electron flood, or neutralization, gun). An example of this type of situation is presented in Figure 2-3.

It is important to note that, whether one is dealing with electrons scattered into a sample from a flood gun or photoelectrons ejected from that sample, they all behave (penetrate through the sample) according to the approximate dictates of the "universal electron attenuation curve",[8] Figure 2-4. It is generally ignored that flood gun electrons operate on the very steep low-energy side of this curve.

Conductive samples, properly lodged in an ESCA, will generally be impervious to the actions of a typical flood gun, whereas the ESCA peaks of an insulator will be displaced. This behavior is thus a way of checking for charging, although certain spectrometer-dependent features may mitigate this test.

Unfortunately, it often goes unrecognized that removal of the charging shift, even when totally successful, generally does not produce a "valid" binding energy by the previously described conventional measurement procedure.[6,9] This occurs because, as

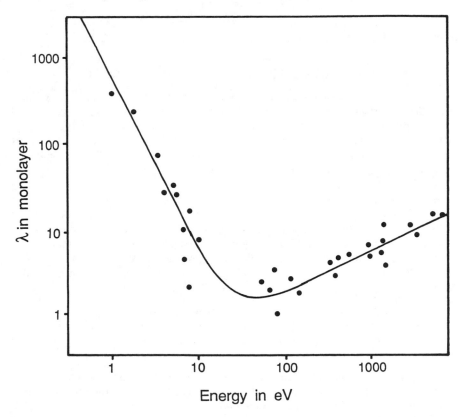

Figure 2-4 The dependence of the escape depth on electron energy. The "universal curve".

explained above, precise binding energy measurements in the conventional ESCA spectrometer require the coupling of the Fermi edge of the sample involved with that of the spectrometer. "Neutralization" of an insulating solid may leave its electronic energy scale still floating with respect to that of the spectrometer. In fact, it may be argued that an insulator does not even have a Fermi edge to use in this coupling process. (It is possible, however, to induce a Fermi edge into some insulators and promote the aforementioned coupling, but the procedure for doing this generally alters the physical and chemical status of the sample.[10])

The most common procedure for producing valid binding energies for nonconductive solids, therefore, has been to reference all binding energies to that for a particular constituent (or dopant) with a purportedly fixed, easily reproduced set of binding energies. Often, in fact, the C(1s) binding energy of the ubiquitous, adventitious carbon, generally deposited on all air-exposed surfaces, is employed for this purpose[11] (see Chapter 6). Although generally utilized (and often correctly so) this "reference" method is fraught with potential problems, since such features as morphology, degree of dispersion, and other variable parameters may significantly alter the basis for these measurement procedures.[12] As a result, a variety of alternative techniques for the removal of charging and establishing the valid Fermi edge have been developed. None of the techniques appear to be a panacea, while most have at least specialized merit. Some of these newer techniques are outlined later in this book.

Interestingly, in our own studies we have discovered that the variable presence of charging and decoupling (particularly of the differential type) actually provides an auxiliary tool for materials analysis by ESCA that, in some cases, produces unique results.[5,6,14,15] Hence these studies are likewise described later in this book.

IV. INTRODUCTION TO THE CHEMICAL SHIFT

The preceding discussion features three basic and related contentions. First, we have stated that the principal attribute of ESCA (or XPS) lies in its ability to provide detailed *chemical* information about the surface-near surface regions of almost any type of solid material. Second, we have pointed out that the principal feature of ESCA that contains this chemical information is the so-called *chemical shift* — a term that is commonly employed to designate the changes in binding energy apparently induced in many core level, photoelectron lines as a result of changes in the chemical environment of the material. Third, we have apparently complicated the former statements by noting that there are serious problems that result from any straightforward simple view of chemical shifts — notably (1) the so-called *charging shift* and (2) the propensity of most systems to suffer final state relaxations during photoelectron ejection in a manner that significantly alters the binding energy position of the core level peaks. All of these cautions were presented without presenting any of the details of the arguments that surround this *chemical shift*. At this point, we correct this omission for two reasons:

1. To clarify the record and present the historical arguments
2. To show that despite its deficiencies the *classical chemical shift* is still thriving and is still the preeminent method employed to generate chemical information in surface analysis

We presented in Equation 2-4 the basic (simplified) prescription for the photoelectron chemical shift, without specifying how this is achieved, i.e., we are still left at this point with the uncertainty of realizing the orbital energies, $\varepsilon_{A(j)}(X)$, for any system, (X), and despite its apparent simplicity this is not an easy task for a complicated solid material.

As usual, in chemistry when problems become difficult, it is common to turn to well-reasoned semi-empirical arguments. The first of these to consider is the charge or electrostatic potential model originated by the Siegbahn group.[16] In this method the chemical shifts experienced by the core level (j) for atom A in the X system is given by

$$\Delta E_{A(j)}(X) = kq_A + V_A + L \qquad (2\text{-}6)$$

where q_A is the "effective" charge experienced by A in the X system, and V_A is the interatomic potential that expresses the influence on the energetics of A of the atoms that surround A in the X system. The k term is related to the interactions experienced by the jth core state as a result of the changes induced into the valence shell of the Ath atom. These features are obviously coulombic in nature and should be expressed in terms of rather complicated quantum mechanical electron-electron interaction integrals, but in the present case the hope is that they can be replaced by an empirical average. The term L is a constant that fixes the energy scale, and thus references the system under consideration to some "standard material". It is common to shift this scale so that L is zero, although for systems formed from metallic elements the neutral metal may be that "zero".[8]

It has also proven to be useful to consider the possible sources for the terms in Equation 2-6, and to use this information to break these terms down into more "traditional" expressions. Thus, the interaction potential V_A is often expressed in terms of specific parameters for the solid structure,[8] i.e.,

$$V_A \rightarrow \sum_{c \neq A}^{N} \frac{q_c}{r_{cA}} \tag{2-7}$$

where $\{q_c\}$ is the effective charge on the set of atoms (except A) in the environment of A and the $\{r_{cA}\}$ are the appropriate interatomic distances. It is not difficult to see that the $\{r_{cA}\}$ terms in Equation 2-7 may be readily extracted from lattice measurements or spectroscopic determinations, and the effective charges, $\{q_c\}$, are directly determined by any argument of the ionic character of the bond, e.g., from the Pauling[17] or Phillips-Van Vechten[18] procedures. It also should be apparent that the second term in Equation 2-6 may be directly related to the Madelung potential.[8]

Even ignoring the problems enumerated above, one should recognize that the model expressed in Equation 2-6 is filled with potential pitfalls. First, it should be noted that the concept of an effective charge, q_A, arises from arguments to achieve a recognition of the partial ionicity known to arise in many key materials, e.g., compound semiconductors, such as GaAs. The basic tenant of this approach is the Coulson concept[19] that the state of an electron (P) in a system may be represented as in a state of partial covalency (C) and partial ionicity (I), where the sum of the two effects represent the whole, i.e.,

$$\psi_P = a\psi_c + b\psi_I \tag{2-8}$$

and

$$a^2 + b^2 = 1 \tag{2-9}$$

As a result the bond description pivots on the determination of these relative (Ψ_C, Ψ_I) states. As described in more detail elsewhere, it has been argued that one may employ the Pauling electronegativity approach for this purpose,[17] but Phillips and Van Vechten have shown[18] from band theory arguments that the former approach is sometimes in error, and unfortunately the degree of this error may be substantial for some of the key semiconductor materials. On the other hand, the Shirley group[20] has employed XPS to examine several other features of the valence band structure of key compound semiconductors. As a result, they have discovered that the splitting between the two deepest subbands in the typical, three-peak valence band structure, realized by a tetrahedral semiconductor, see Figure 2-5, is a feature that seems to vary with changes in chemistry. Kowalczyk et al.[20] (S) have demonstrated that this splitting is directly related to the degree of ionicity in a typical A_nB_{8-n} compound semiconductor, i.e., the splitting grows directly with the ionicity, see Table 2-2. The key to these results is that they correspond more closely to the Phillips-Van Vechten (PV)[18] values, than those of Pauling,[17] but there are also progressive differences between S and PV. Kowalczyk et al.[20] argue that these differences arise from the fact that their method more correctly realizes the status that results when materials become increasingly ionic. It should also be noted that Levine[21] has enlarged on the concepts of PV, and developed the equivalent arguments for many additional binary compounds (including oxides), and even more complex, polyatomic species. The author's research group has also utilized modified versions of these Levine arguments, in combination with detailed valence band spectra, to develop an explanation for the progressive variations in ionicity for the oxides found throughout the periodic table.[22] These arguments are presented below and in Chapter 7.

Despite these variances and problems, the charge potential model has often been applied in total or in partial form with significant success. Many of the early examples

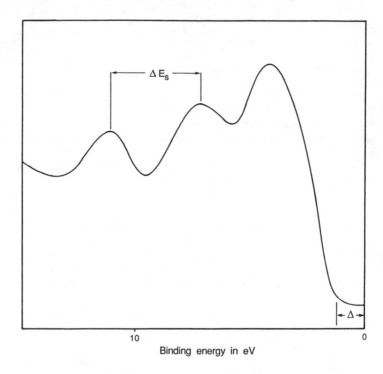

Binding energy in eV

Figure 2-5 Representative three-peaked valence band spectrum for typical tetrahedral (sp³) bonded compounds.

Table 2-2 Binding energies (in eV ± 0.1) of principal valence-band feature for key semiconductors

		GE	GAAS	GASB	INSB
P_I	L	2.6	2.4	2.1	2.5
	B	—	3.2	2.8	2.4
P_{II}	L	7.4	6.6	6.4	5.9
	B	7.2	7.4	6.9	6.0
P_{III}	L	10.5	11.4	10.0	10.0
	B	10.3	12.2	9.25(P)–10.0(M)	10.0
$\Delta(P_{III}\text{-}P_{II})$	L	3.1	4.8	3.6	4.1
	B	3.1	4.8	2.35	4.0

Note: L = Ley, L., *Photoemission in Solids,* Cadona, M. and Ley, L., Eds., Springer-Verlag, Berlin, 1978, chap. 1. B = Barr, T. L., et al., *Mat. Res. Soc. Proc.,* 47, 205, 1985. P = primary, and M = minor shoulder.

of its use were comparative analyses of the C(1s) shifts induced in various organic systems with progressive variations in hetero groups.[23] Recently, the author's research group has also evolved some chemical shift results for group A oxides that seem to exhibit progressions related to the arguments of the charge potential model.[16,22,24,25] The essential feature of these arguments is presented later. A related consideration for group B oxides has been recently described by Minachev and Shapiro.[26]

In a few cases in the past, researchers have tried to expand on these arguments to determine and explain the "collective" chemical shifts apparently realized by particular functional groups when the latter are attached to a variety of sites. The simple theoretical argument for this situation may be expressed[23] by noting that the electroneutrality of an uncharged complex compound requires that

$$q_A = -\sum_{d \neq A} q_d \tag{2-10}$$

$$\therefore \quad \delta E_A = \sum_{d \neq A} \left(\frac{1}{R_{dA}} - k_A \right) q_d \tag{2-11}$$

Now the structure of the sum in Equation 2-11 may be substantially simplified by noting that a series of common "groups" influence atom A, such that

$$\Delta E_A = \sum_{G_i} \Delta E_{G_i} \tag{2-12}$$

and

$$\Delta E_{G_i} = \sum_{\substack{d \neq A \\ \text{in group} \\ G_i}} \left(\frac{1}{R_{Ad}} - k_A \right) q_{dG_i} \tag{2-13}$$

where ΔE_{G_i} is the group chemical shift induced by group G_i, e.g., $G_i \equiv \{CH_3, CBr_3, CF_3, COH, \text{etc.}\}$.

Later in this section, we will consider a form of this group shift argument employed in a description of the use of ESCA as a means to monitor the differences in bonding experienced by an element in a variety of "complex" compounds, e.g., Al in Al_2O_3 and $Na_2Al_2O_4$, etc.

All of the factors considered so far to describe the chemical shift are encapsulated inside an initial state $[\psi_I(N)]$ argument when, in fact, it has been correctly recognized in the arguments outlined above that the true detailed nature of the measurement process in photoelectron spectroscopy involves a transformation to a state of $\psi'_F(N-1)$ before detection of the photoelectron, and the latter generally has to experience some form of relaxation. This fact will be considered theoretically in much more detail in Chapter 4. Suffice it to say at present that it has been common to represent this relaxation as a (reasonable, but not exact) linear correction, ε^R, to the (measured) initial state orbital energy as described in Equation 2-3, i.e.,

$$E_j^m(1) = \varepsilon_j(1) - \varepsilon_j^R(1) \tag{2-14}$$

Unfortunately, it has been found that this relaxation correction is often as large as (and sometimes larger than) a typical chemical shift.[23,25] In fact, we may loosely represent the situation as that depicted in Figure 2-6. It should be noted that there are several additional

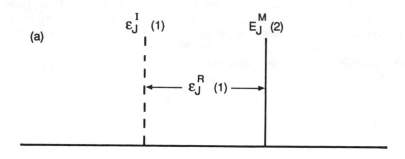

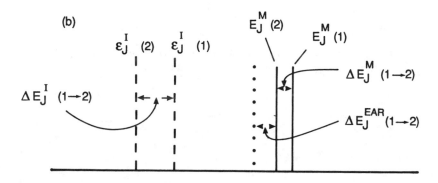

Figure 2-6 Representative relaxation shifts (E^R) for E_m and extraatomic relaxation (E^{EAR}) for E_m.

problems inherent in trying to employ Equation 2-14 to represent a measured binding energy. As indicated above, also still missing from the complete argument for an atom in a solid are the electron correlation, proper treatment of the Madelung potential, and a variety of repulsion terms. Recent consideration has been given by Bagus[27,32] to a detailed theoretical development of the binding energies of select simple oxides employing many of these "addendum" features. The results are encouraging, but their extension to larger systems will, at present, require a substantial, perhaps prohibitive, effort.

At first sight it may seem that the relative size of the relaxation effect would not affect the chemical shift because the latter are differences in binding energies for a particular element (or group of elements), i.e.,

$$\Delta E^m_{jA}(1 \rightarrow 2) = \Delta \varepsilon^I_{jA}(1 \rightarrow 2) - \Delta E^R_{jA}(1 \rightarrow 2) \qquad (2\text{-}15)$$

and the (difference) relaxation term, ΔE^R, might be suspected to cancel. Indeed, a principal contribution to this relaxation, i.e., that due to the influence of the atom A itself (labeled as *intra-atomic*), does cancel. However, as Davis and Shirley have shown[28] there is also a contribution due to the atomic units that surround A that contribute an *extra-atomic* relaxation, i.e.,

$$\Delta E^R_{jA}(1 \rightarrow 2) = \Delta E^{EAR}_{jA}(1 \rightarrow 2) \qquad (2\text{-}16)$$

and as depicted in Figure 2-6b, this component may be finite, and it is not accounted for in any initial state arguments, such as those implied in employing Equation 2-16.

It has been pointed out by a number of investigators that inclusion of any reasonable form of ΔE^{EAR} generally provides a significant improvement in the calculated chemical shift values.[23,26,29] The direct calculation of relaxation terms, however, is generally very difficult (see Chapter 4), and this has naturally led to various forms of empirical approximation. One form of that approximation involves the incorporation of thermo-dynamic data and the N + 1 or equivalent core method of Jolly and Hendrickson.[30] This approach recognizes the fact that, because of enhanced nuclear charge, the next larger (N + 1 electron) atom in the periodic table has, in effect, a core of electrons that are somewhat collapsed onto their nucleus (compared to the N electron atom). When properly incorporated, this collapse somewhat mimics a relaxed photoelectron case for the N electron atom. The equivalent cone method has been successfully used in a number of cases, and has recently been substantially exploited by Johansson and Martensson to describe surface and interfacial chemical shifts[31] (see Chapter 9).

Despite the previously described omissions and approximations, it is still apparent that one may often argue that there is a direct qualitative correspondence between measured chemical shifts, ΔE^m, and the regular progressive changes in initial state characteristics, ΔE^I. Essentially, one is arguing, that although individual cases often have finite ΔE^{EAR}, for any progressive group of "chemical" changes, the differences in ΔE^{EAR} are either small or also modest, progressive changes. Some examples of the application of this argument are presented below and throughout this text.

We have mentioned that the research team led by Bagus[27] has recently tried to partition the factors leading to the chemical shifts for oxides in a manner that removed much of the aforementioned empirical constraints (see also Chapters 7 and 8). Recently Bagus and colleagues[32] have attempted to expand upon some arguments originally presented by Mason[33] and develop a similar first-principals argument for the chemical shifts experienced by an atom when it moves from the free state to that realized in the bulk (or surface) of a small cluster of atoms. The principal purposes of these arguments are to try to evolve a theoretical picture for the surface core level shift for metals (Chapter 9), but the initial aspects of the Bagus arguments[34] are of importance in trying to understand the factors that contribute, in general, to core level shifts. In this manner, they propose that initial state binding energy (BE) situations such as

$$\Delta E_B(\text{bulk}) = E_B(\text{bulk atom}) - E_B(\text{free atom}) \qquad (2\text{-}17)$$

are dependent upon two general effects, an environmental shift and a hybridization shift. In the environmental effect, the changes that occur in the valence region (valence orbital to valence band) are taken into account. Thus, this feature effectively takes into account the delocalization of electrons, with the corresponding growth in electron density. As a result, it is easier to eject the photoelectron and the $\Delta E_B < 0$. In the case of the hybridization effect, electron promotion and directed localization occurs. This makes a positive contribution to ΔE_B. Thus, as amply demonstrated by Bagus et al.[32,34] in calculations on Al and Cu clusters, the two principal initial state terms may be (individually) quite large, but they tend to cancel one another.

In addition, Bagus et al.[32] find that the final state relaxation effects also tend to cancel so that (for the surface-to-bulk shift, in particular) the principal contributions come from the aforementioned initial state features. Effects such as surface core level shifts are described in much more detail in Chapter 9.

V. RECENT EXAMPLES OF THE USE OF CHEMICAL SHIFTS

In the following it is intended to employ ESCA core level chemical shifts to provide a general (but still incomplete) explanation for the bonding patterns that occur at (or near) the surfaces of *oxides*. The primary consideration is on the behavior of the group A oxides in both their simple (M_xO_y) and complex ($A_zM_sO_t$) form, where the former refers to such species as SiO_2 and the latter to compounds such as $Na_2Al_2O_4$. While concentrating on these systems we also make reference to the extensions of these arguments to include the group B oxides (e.g., CuO), and even some of the mixed group oxides (e.g., $NiSiO_3$ and $CuSO_4$).

Considering all of the complexities in these arguments it should be apparent that ESCA studies that are restricted to the principal core level results often may be incomplete or misleading. In many of the cases described below eventual interconnections need to be made with all of the other forms of data as described elsewhere in this book.

The present discussion is limited to a consideration of only a few of the numerous subfeatures of the oxide bonding being analyzed. In this case we concentrate upon

1. The recognition and explanation for the generalized core level shifting patterns and their significance for the simple (M_xO_y) oxides
2. The concept of a possible polarity in the oxide (O) component in these systems
3. The shifting patterns exhibited by the complex oxides ($A_zM_sO_t$) and how and why the latter shifts are generally different from the bonding patterns produced by the corresponding simple oxides

A. BINDING ENERGY PROGRESSIONS FOR SIMPLE OXIDES AND THEIR SIGNIFICANCE

The first features to consider are the progressive patterns, if any, that are present in the different principal core level binding energies for different M_xO_y systems, and what these progressions signify, particularly as M "marches" through the periodic table. As stated, we will concentrate in this examination on the group A oxides. For the present discussion we will further restrict ourselves to a consideration of the common maximum valent species, primarily for those oxides formed by group IIIA (e.g., Al) through VIA (e.g., Te) elements. For those who are interested in the suboxides, we refer them to the investigations of other groups (e.g., Lau and Wertheim for the tin oxides)[35] and several of our previous publications[15,25,36] (e.g., GeO).

Armed with these "tools" we have generated the results presented in Table 2-3. In this regard, a useful tabulation of measured binding energies and possible chemical shifts are presented for both O(1s) and the principal core states for the corresponding metal (or metalloid) M in M_xO_y.

A feature that is not apparent is that, although inclusion of E_R could significantly affect the absolute, numerical values of the measured "chemical" shifts, the *general* trends and progressions expressed in rows 1 and 2 would generally be unchanged by E_R and, therefore, one may ignore E_R and use the measured results in these general chemical arguments.[15,22,32]

Several other features should be immediately apparent from Table 2-3. First, since we have no appropriate state for "unreacted" oxygen, we have listed the absolute O(1s) binding energies E_M for the oxides, whereas in the case of the metals we have employed the generally well-produced $M° \rightarrow M_xO_y$ chemical shifts, ΔE_M. In the case of oxygen we could have chosen some arbitrary, but realistic, value for $E°$ ($O°_2$ ads), such as 534.0 eV, and tabulated the ΔE, "shifts" with respect to that value, but that seemed too capricious. Therefore, we have included the absolute O(1s) binding energies, and everyone should

Table 2-3 'Key' XPS binding energies in eV for group A oxides
1. (E^m (O(1s)) 2. E^m (M_xO_y) – E^m ($M°$)

	IA	IIA	IIIA	IVA	VA	VIA
	Na	Mg	Al	Si	P	S
1.	~529.7	~530.9	(531.2)	532.8	—	—
2.	0.8	0.9	(1.3–2.7)	3.9	5.4	6
	K	Ca	Ga	Ge	As	Se
1.	—	~529.8	530.6	531.3	531.4	—
2.	—	0.6	1.9	2.8	4.4	5
	Rb	Sr	In	Sn	Sb	Te
1.	~529.0	~529.0	530.1	530.1	529.8	—
2.	—	—	1.1	1.4	2.4	4.0
	Cs	Ba	Tl	Pb	Bi	Po
1.	~529.4	~528.2	—	(531.4)	—	—
2.	—	0.4	—	0.8	—	—

Note: Highest (common) valence oxide; $\delta E = \pm 0.2$ eV. Binding energies of insulators fixed by several methods including $E(C_{1s}) = 284.6$ eV for adventitious carbon. Numbers in parentheses suggest values in question. ~ indicates result achieved from impure sample. $E(0^0_{1s})$ chosen as 534.0 eV.

realize that these represent (to varying degrees) the negative chemical shifts from adsorbed O_2 induced by the formation of oxide systems, with differing degrees of (negative) ionicities.

A second feature is to note that we may be trying to produce too much from too little data if we attempt to reach chemical conclusions based solely on the data in Table 2-3. Therefore, in many cases it is also very informative to measure the corresponding valence band spectra of these oxides (see Chapter 7).

The results described above demonstrate that, with few exceptions, the core level binding energies for the (M_xO_y) oxides under consideration shift and change size in a regular fashion as M progresses through the columns and rows of the periodic table. Several related properties of the M–O unit also should change in a similar progressive fashion. Two of the most important features that do shift in this manner for the oxides under consideration are the size and number of the M bonding units involved and the degree of covalency/ionicity in the MO bond.

1. The Progressive Changes in the Chemical Shifts of the M Species

It is apparent in Table 2-3 that the core level chemical shifts of the M units under consideration increase (with few exceptions) in a regular, progressive fashion as we move to the right and up in the periodic table. It may seem intriguing to consider that some aspect of the (electron sharing) bonding character is changing for M in these cases, but it is far more obvious to attribute most, if not all, of the progressive differences to two obvious features: (1) the number of oxide units attached to each of these M species and (2) the regular changes in the size of the M unit. Thus, as different M species "neutralize" more and more oxide units, the degree of chemical shift experienced by each M unit increases in an almost linear fashion. Indeed, this is what we find in Table 2-3, where, as an example, the chemical shifts for the sequence MgO, Al_2O_3, SiO_2, P_2O_5(s), and $-SO_3$(s) are presented. In fact, while adsorbed SO_3 exhibits an approximate 6-eV chemical shift (for S with respect to S°), adsorbed SO_2 produces a shift of approximately 4 eV, i.e., almost identical to that for the silicon in SiO_2. The number of oxide units being accommodated is obviously not the only major effect creating the size of the resulting chemical shifts. As can be seen in Table 2-3, the chemical shifts decrease almost inversely with the size of the M units in each (A) column of the periodic table. This suggests that the effective chemical shift must begin at a certain maximum value when the metal (metalloid) in M–O may be described as a point charge, and the effect on each M must decrease as that effect is spread over a larger and larger area. Thus, we are simply restating the previously established charge potential model,[16,22,24] Equation 2-6.

2. The Chemical Shifts Experienced by O(1s) for the Different Oxides

As we have indicated, the shifts experienced by the different M species in these simple oxide systems are not very informative about the differences in bonding chemistry. The binding energy shifts experienced by O(1s), however, are quite different. In this regard, we assume that the oxygen bonding in the O_2 molecule is 100% covalent. As we substitute different M's for one of the oxygens in this molecule we will introduce differing degrees of ionicity into the system, and this degree is, in part, suggested by the relative negative shifts of the O(1s) binding energy. We have employed this argument in other descriptions in this book in analyses of oxide systems with reasonable success. Thus, in this simplified argument, only covalent and ionic "forces" contribute to bonding. Employing this model, and our assumption that final state effects may be ignored, we suggest that the *different measured O(1s) binding energies of different, (M_xO_y) oxides are direct (qualitative and perhaps even semiquantitative) registrations of the degree of ionicity in the corresponding MO bonds*.[15,22,24,25,36,37] This is, of course, nothing more than an explicit example of the original "Siegbahn" concept for chemical shifts.

As suggested above, this argument seems to hold for the highest valent (common) oxide of each group A element as indicated in Table 2-3. The O(1s) binding energies exhibited by certain lower valent, group A oxides, and those of group B, are somewhat more involved but, as we argue elsewhere (Chapter 8), seem to fall into a consistent pattern that may be argued using a slightly modified version of the previous statement.

The oxides of the metals found near the bottom of groups IA to VA form a special set that we have labeled in a companion study as very ionic oxides (VIO).[22,24,37] The O(1s) values for these systems seem to follow the qualitative concept of the presently suggested progressive shifts, but in a much more dramatic fashion. The special reason for these differences are suggested elsewhere in this text (Section III.A in Chapter 8).

B. MIXED OXIDE SYSTEMS

The chemical mixing of several types of metal and/or metalloid oxides to create a composite oxide is obviously a complex process that involves a variety of sensitive steps that must be properly activated to ensure that the end product will be the desired compound. Based upon the spectroscopic results achieved in our laboratory, and those of numerous other scientists, the formation of these mixed oxide (symbolized herein as $A_zM_sO_t$) is almost always competing with other processes such as the creation of a mixture of simple oxides, e.g., A_mO_n and M_xO_y.[22,24] One also must be concerned that while production of one form of these species may be relatively easy in the bulk (or conversely, on the surface) the opposite site may favor the reverse reaction. Thus, a great deal of care must be employed in trying to generalize surface observations (such as ESCA) to the bulk and vice versa.

With these provisos in mind, we should make note that, in a number of key cases of this type, XPS data have already been produced in our laboratory[22,38] and many others.[8]

Further, the validity of the resulting data can be established because of certain characteristic features that persist throughout most of these systems. If, as indicated above, we
concentrate solely on those oxide systems that result from mixtures of species involving
two metals (or a metal and a metalloid), such that one (M) has a larger electronegativity
than the other (A), then we find that when examined by ESCA if a complex oxide is truly
formed then all of the elements involved (A, M, and O) will generally exhibit binding
energies that are shifted from those produced by either of the simple oxides, whereas if
this is not the case, and only a mixture of simple oxides results, then the XPS results will
reflect this in a lack of chemical shifts.

1. Core Level Shifts for the Metallic Species

The chemical shifts displayed in Table 2-4 for most of the complex oxides project an
obvious pattern, both compared to their corresponding simple oxides and also in comparison to one another. In this regard, we may paraphrase the results with the basic arguments
described elsewhere[22,24,36-38] and repeated herein. If one considers the influence of a metal,
A, in two types of systems A_mO_n and $A_zM_sO_t$, and another metal, M, in the latter and in
M_xO_y, where A is obviously more basic than M (i.e., an A_mO_n oxide will be substantially
more ionic than M_xO_y), then the relative covalency of M will make A in $A_zM_sO_t$ more
ionic than that for A in A_mO_n. As a result, the binding energy of A in $A_zM_sO_t$ will usually
be larger than that for A in A_mO_n. Further, we will find that the M–O bond in $A_zM_sO_t$ is
generally more covalent than that for the M–O bond in M_xO_y.

This latter result is the effect of the enhanced ionicity of A–O (compared to M–O).
Thus, as one interjects A into an M–O lattice, the formation of ionic A–O bonds on one
side of the oxide further enhances the covalency of the M–O bond (compared to its status
when there are only M–O–M-type units). Conversely, one may argue that the relatively
covalent M–O unit interjected into ionic A–O–A-type bonding will tend to pull electrons
away from A, thus making the A–O bond even more ionic.

Representative binding energy shifts are cataloged in Table 2-4 that demonstrate the
general, chemical effects realized by both the A and M elements as the systems under
consideration are changed from M_xO_y to $A_zM_sO_t$ or, in some cases, even when the change
is from $A_zM_sO_t$ to $A'_sM_sO_t$ or to $A_zM'_sO_t$.

We may generalize the resulting process by noting that *given the mixing of two (or
more) oxides (A_mO_n and M_xO_y) to form a complex mixed oxide ($A_zM_sO_t$), the metal (or
metalloid) oxide with the most ionic bond (e.g., A_mO_n) will be made even more ionic
following formation of the complex oxide, whereas the other more covalent oxide will
experience a corresponding increase in covalency.*

Based upon the results analyzed so far we believe that these statements are reasonable
generalities. Obviously, a few cases exist that will not follow these patterns, since we are
employing a very simple model for these arguments that disguises a variety of potential
difficulties. Perhaps the two most important possible problems with our simple model are
the aforementioned final state effects and the fact that as one progresses through the
periodic table for both simple and complex oxides there are progressive changes in the
structures of these systems that accompany the changes in bonding.

2. Chemical Shifts of O(1s) for Complex Oxides

Most of the suppositions suggested above for an XPS argument concerning the
bonding of simple oxides were based upon the effects experienced by the O(1s) core level
peaks. Our analysis of the chemical consequences of the XPS results for complex oxides,
on the other hand, has been rendered to this point without considering the O(1s) results.
The reason for this should be obvious, that is, whereas the change induced in O(1s) by
changing M to M′ may be the critical result, when one substitutes A_mO_n into M_xO_y, to
form the mixed complex oxide, one finds it difficult initially to determine which oxygen

Table 2-4　**Chemical shifts reflecting covalency/ionicity changes in complex oxides**

Element	Compound	Peak B.E. in eV	Comment	Ref.

(A)

$$M_xO_y \longleftrightarrow A_zM_sO_t$$

Enhanced covalency (decreased E_B) of representative M due to interjection of ionic A

Element	Compound	Peak B.E. in eV	Comment	Ref.
		Al(2p)		
Al	Al_2O_3	73.8		1
	$Na_2Al_2O_4$	72.9		1
		Si(2p)		
Si	SiO_2	103.4		1
	Talc	103.1		2
	Mica	102.6		2
	Soda glass	102.9		2
		P(2p)		
P	Na_2HPO_4	133.1 (i)	(i) has more ionic units	3
	$NaPO_3$	134.7 (ii)	than (ii)	4
		Se $(3d_{5/2})$		
Se	$(NH_4)_2SeO_4$	59.2	Variable ionicity of A:	5
	Na_2SeO_4	60.6	$NH_4^+ > Na^+ > H^+$	5
	H_2SeO_4	61.2		5

(B)

$$A_mO_n \longleftrightarrow A_zM_sO_t$$

Enhanced ionicity (increased E_B) of representative Group A element due to interjection of relatively covalent M

Element	Compound	Peak B.E. in eV	Comment	Ref.
		Mg(2p)		
Mg	MgO	50.6	Enhanced covalency of	1
	$MgSeO_4$	51.1	M–O units; S–O more	5
	$MgSO_4$–$7H_2O$	51.6	covalent than Se–O	5
		Ca$(2p_{3/2})$		
Ca	CaO	346.5		6
	$CaCO_3$	347.0	S–O more covalent than C–O	7
	$CaSO_4$	347.6		6
		Ba$(3d_{5/2})$		
Ba	BaO	777.9		1
	$Ba(OH)_2$	779.5		1
	$BaSO_4$	780.8		8
		In$(3d_{5/2})$		
In[a]	In_2O_3	444.7	Small shifts due to	1
	$In(OH)_3$	445.0	amphoteric behavior of H	1
		Pb$(4f_{7/2})$		
Pb[a]	PbO_2	137.6	Enhanced covalency of	1
	$PbSiO_3$	138.65	S–O compared to Si–O	9
	$PbSO_4$	140.0		9

(C)

$$A_mO_n \longleftrightarrow A_zM_sO_t$$

Enhanced ionicity (increased E_B) for A from group B elements due to interjection of relatively covalent M

		Co(2p$_{3/2}$)		
Co	Co$_3$O$_4$	780.3		6
	CoSiO$_4$	781.3		6
		Ni(2p$_{3/2}$)		
Ni	NiO	854.6		1
	NiSiO$_3$	856.9		5
		Cu(2p$_{3/2}$)		
Cu	CuO	933.7	Order of increased	1
	Cu(OH)$_2$	934.7	covalency:	1
	CuSiO$_3$	934.9	O–H < O-Si < O–S	5
	CuSO$_4$	935.5		6
		Pd(3d$_{5/2}$)		
Pd	PdO	336.9		1
	Pd(NO$_3$)$_2$	338.2	S–O more covalent than	5
	PdSO$_4$	338.7	N–O (?)	5
		Cd(3d$_{5/2}$)		
Cd	CdO	404.2		1
	Cd(OH)$_2$	405.1		5

[a] Borderline A or M.

1. This study.
2. Wagner, C. D., Passoja, D. E., Hillary, H. F., Kinisky, T. G., Six, H. A., Jensen, W. T., and Taylor, J. A., *J. Vac. Sci. Technol.*, 21, 933, 1982.
3. Wagner, C. D., *J. Vac. Sci. Technol.*, 15, 518, 1978.
4. Wagner, C. D. and Taylor, J. A., *J. Electron Spectrosc. Rel. Phenomena*, 20, 83, 1980.
5. Wagner, C. D., Gale, L. H., and Raymond, R. H., *Anal. Chem.*, 51, 466, 1979.
6. Wagner, C. D., unpublished results.
7. Wagner, C. D., Riggs, W. M., Davis, L. E., Moulder, J. F., and Muilenberg, G. E., Handbook of XPS, Perkin-Elsneo Co., Eden Prarie, MN, 1979.
8. Hoogewigs, R., Fiemans, L., and Vennik, J., *J. Microsc. Spec. Electron,* 1, 109, 1977.
9. Pederson, L., *J. Electron Spectrosc.*, 28, 203, 1982.

is experiencing any resulting change in O(1s) binding energy values. Perhaps this is the reason why many ESCA scientists fail to report any O(1s) results when examining complex oxides. Thus, for example, the tabulated results in the two most commonly employed general sources — the *PHI XPS Handbook*[39] and *Practical Surface Analysis* by Briggs and Seah[8] report M and/or A binding energies and no O(1s) values. This is unfortunate because, as we shall point out below, several of the key unanswered questions, and incorrect assumptions, in oxide bonding may be substantially addressed through an analysis of the O(1s) results.

Because of this lack of information, there are only a few cases in which detailed O(1s) results exist for progressive series of different $A_zM_sO_t$ systems. In fact, the most detailed would appear to be the results reported by the authors group on the ESCA analysis of the key binding energies of silicas, zeolites, clays, aluminas, and aluminates.[6,40-44] Fortunately, once again, we generally seem to need only the trends in the O(1s) values to argue many of the key chemical effects and changes.

Based upon the assumption made previously that the degree of ionicity in these systems may be keyed by the resulting O(1s) binding energies, we see from Table 2-5 that, of the compounds reported, α-SiO$_2$ is the most covalently bonded [largest O(1s) binding energy], whereas Na$_2$Al$_2$O$_4$ is the most ionic, although it should be noted that we have not, as yet, identified which bonds in the latter are the ionic ones.[22,42,43] Further, we note that, based upon the O(1s) results, changing the ratios of ingredients (e.g., xSiO$_2$ to yNa$_2$Al$_2$O$_4$) results in significant changes in O(1s) that seem to consistently track the x/y ratio.[6,40-44]

Even without considering the specific numbers it is possible to use these facts to craft several consistent, and potentially general, suppositions.

1. The formation of true (chemical) complex (A$_z$M$_s$O$_t$) oxides produces binding energy changes that suggest that both the A–O and M–O bonds are altered (often substantially) from those in the corresponding simple oxides as a result of the formation of the complex oxide.[22,43]

2. This alteration is such that one (rather singular) type of oxygen results (argued based on the production of a single peak with relatively narrow line width).[22,35,38,43] Further, *the complex bonding experienced by this oxygen is not as covalent as was the case for M$_x$O$_y$, nor is it as ionic as in A$_m$O$_n$, but rather it is shifted somewhere in between, and the size of this shift seems to be directly related to the ratio of the two simple oxides found in the complex oxide.*

3. There is some indication that oxygen prefers an ionic environment in that, following complex oxide formation, the negative O(1s) shifts experienced by the relatively covalent M$_x$O$_y$ seem to be larger than the corresponding positive shifts in the O(1s) for A$_m$O$_n$.

The chemical significance of point (1) may stretch far beyond the boundaries of the analysis of oxide surfaces. It is a common practice in quantum chemistry, certain types of spectroscopies, and fundamental chemical bonding arguments to refer to the M–O bond environment for most M$_x$O$_y$ oxides as identical to that of the corresponding metal(ate) or metal(ite) system, e.g., SiO$_2$ is equated to SiO$_4^{-4}$ and Al$_2$O$_3$ to AlO$_6^{-6}$, etc.[45] The implied (and often used) assumption is that, for example, the Si–O bond in SiO$_2$ (1) is the same chemically as that in, for example, Ca$_2$SiO$_3$ (2). Our XPS results, and those of others, demonstrate that this is not generally true. Thus, as indicated above, the silicon in (2) is much more covalently bonded, whereas the complex bonding makes the oxygen much more ionic. [Pictorially, this suggests that the valence shell of electrons about the oxygen in (2) is "held" much tighter than in (1).]

Further, one should note that these results argue against the implied (or, in some cases, used) assumption that a complex oxide is merely a mixture or quasi solution of its corresponding simple oxides. Thus, it is inappropriate to consider that any aspect of Mg$_2$SiO$_3$ other than elemental composition can be represented as 2MgO · SiO$_2$. The chemical changes experienced by the elements in forming the former have substantially altered the bonding features present in the latter.

Based upon some of the data in Table 2-5, supposition (2) above would appear to be somewhat inaccurate, because we find that the O(1s) in Na$_2$Al$_2$O$_4$ is noticeably more ionic than Al$_2$O$_3$ or any of the other systems in that table. Thus, the concept of a correspondence between binding energy shifts and degree of ionicity would seem compromised. The problem is rectified by noting that the ultimate A$_m$O$_n$ for the complex systems in Table 2-5 is not specified. The appropriate A$_m$O$_n$ is Na$_2$O, for which we have found an O(1s) of 529.6 eV.[22] Thus, the shift to the O(1s) for Na$_2$Al$_2$O$_4$ is a slight "relaxation" of the ionicity (negativeness) in Na$_2$O, while as suggested in point (3) the change of the O(1s) bonding energy from that for Al$_2$O$_3$ (to that for Na$_2$Al$_2$O$_4$) is much more substantial.

C. OXIDE ION SIZE AND SHAPE "POLARIZATION"

The feature which has not been analyzed in all of this concerns the implication of these results on the conventional picture of the oxide ion. Literally thousands of calculations,

Table 2-5 **ESCA binding energies in eV ± 0.2 eV for zeolites and related materials (based upon C(1s) = 284.8 eV)**

	SiO_2	REY[a]	NaY	CaX	NaX	$\gamma=Al_2O_3$	NaA	$Na_2Al_2O_4$
Si(2p)	103.5	102.9	102.55	102.45	102.0	—	101.1	—
Al(2p)	—	74.65	74.2	74.35	73.95	73.9	73.5	73.4
O(1s)	532.9	532.2	531.8	531.6	531.1	530.6	530.5	530.3
Na(2s)	—	—	64.0	—	63.75	—	63.2	?
Valence band width (half maximum)	10.4	10.3	10.4	9.8	9.8	9.1	9.1	?

[a] RE = rare earth mixture predominantly La and Ce.

lattice formulations, and depictions of reactions exist in the literature and textbooks in which the concept of a fixed spherical size is promoted for select ions, including the oxide ion.[17] The concept is known to be approximate, and is generally specified as such, but this is not the problem. The problem is that the concept is often wrong.

The radius of an ion is by definition the distance from the center of the nucleus of an elemental system (that is assumed to have gained or surrendered a certain number of valence electrons) to some average separation distance of the outer electron (valence) shell of the ion. Further, this ion is assumed to be spherical. The photoelectron spectrometer registers, in part, the effect of this outer shell of electrons on all of the electrons in the ion. These effects are felt through the combined influence of separation and charge. In the depictions of simple oxides it is often common to assume the presence of the same sized, $O^=$ ion in all oxide systems with the lattice of different oxides induced by assuming that the latter is slightly disturbed by the (interstitial) presence of different sized, and sometimes different charged, M^{+x} units. Unfortunately, this model loses validity when one considers the covalency of oxides, and it becomes necessary to realize that *different M species will, in effect, alter the effective size of the oxide ion!*

Even more problems occur in trying to rationalize a constant spherical shape for the oxide ion for most $A_zM_sO_t$, complex oxides. In these cases we have bonding situations where, effectively, one or more very ionic A^{+m} unit is located on one side of the oxide unit, whereas one or more much less ionic M^{+n} unit is located on the other side. No matter how one argues the results of this situation it is apparent that, not only the size, but the effective shape of the oxide unit must be altered by this situation. It seems apparent that this alteration may be reasonably approximated by assuming that *in $A_zM_sO_t$-type systems the oxide ion is polarized.*[22,42,43] The effect (or degree) of this polarization is reflected in the unique chemical shifts realized by the A and M species, as depicted in Figure 2-7. It is important to repeat that our limited results in this area suggest that the relative effect on (or distortion of) the binding energies of the oxide unit is generally smaller for the more ionic A than for the relatively covalent M species.[22,25,43] The influence of all these features on the structure of the basic charge potential model should be obvious.

D. CHEMICAL SHIFTS AND ZEOLITE STRUCTURE AND ACIDITY

Another example of the use of measured chemical shifts to correlate with other experimental features realized for complex materials systems involves the previously mentioned extensive studies of zeolite systems. (Zeolites are a form of crystalline aluminosilicates.) In our case we have discovered that changes in the ratio of Al to Si in the

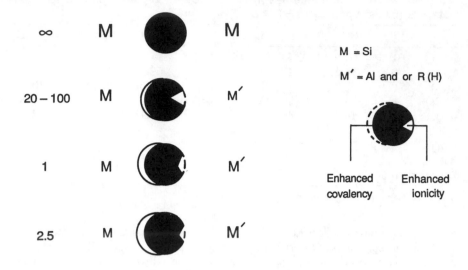

Figure 2-7 Representative polarizations of the oxygen in select mixed (M–O–M′) oxides where M and M′ differ substantially in covalency.

complex zeolitic oxide lattice will produce a corresponding significant change in the core level binding energies.[6,40-44] We have used these progressive shifting results in correlation with a variety of effects. In Figure 2-8 and Table 2-5, for example, we plot the change in O(1s) binding energy vs. the variation in Sanderson electronegativity. We also find progressive changes in binding energies that directly mirror changes in the bonding ionicity, i.e., the acidity of any OH-containing $\underset{Si \diagup \diagdown Al}{O\overset{H}{\diagup}}$ bond.[42-44] The latter is crucial to the use of zeolites.[46]

VI. SUMMARY OF THE PROS AND CONS OF CONVENTIONAL ESCA

In summary, we have noted that the basic ESCA process provides a rather straightforward measurement procedure that generally produces readily identified peaks, particularly from core level emissions. Shifting patterns in the binding energies of these peaks were originally determined by the Siegbahn group to indicate changes in chemistry. In order to fix a common reproducible scale for these binding energies, the Siegbahn group recommended Fermi edge coupling between the (conductive) sample and the spectrometer.[16]

Unfortunately, several problems interfere with the total ready use of these arguments and methods for all systems. Thus, we have suggested that the measurement process itself interferes with the actual measured binding energy values by mixing relaxation and/or satellite peak effects into the chemical (initial state) arguments. The latter would be preferred by those seeking a simple "chemical" shift. We also have shown that nonconductive materials may be subjected to Fermi edge decoupling and a charging shift following photoelectron ejection. As stated earlier, these effects tend to complicate the measured results.

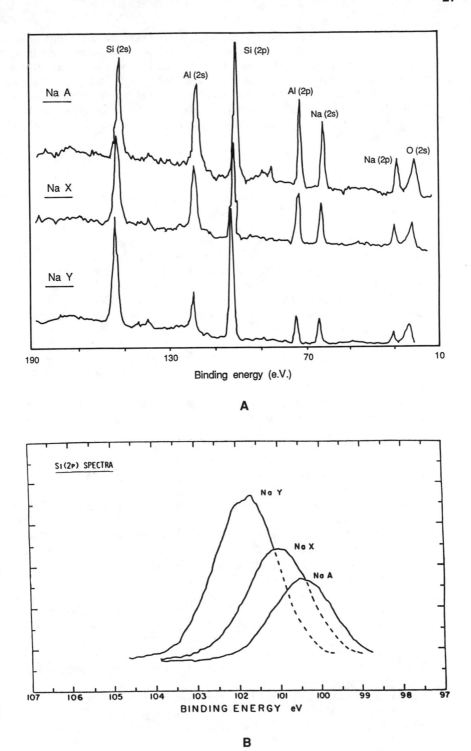

Figure 2-8 Examples of core level shifts induced in zeolites by changes in [Si/Al] ratio: (A) relative quantification; (B) Si(2p) shifts for NaA, NaX, and NaY.

Fortunately, as we have seen, one may generally ignore or remove these difficulties. Thus, it is still possible to combine measured shifts and chemical arguments to achieve useful data. In addition, we will describe in some detail how various aspects of the "problem" phenomena may be used as important auxiliary tools. In addition, as we will also explain, many features of a detailed ESCA experiment may be exploited that are not affected by these problems. First, however, it will be useful to describe the basic rudiments of the experimental systems, and the quantum concepts, that are the basis of the ESCA process.

REFERENCES

1. **Barr, T. L. and Davidson, E. R.,** Nature of the configuration-interaction method in ab-initio calculations, *Phys. Rev. A,* 1, 644, 1970.
2. **Martin, R. L. and Shirley, D. A.,** *J. Chem. Phys.,* 64, 3685, 1976.
3. **Szabo, A. and Ostland, N. S.,** *Modern Quantum Chemistry, Introduction to Advanced Electronic Structure Theory,* MacMillan, New York, 1982, chap. 3.
4. **Koopmans, T. A.,** Uber die Zuordnüng von Welten Fuhktionen and Eigenwerten zu den ëinzelnen Elektronen eines Atoms, *Physica,* 1, 104, 1933.
5. **Barr, T. L.,** Studies in differential charging, *J. Vac. Sci. Tech.,* A7, 1677, 1989.
6. **Barr, T. L.,** Applications of electron spectroscopy to heterogeneous catalysis, in *Practical Surface Analysis,* 2nd ed., Briggs, D. and Seah, M. P., Eds., John Wiley & Sons, Chichester, England, 1990, chap. 8.
7. **Lewis, R. T. and Kelly, M. A.,** *J. Electron Spectrosc. Rel. Phenomena,* 20, 105, 1980.
8. **Briggs, D. and Seah, M. P.,** *Practical Surface Analysis,* 2nd ed., John Wiley & Sons, Chichester, England, 1990.
9. **Barr, T. L.,** Applications of ESCA in industrial research, *Amer. Lab.,* 10, 40, 65, 1978.
10. **Stephenson, D. A. and Binkowski, H. J.,** *J. Noncrystalline Solids,* 22, 399, 1976.
11. **Wagner, C. D.,** *J. Electron Spectrosc. Rel. Phenomena,* 18, 345, 1980.
12. **Hicks, R. F.,** Characterization of supported metal catalysts by X-ray photoelectron spectroscopy, in *Characterization and Catalyst Development: An Interactive Approach,* Bradley, S. A., Gattuso, M. J., and Bertolacini, R. J., Eds., (ACS Symp. Ser. Vol. 411), American Chemical Society, Washington, DC, 1989, chap. 20.
13. **Barr, T. L.,** XPS as a method for monitoring segregation in alloys: Cu-Be, *Chem. Phys. Lett.,* 43, 89, 1976.
14. **Barr, T. L. and Yin, M. P.,** Studies of Pt-metal catalysis by high resolution ESCA, in *Characterization and Catalyst Developments: An Interactive Approach,* Bradley, S. A., Gattuso, M. J., and Bertolacini, R. J., Eds., (ACS Symp. Ser. Vol. 411), American Chemical Society, Washington, DC, 1989, chap. 19.
15. **Barr, T. L.,** *Crit. Rev. Anal. Chem.,* 22, 224, 1991.
16. **Siegbahn, K., Nordling C., Fahlman, A., Nordbeng, R., Hamrin, K., Hedman, J., Johansson, G., Bergmark, T., Karlsson, S. E., Lindgren, I., and Lindberg, B.,** ESCA, atomic, molecular and solid structure studied by means of electron spectroscopy, *Nova Acta Regiae Soc. Sci., Ups.,* 4, 20, 1967.
17. **Pauling, L.,** *The Nature of the Chemical Bond,* 3rd ed., Cornell University Press, Ithaca, NY, 1960, chap. 3.
18. **Phillips, J. C. and Van Vechten, J. A.,** *Phys. Rev. B,* 2, 2147, 1970.
19. **Coulson, C. A.,** *Valence,* 2nd ed., Clarendon Press, Oxford, 1961.
20. **Kowalczyk, S. P., Ley, L., McFeeley, F. R., and Shirley, D. A.,** *J. Chem. Phys.,* 61, 2850, 1974.
21. **Levine, B. F.,** *J. Chem. Phys.,* 59, 14, 53, 1973.
22. **Barr, T. L.,** *J. Vac. Sci. Technol. A,* 9, 1793, 1991.

23. **Ghosh, P. K.,** *Introduction to Photoelectron Spectroscopy*, John Wiley & Sons, New York, 1983.
24. **Barr T. L. and Brundle, C. R.,** *Phys. Rev. B,* 46, 9199, 1992.
25. **Barr, T. L.,** *Multicomponent and Multilayered Thin Films For Advanced Microtechnologies: Techniques, Fundamentals, and Devices*, Auciello, O. and Engemann, J., Eds., (NATO, ASI Ser. E, Vol. 10), Kluwer Academic, Amsterdam, 1993.
26. **Minachev, K. M. and Shapiro, E. S.,** *Catalyst Surface: Physical Methods of Studying*, Mir Publishers, Moscow, 1990, chap. 1 (republished in English by CRC Press, Boca Raton, FL).
27. **Broughton J. Q. and Bagus, P. S.,** *Phys. Rev. B,* 36, 2813, 1987.
28. **Davis, D. W. and Shirley, D. A.,** *J. Electron Spectrosc. Rel. Phenomena,* 3, 137, 1974.
29. **Waddington, S. D.,** in *Practical Surface Analysis*, 2nd ed., Briggs, D. and Seah, M. P., Eds., John Wiley & Sons, Chichester, England, 1990, app. 4.
30. **Jolly, W. and Hendrickson, D. N.,** *J. Am. Chem. Soc.,* 92, 1863, 1970.
31. **Johansson, B. and Martensson, N.,** *Helv. Phys. Acta,* 56, 405, 1983.
32. **Bagus, P. S., Brundle, C. R., Pacchioni, G., and Parmigiani, F.,** *Reports on Surface Science*, Barr, T. L. and Saldin, D., Eds., 1993, in press.
33. **Mason, M. G.,** *Phys. Rev. B,* 27, 748, 1983.
34. **Bagus, P. S., Pacchioni, G., and Parmigiani, F.,** *Phys. Rev. B,* 43, 5172, 1991; **Bagus, P. S., Pacchioni, G., Sousa, C., Minerva, T., and Parmigiani, F.,** *Chem. Phys. Lett.,* 196, 641, 1992.
35. **Lau, C. L. and Wertheim, G. K.,** *J. Vac. Sci. Technol.,* 15, 622, 1976.
36. **Barr, T. L., Mohsenian, M., and Chen, L. M.,** *Appl. Surface Sci.,* 51, 71, 1991.
37. **Barr T. L. and Liu, Y. L.,** *J. Phys. Chem. Solids,* 50, 657, 1989.
38. **Barr, T. L.,** to be published.
39. **Wagner, C. D., Riggs, W. M., Davis, L. E., Moulder, J. F., and Mullenberg, G. E.,** Eds., *Handbook of X-ray Photoelectron Spectroscopy*, Perkin-Elmer Co., Eden Prairie, MN, 1979.
40. **Barr, T. L.,** *Appl. Surface Sci.,* 15, 1, 1983.
41. **Barr, T. L. and Lishka, M. A.,** *J. Amer. Chem. Soc.,* 108, 3178, 1986.
42. **Barr, T. L.,** *Zeolites,* 10, 760, 1990.
43. **Barr, T. L.,** to be published.
44. **He, H., Alberti, K., Barr, T. L., and Klinowski, J.,** *Nature,* 365, 429, 1993.
45. **van Santen, R. A.,** *Theoretical Heterogenous Catalysis*, World Scientific, Singapore, 1991, chap. 4.
46. **Breck, D. W.,** *Zeolite Molecular Sieves*, John Wiley & Sons, New York, 1974.

Chapter 3

The ESCA Spectrometer and Its Use

I. INTRODUCTION TO THE BASIC SYSTEM

The conventional commercial ESCA experiment generally employs either $Al_{k\alpha}$ (1486-eV) or $Mg_{k\alpha}$ (1254-eV) X-radiation. Whether utilized with direct "line of sight" from the X-ray anode, or a crystal monochromator, the incidence angle of the X-rays are typically in the range of 33 to 90°, with respect to the sample normal. The ejected photoelectrons are often collected along this sample normal, although almost all angles down to the specular are employed. Initially, a variety of electron (or β) analyzers were employed to collect and energy sector the photoelectrons, with particular use of either hemispherical or cylindrical mirror systems. Many spectrometer manufacturers originally employed various arrangements of different-sized slits to introduce and remove the photoelectrons from these analyzers. The lack of consistently satisfactory results also has generally led to the use of a variety of lens systems (electrostatic or magnetic) that are employed (often with slits) to focus the photoelectrons properly inside the analyzer. (Recently most manufacturers have realized the superiority of the hemispherical analyzer for the ESCA process.[1] A typical arrangement of this type is displayed in Figure 3-1.) A variety of detectors, often arranged into particular geometries, are employed to collect the final electron emissions. Successful detectors have included channeltrons, microchannel plates, and resistive anode plates, often coupled with various arrangements of phosphor screens and high-sensitivity TV cameras.

II. EXTENSIONS AND IMPROVEMENTS

A. MONOCHROMATORS — HIGH RESOLUTION AND ACUTE SENSITIVITY

As mentioned earlier, conventional ESCA is still often accomplished employing direct impingement of the relatively soft X-rays from their anode source onto the sample of interest. This approach is generally simple to operate and permits the maximum use of the photon density, and thus the maximum "flexibility" with the *total* electron output. Early in the use of ESCA, however, the University of Uppsala group under Kai Siegbahn recognized that this simple approach often defeated some of the purported attributes of the ESCA technique.[2] To illustrate: (1) direct X-ray impingement with its accompanying thermal effects and additional "stray" particles (e.g., Bremsstrahlung, etc.) was sometimes found not to provide a "nondestructive" analytical method, as generally advertised (e.g., direct impingement ESCA cases were reported that beam reduced some samples [e.g., Cu^{++} to Cu^+],[3] and seriously chemically damaged various polymers);[4] (2) the secondary electron continuum mentioned above was often so large that it produced a signal to background that sometimes interfered with the detection of weak peaks; (3) concepts of optics dictated that the maximum electron output resolution (minimum line width) could not exceed that of the input X-ray line; and (4) all of the significant satellites exhibited by the impinging X-ray line also produce photoelectron peaks, sometimes in binding energy regions that screened the peaks of sparse species or interfered with chemical shift registration.

Faced with these difficulties, the Siegbahn group proposed,[2] demonstrated,[5] and brought to general practice the use of monochromatically scattered X-rays. This led to the development and demonstration[6] of the use of geometrically arranged silica (quartz) crystals, bent to produce the proper Bragg angle, such that a Rowland circle arrangement

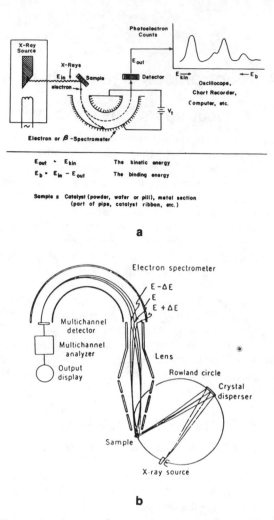

Figure 3-1 (a) Generalized schematic for ESCA spectrometer and typical results. (b) Figurative rendition of older monochromator based ESCA.

will permit the focus on the sample of a diffracted segment of the principal $Al_{k\alpha}$ X-ray peak. As a result, the above-mentioned beam damage is substantially curtailed,[6] line widths may be reduced (often by several times), and the general background may be sufficiently reduced to realize an improvement in signal to background of several orders of magnitude.

Unfortunately, however, the carefully worked out details of the Uppsala arguments were often not totally heeded. Thus, many manufacturers and practitioners failed to note that successful implementation of an X-ray monochromator often depends upon the care employed to couple it with the *proper* electron spectrometer (particularly, the right arrangement of lenses and hemispherical analyzer). Originally, for example, the Siegbahn group suggested a dispersion compensation arrangement for X-ray monochromator and electron lens, which could, theoretically, completely decouple the output resolution from that of the input.[5] This arrangement was sucessfully implemented in the Hewlett-Packard ESCA system marketed from 1971 to 1976.[6]

Most manufacturers have finally realized the importance of providing a satisfactory monochromatic option along with a conventional direct impingement system. It is now not uncommon for ESCA systems to be equipped with both procedures, permitting the use of the faster, high-intensity, moderate resolution in cases where sensitivity or resolution is not critical, with relatively easy conversion to a high-resolution/acute sensitivity monochromatic mode when these features are needed.

B. OTHER DEVELOPMENTS — SPATIAL RESOLUTION

Several other recent improvements in ESCA technology deserve special mention. It is now possible to fine focus the X-ray beam before monochromatization, thus permitting enhanced spatial resolution with further enhancement in energy resolution.[7] It is also possible to employ adroit arrangements of the lens and slit system to segment the spatial aspects of the photoelectron output. Focusing of the latter output would appear to permit even greater spatial resolution (1 μm or less is predicted) with rastering.[8] Fine focus of the X-rays, however, also permits a much more intense (faster) small spot analysis. It should be apparent that all of these improvements have ushered in a period of intensive effort in spatial analysis, thus effectively removing one of the often-mentioned shortcomings of ESCA — the statistical (relatively large x-y) nature of its analysis.[9] With recent alterations, one may now routinely switch many systems from "large spot" (analysis generally of the order of 1 to 5 mm in diameter) to small spot inside the same experiment, thus permitting the sequential realization of the advantages of both approaches. These small spot capabilities have naturally led to the successful beginnings of a form of "photoelectron microscopy" (see Section V in Chapter 9) with the obvious goal of providing a replacement (or at least supplement) for modest electron microscopy and/or Auger microscopy in a mode that will simultaneously provide detailed chemical, as well as qualitative, analysis.

III. THE PRESENT PERFORMANCE FRONTIER

Subsequent developments in resolution and sensitivity have recognized not only the need to optimize the separate features of ESCA systems (e.g., source, monochromator, sample presentation and treatment capability, lenses, slits, hemispherical analyzer, detector, and data system), but have also noted that this improvement will depend particularly upon the ability of the spectrometer manufacturer to optimize most (if not all) of these features into a simultaneous, cohesive package.[10] On the commercial front this motivation is presently best represented by the new system being produced in Uppsala, Sweden, by Scienta Instruments AB.[11] This system (depicted in Figure 3-2) begins with a fine focus electron gun that is employed to eject $Al_{k\alpha}$ X-rays from an anode that is electroplated onto the circumference of a high-speed (~10,000 rpm) rotor (rotating anode). These X-rays are then focused onto 7+ strategically bent and placed quartz crystals, aligned on an adjustable backing plate. The monochromatic, diffraction scattered output of the latter is further focused onto a sample strategically placed to form the optimal, 650 mm, Rowland circle geometry with both the anode and silica crystals. A multistaged, high-resolution, electrostatic lens is employed to collect the photoelectron output and channel it through a 300-mm diameter hemispherical analyzer using various intermediate slits. Final detection is achieved with a microchannel plate detector, phosphor, and a computer-interfaced CCD-TV camera. The preliminary resolution achieved by this system (see Figure 3-3) exceeds that of any other commercial instrument, and the sensitivity enhancement appears to be more than an order of magnitude better (see Figure 3-4). All of this is achieved with minimal sample damage, while collecting data at a rate that not only exceeds that of all other commercial systems but also most synchrotron radiation sites. Such improvements are not cheap, of course, and in the view of some practitioners may not be necessary, particularly if the desired results fall into an area loosely labeled as "conventional" ESCA. It should be noted, however, that many of the more sophisticated techniques to be

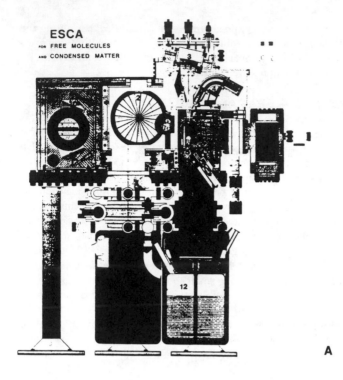

A

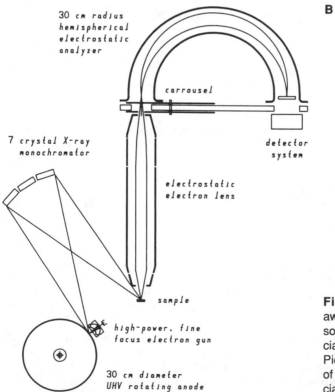

B

30 cm radius
hemispherical
electrostatic
analyzer

carrousel

7 crystal X-ray
monochromator

detector
system

electrostatic
electron lens

sample

high-power, fine
focus electron gun

30 cm diameter
UHV rotating anode

Figure 3-2 (A) Cutaway rendition of precursor to present commercial Scienta ESCA. (B) Pictorial representation of present day commercial Scienta ESCA.

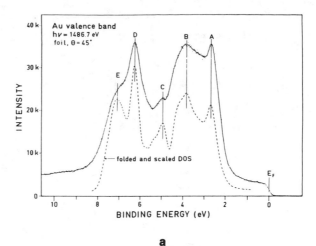

a

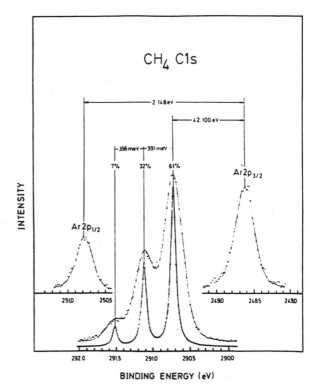

Figure 3-3 Typical set of high resolution valence band and gas phase core level result achieved with Scienta precursor.

b

described later in this book (e.g., Chapter 9) are substantially enhanced by technology, like that employed in the Scienta 300. Thus, the fact that both photons and electrons are focused in the Scienta system produces the resolution and geometry needed for some applications in surface-to-bulk photoemission, photoelectron diffraction, and photoelectron microscopy (see Section V in Chapter 9). It is also possible with adroit plating or rotor wheel choice to achieve reasonable monochromatic output from several X-ray

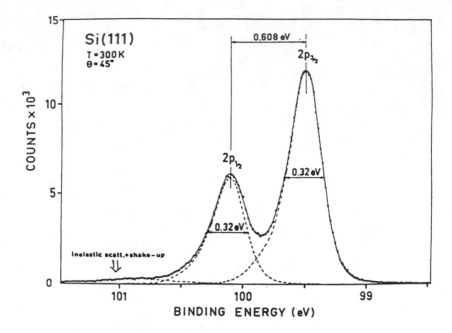

Figure 3-4 ESCA results from Scienta precursor demonstrating extreme resolution and sensitivity; SiO_2.

sources at substantially different energies (e.g., $Al_{k\alpha}$ and $Cr_{k\beta}$). Thus, a system configured similar to the Scienta may, for some purposes, function as a reasonable "in-house" alternative to synchrotron radiation.

IV. STANDARD USAGE OF THE ESCA SYSTEM

A. INTRODUCTION

In a book that concentrates on the theory and new methodology of ESCA, it may seem strange to include a chapter on the practical aspects of the science as they have been practiced since the 1970s, but it should be recalled that a primary goal of this book is to be helpful to those scientists and engineers who want to use ESCA as well as understand it. In this regard, the author has learned that there are important experimental factors that lie beyond the formalities of the science and that these factors can substantially inhibit proper use of ESCA if they are not understood.

In order to discuss these factors we will return to the beginning and describe the basic ESCA experiment in a stepwise fashion. Much of this discussion will be too elementary to be of great value to the experienced ESCA scientist, but it should be helpful for the novice or less-experienced user. In addition, it should be noted that several points and procedures will be raised that may be handled differently by some of the author's experienced colleagues. The uninitiated should be well aware that there are often several "right" ways to proceed to a common objective, and it is hoped that with experience they will develop their own "method of practice".

In this book the meaning of the word calibration will be widened to include "readying" the ESCA user, as well as the spectrometer. For this reason a much more expansive list of procedures is described than generally appear in other texts. Some, of course, may be omitted as outside of the areas of interest or need of the user or as redundant, but the serious ESCA user may find careful exercise of a number of these good practices in developing his/her skills before (or while) tackling the "real thing".

Remember — never accept the word of a person inexperienced in ultra-high vacuum (UHV) technology as to the surface cleanliness of gas lines, materials, containers, chambers, gaskets, etc. There are brilliant scientists and engineers throughout the world who still think that their eye is a good surface detector and that, for example, a shiny piece of copper (e.g., a penny) sitting on a laboratory bench has an outer surface composed of elemental copper.

B. VACUUM AND SYSTEM MAINTENANCE[1]

It is vital for the serious ESCA user to understand the UHV system in his/her spectrometer. First, it is important to understand how one's UHV conditions are achieved and maintained. The intimate interplay between the pumps and the "vacuum tight" sealing of the system should be studied. The user should experiment with what happens to his or her system when the pumps are turned off. In general it will be noted that it is the seals that maintain vacuum, whereas it is the pumps that achieve and maintain UHV. It is very important to understand how much vacuum capability your system has and how fast it can "recover" under the working conditions that it will commonly experience. It should be noted that pumping capacity, speed, and the eventually pressure achievable all depend on the type and amount of the gases that the pump will be called upon to remove. The user should experience several "pump downs" from atmospheric conditions before he/she seriously uses the spectrometer. It should be noted that the generation of "conventional" vacuum ($\sim 10^{-3}$ to 10^{-4} torr) is almost never achieved with the pumps that will be employed to "maintain" UHV. Thus, initial pumping is usually accomplished with a type of mechanical "roughing" pump (often several of these). Since these are generally oil-based pumps, they must be carefully trapped and vented to prevent (as much as possible) the encroachment of oil vapor into the UHV chambers. In particularly sensitive areas it is common to resort to non-oil-based pumps, such as molecular sieve-activated, vacsorb pumps.

It should also be noted that, during moderate shut-down times (perhaps even days) for adjustments, checks, and simple repairs, it is advisable to isolate the unexposed parts of the system under vacuum and keep the "exposed" UHV parts of the spectrometer system under a positive (flowing) atmosphere of some inert gas, e.g., Ar or, at least, dry (oil free) nitrogen. All openings that are not sealed should be covered tightly with aluminum foil. Dust and dirt are unbelievable adsorbers of O_2 and H_2O, and will also bleed other chemical contaminants for years, if not painstakingly removed. (Remember also that everything that comes in contact with a surface leaves behind some kind of residue. The art of UHV technology is to minimize that residue.) The use of a positive pressure of inert gas is not necessarily intended to shorten the subsequent pumpdown time. Many UHV pumping systems, for example, are less efficient for Ar than O_2. However, any O_2 or H_2O introduced to the system is preferentially adsorbed on all surfaces, and immediately begins to oxidize that surface. Titanium and aluminum are essentially O_2 getters at room temperature. This means that any subsequent pumping action must be operated against these adsorptions.

Most of the gauges and "active" units in the UHV system have surfaces that are subject to electrical and/or thermal pulses that if activated too rapidly following air or H_2O exposure will "burp" and may do other things that can seriously damage the system. Therefore, where possible they should be gradually activated with continuous monitoring of the pressure.

If any of the aforementioned problems are severe then the system *must* be baked out. The baking process is unique to each individual type of system and one should not try it without detailed instructions from the manufacturer. Be particularly careful about instructions designed for one set of components when one's system is equipped with another. Different types of pumps, for example, require completely different baking and cleaning procedures. Many units must be left outside the baking region or baked separately, and if the inexperienced is unsure he/she should seek advice. In any case, even if instructions are followed exactly, the first few "bakes" should be monitored on as tight a basis as possible. The author is personally aware of numerous major accidents during baking. Establishing a regular, periodic baking schedule is often good practice, but, as with everything else, one can overdo it. In practice, one should let the use of the system dictate the baking schedule.

It must be remembered that many severe bleeding materials, such as high surface area catalysts and ceramics are also often active gas getters. Thus, as one examines these materials in UHV they are constantly "breathing" gases in and out. This means that, even when the pressure gauges suggest that a good vacuum has been reestablished, one should realize that with continuous introduction of these types of materials the vacuum integrity of the system is being gradually compromised. After a certain point, full restoration may only be achieved with an extensive clean and bake exercise.

C. BASIC CALIBRATIONS[1]

Spectrometer calibration may be more appropriately described as "know your spectrometer". Before extensively employing one's ESCA to try to generate useful data, one should use the services of experienced practitioners to adjudge what type of periodic calibrations to employ and how often. Generally four types of calibrations need to be examined with some regularity. First, there are the binding energy positions and linewidths of a nonreacted, clean, smooth metal surface, especially $Au^0(4f_{7/2})$ or $Ag^0(3d_{5/2})$. It is good practice to also determine the spin-orbit split for these peaks, e.g., $Au^0(4f_{5/2})$ to $Au^0(4f_{7/2})$. One should also take the time to determine the appropriate peak-to-peak intensity ratios, e.g., $[Au^0(4f_{7/2})]/[Au^0(4f_{5/2})]$. The latter should be achieved by several means, e.g., ratios of the peak heights and peak areas. It is also useful to generate these by hand and by using any computer packages provided with the spectrometer. Note how close each method comes to the correct $(2J + 1)$ ratio (e.g., 4/3 for Au^0).

Lastly, one should always try to calibrate the signal to noise or signal to background. Perhaps one of the most useful procedures intended for these types of analyses is to produce an offset C(1s) peak for a sample of graphite that simultaneously exhibits the background ~10 eV down field (to lower binding energy).[6] Since there is almost no photoelectron density below the C(1s) peak of pure graphite this background is almost totally that of the spectrometer. The maximization of the signal to background for this C(1s) is a useful signpost of the ability of any spectrometer to extract information density from modest peaks for constituents that constitute less than 0.5% of any mixed system. The success realized by many of the HP-5950 ESCAs in catalysis and composite studies (see, for example, Section VI of Chapter 8) in the period between 1975 and the present was often predicated on their ability to achieve a 200+ value for the aforementioned ratio![6]

D. MORE EXACTING CALIBRATIONS

Further, more exacting, calibrations of your ESCA may be accomplished in the valence band region. (These procedures are very time consuming and, therefore, unless the user is conducting detailed research in this region, they need to be performed only following extensive down times.) The materials needed for these studies include clean, flat, pure (polycrystalline) foils of Ni, Ag, and Zn, plus a thin film of pure Si^0, a clean, pure wafer of SiO_2, and a basal oriented piece of graphite.

The metal foils — These can be employed to examine resolution, establish the Fermi edge (E_F) of the spectrometer, and gauge the background produced by the spectrometer operation above E_F.

The easiest sample to examine is Ni^0. This should produce a characteristic near E_F peak structure, due primarily to the $3d^8$ bands of the metal. All spectrometers should, at least, reveal the apparent asymmetry of these bands, and high-resolution systems should

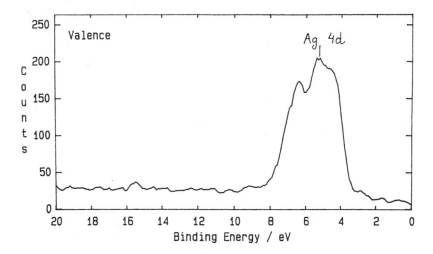

Figure 3-5 XPS generated band structure for silver exposed to O_2. Note that G monochromator was employed.

generate some of the d band structure readily detected with the Scienta ESCA and on synchrotron radiation systems.[12] *All ESCA systems* should also yield a smaller shoulder protrusion to lower binding energy. This structure is predominantly produced by the s^2 band of the Ni^0, and, for measurements at ~25°C, its leading edge should cross the 0.0-eV point of the spectrometer at ~1/3 of its peak height. This process establishes the Fermi edge (0.0-eV point) of the spectrometer and connotes proper Fermi edge coupling between a conductive sample and the spectrometer. If this does not occur it suggests one, or more, of several possible problems: (1) the Ni^0 foil may be improperly lodged on the sample probe negating good coupling; (2) the probe itself is improperly coupled to the spectrometer; (3) the energy zero built into the spectrometer has been improperly established, e.g., when the aforementioned Au^0 sample was inserted, $Au(4f_{7/2}) \neq 83.95$ eV, that is, the spectrometer work function has not been properly registered; and (4) the nickel sample is significantly oxidized and the ESCA registration is actually that of nonconductive NiO, rather than Ni^0.

The valence band of the Ag^0 sample is also a useful measurement. In this case the already established binding energy of the system should be offset several electronvolts so that one can also read the above E_F output. Once again the character of the d band (4d) should be monitored and judged against the excellent results reported by Siegbahn[13] and Gelius et al.[11] and reproduced herein in Chapter 9 in Figure 9-15. In addition, the user should examine the character of the relatively weak s band.[14] This should be a nearly flat plateau, evolving out of the d band at low binding energy, falling off rather abruptly just before the Fermi edge. This exercise is particularly demonstrative of the capability of an ESCA system to extract an extremely weak peak density (s band) out of the general spectrometer background, against the adjacent presence of a much more intense peak structure. It should be noted that there is no "true" photoelectron density above E_F (see Section II in Chapter 9 on Inverse Photoemission) and therefore the density that one finds in this region is indicative of this background. This is strictly true, however, only for monochromator-based systems. ESCA using conventional Al or Mg anodes will, of course, exhibit weak X-ray satellite lines from the $Ag°(4d)$ band peaked at ~⊕4 to 6 eV.

It is appropriate to ask at this point why we are placing so much stress on the seemably insignificant s band character. As a partial answer one should note in Figure 3-5 that a Ag valence band spectrum has been generated for another silver sample that seems, at first

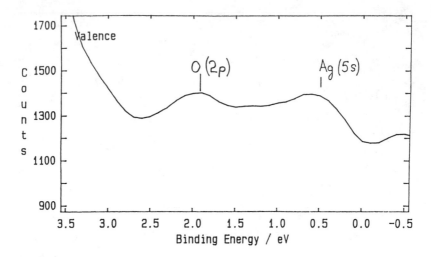

Figure 3-6 Expanded version of near Fermi-edge densities of states for example in Figure 3-5.

sight, to replicate the d band in Figure 9-15b (except for the obvious loss in resolution compared to the Scienta-generated result). Close examination of the s band region of this second Ag sample in Figure 3-6, however, demonstrates a ready presence of the O(2p) of an oxide apparently formed at the expense of some of the Ag(5s). (This should remind the reader that the noncomplex-forming chemistry of the group B metals does not involve the d electrons.)

The valence band structure of the Zn^0 system, Figure 3-7, is similar for Ag^0 in that there is a distinct d band density [in this case Zn(3d) occurs at ~10.0 eV]. This band is somewhat core-like, and therefore is quite narrow and exhibits little structure. (Note that the chemical shift of this band for the reaction $Zn^0 + \frac{1}{2}O_2 \rightarrow ZnO$ is very small, but should involve noticeable broadening.) Once again the principal reason for this measurement is to gauge the character of the background above E_F and the s band. (In this case a long flat s structure occurs from ~8 eV to the Fermi edge.) When a monochromator is used the Fermi edge for Zn^0 should be easily discerned, see Figure 3-7B.[14] If a conventional anode is used the X-ray satellite will bracket the Fermi edge and totally hide it. This measurement then becomes a useful signpost to give the user an accurate view of the nature of any satellite subtraction routine provided in his/her data analysis package. The overt action of one such package is demonstrated in Figure 3-8.

Nonmetallic-nonconductors — The Si^0 film is a useful material to examine because it "exposes" (and thus checks) several features of ESCA not readily apparent with the metal foils. Before the metal foils were examined with ESCA, all must be ion etched to remove obvious oxide films formed on their surfaces due to air exposure. Many scientists are surprised to discover that the same process must be employed to clean Si^0 before surface characterization. As described elsewhere in this book, it was a shock to many early semiconductor technologists to discover that all their surfaces suffered from the same type of oxide layer "passivation" as do metals.[15] Removal of that "tarnish"[16] from well-formed Si^0 should reveal a distinct three-peaked valence band pattern,[17] see Figure 3-9. The splittings and binding energy positions of these peaks should be fixed as described by Ley;[18] however, optimal resolution of those peaks may require some annealing. It is also true that the leading edge of the peak closest to E_F should exhibit a separation from E_F of ~0.55 eV. This $\frac{1}{2}E_F$ will be true only if the Si^0 is a totally intrinsic semiconductor and only if its Fermi edge (located at the midgap) is coupled to that of the

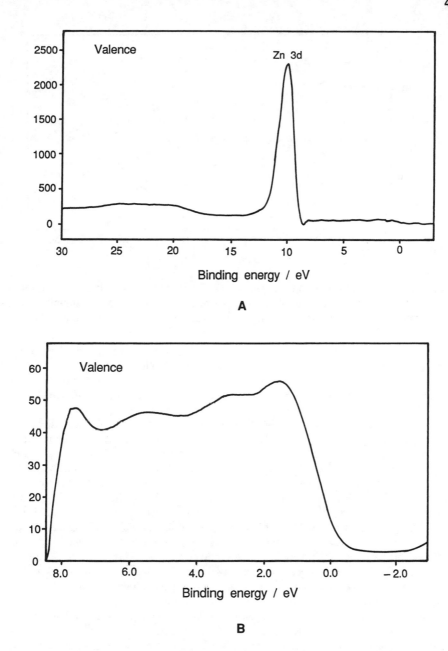

Figure 3-7 XPS valence band (V.B.) spectrum for Zn. (A) Structure from 0 to 30 eV. Note character of Zn⁰ (3d) near-core line. (B) High-resolution scan showing only s band region. Note width and structure of the band and the Fermi edge at 0 eV.

spectrometer. This is a good exercise in charge shift phenomena, as the room temperature electron flow for a material with a band gap of ~1.1 eV might be construed to be small enough for the material to act as a surface insulator and produce a charging shift. So far the author has not found any cases where this is true.

The (SiO_2) silica case is quite different from any of the preceding descriptions. The chemistry and comparative oxide chemistry in the valence band pattern for SiO_2 is presented

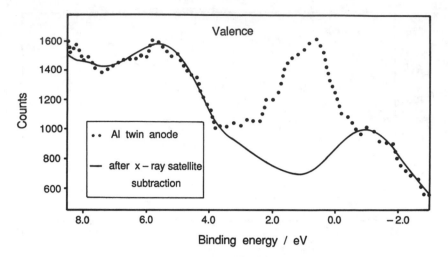

Figure 3-8 XPS valence band for mixed Zn/ZnO system. Generated with conventional (direct focus) $Al_{k\alpha}$ anode. Note X-ray satellite of Zn (three-dimensional) covers the Fermi edge. Note further the results of computer subtraction of this satellite.

in detail in Chapter 7. The problem with employing this material as a calibrant rests on the difficulties involved with trying to clean the SiO_2 surface following exposure to STP air. The latter process will chemisorb some oxygen and cause a significant compromise of the surface by H_2O and adventitious carbon. Removal of these species can be affected by Ar^+ ion etching, but as depicted in Figure 3-10 this process will also alter the surface structure of the SiO_2 and may introduce selective defects. It is possible to return this SiO_2 surface to its maximum integrity through certain selective annealing-type processes. If the material is then maintained in UHV, a reasonable rendition of the "proper" SiO_2 valence band spectrum *may* be generated, as depicted in Figure 3-10. The difficulty here in establishing total calibration, however, rests more on the lack of conductivity of SiO_2 than its cleanliness. Thus, if a wafer of silica is examined, the nearly 10-eV band gap will introduce significant charge buildup following photoelectron ejection that if not rectified will produce a substantial charging shift upfield (see Chapter 7). The extent of this shift should be essentially the same for all photoelectron peak structures of the SiO_2, including the valence bands. The extent of this charging in the recorded spectra may be as large as 100 eV. In this case, the resulting SiO_2 photoelectron peak structures will usually exhibit extensive distortion, as well as shifts. If this occurs, it is necessary to apply some experimental procedure for charge shift removal. This can generally be achieved with application of some type of electron flood gun device (or related system). As described in Chapter 7, it is difficult to remove all of the charging and also achieve the type of Fermi edge coupling needed to realize universally true binding energies. At this point it is important to examine if adjustments may be made to the SiO_2 spectra to achieve reasonably good binding energies and peak shapes. As with the aforementioned conductive samples, the precision of the ESCA measurement in the SiO_2 case can be gauged, in part, by the pattern of its valence band (V.B.). Thus, the structure should closely resemble Figure 3-10, and the leading edge should start at ~5 eV (indicative of a $\Delta E_g \cong 10$ eV, that is, typical of a totally intrinsic semiconductor or, more appropriate, an insulator). The total width at half maximum of the SiO_2 V.B. should be ~10.2 eV. The region of binding energy between the leading edge of the V.B. and the pseudo-Fermi edge should be reasonably flat and devoid of electron density. (The appearance of any finite density in this region may suggest the presence of some type of defect structure.)

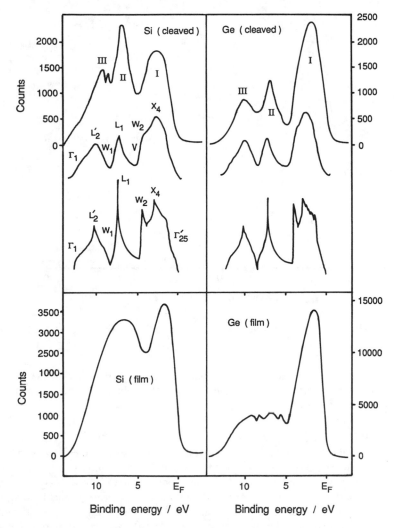

Figure 3-9 XPS experimental and theoretical valence band spectra for select elemental species configured in tetrahedral structure. (a) Si; (b) Ge.

The valence band of graphite provides a different set of problems from those previously described. If the surface of graphite is properly cleaned, it will be a reasonable conductor even if examined as an unoriented powder. The two-dimensional conductivity arises, of course, from the delocalized π density in the basal plane. The XPS pattern of the valence band for graphite, even under superior resolution, is dominated by its σ structure.[19] The π density is shown by McFeeley et al.[19] to overlap the Fermi edge, but very weakly [due to the extremely small size of the relative π cross section in XPS. Note that this π density is much more intense in ultraviolet photoelectron spectroscopy (UPS)].[20] The detection of this weak structure is a challenge that an experienced ESCA user, particularly one dealing with carbonaceous systems, may wish to tackle.

The previously described valence band results are suggested calibration addendums to the common core level peak examinations (see Reference 1 for a detailed discussion). Obviously, one can calibrate one's ESCA without the use of any of these valence band examples and, even if these procedures are utilized, only a few select cases need to be

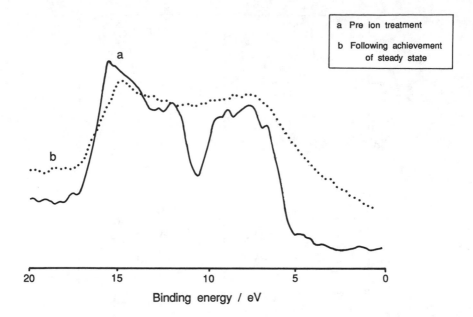

Figure 3-10 XPS valence band for SiO_2. Note "true" structure and effects of ion etching.

considered. It should be noted, however, that in some cases these valence band analyses are more informative than the simpler core level calibrations and sometimes more exacting than the latter.

E. SURVEY SCANS

In view of the uncertainty that surrounds the status of surface cleanliness and composition, it is *always* good form to begin any XPS study with a good survey scan. This can generally be accomplished by generating a spectrum from 0 to 1000 eV in binding energy, since essentially all common elements produce significant peaks in this binding energy region. It should be remembered, however, that due to conflicting peak positions it may be necessary to extend this scale to as high as 1250 eV to verify the presence of such common elements as Na, Zn, and Ge. It should be noted that the presence of even a modest amount of some element must generate its entire photoelectron signature, not just its most intense line. Thus, when the user employs a reference document to identify a particular element she/he should note the entire photoelectron *pattern* of that element, not just the position of the element's most intense core line.[21] The author chooses the word "pattern", rather than binding energy "position" because the relative cross-sectional intensity pattern is as important as the peak binding energies. The user should obtain a reference source, however, that clearly labels the most intense line and also the relative "pecking order" of the intensities of the rest of the photoelectron peaks.[1] One should then get in the habit of always looking for this pattern before suspecting the surface presence of a particular element. Thus, for example, a spectrum similar to Figure 3-11 should be produced by the presence of elemental Cd and that in Figure 3-12 for a typical zeolite (crystalline silica aluminate).

The author has once again purposefully avoided making a series of absolute statements here, because the reader will soon become aware that there are a number of problems that may arise at this point to complicate and even negate qualitative judgments.

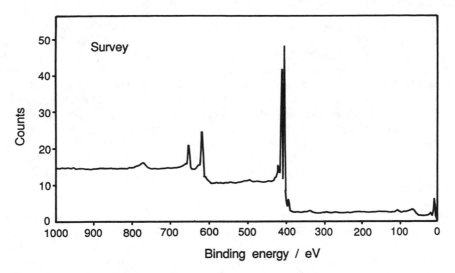

Figure 3-11 Representative 0 to 1000-eV survey scan, in this case for Cd. Note appearance of general background, principal peaks, and loss lines.

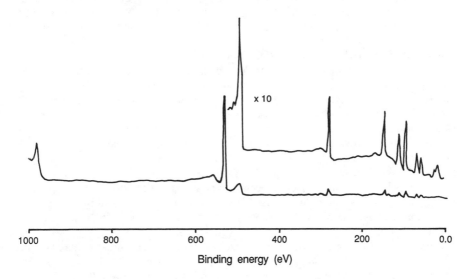

Figure 3-12 0 to 1000-eV survey scan for Na-X zeolite. Note indication of a pure air exposed material, i.e., only C(1s) impurity. Note also the distribution of the peaks for elements of Na-X.

F. NARROW SCANS

The generation of ESCA spectra designed to distinguish key features of individual peaks must always be accomplished with great care, even by the most experienced ESCA users. It is important to remember that "fragile" peaks protruding from the background, characterized by inappropriate (non-Gaussian-Lorentzian) shapes, and line widths, that are thinner than the best resolved significant peaks, are generally just an unfortunate part of the aforementioned background. Rolling shoulders and protrusions (often covered with extensive background spikes) are, on the other hand, often the "half-hearted" signatures of real components, sometimes — the critical species. When these various uncertain effects occur and they are deemed to be of potential importance, it is a good practice to try to rely on the (hoped for) long-time stability of the spectrometer and the eventual rewards of statistics and run extremely long (after hours?) scans. Even overnight runs should be considered. (There seems to be an apparent magic based around the classic saying "a watched pot ....".) Often, in the case of repetition, if this procedure is employed for one or two samples, computer enhancement methods may be used on the balance.

Before performing these extremely long scans the spectroscopist must give due consideration to the potential for alteration of the material due to extensive time in the vacuum and under the X-ray beam. All things considered the author often chooses the monochromator source (if available) during these types of "peak extraction" studies, even if enhanced resolution is unnecessary.

Even if the aforementioned practices are followed there are still sometimes cases in which weak (suspected) peaks are rivaled by the background and the user may find it difficult to impress his/her colleagues (and sometimes even himself) as to the efficacy of his/her judgment. When this occurs it is often helpful to generate two completely different scan regions covering the protrusion in question. Thus, a series of 20-eV scans may be run accompanied by a series of, for example, 30-eV scans that are slightly offset from the former. It is virtually impossible for the spectrometer background to duplicate itself in these two independent regions. Thus, reproduction of specific peak structures generally signifies that the protrusion is indicative of a true peak for the material being examined. The correct interpretation of the latter, however, often requires skill and experience.

G. NARROW SCANS — ADDENDUM

To a certain extent the scan width employed in narrow scans is a matter of choice. Most spectrometer data systems are pregraduated on a ten-unit scale so that many practitioners automatically choose 10- or 20-eV narrow scans as a matter of habit. Although one can generally play with the key parameters, many spectrometers are designed to achieve their maximum resolution during 10-eV scans. A few can be programmed to even narrower windows and this may on occasion be of some value. In many cases, however, ranges of 5 eV (or even less) have been exploited in attempts to try to produce line widths that "rival the best" (e.g., the Scienta and/or synchrotron radiation results) and, if present, the slits and other features have been constricted to the point of making (unfortunately, sometimes useless) operations under these conditions. Typically, as mentioned above, scans of 10 and even 20 eV are employed for practical high-resolution analysis.

There is no absolute rule of thumb here, but typically one tries to maintain the linewidth of the $Ag(3d_{5/2})$ and $Au(4f_{7/2})$ peaks below (or as close to) 1 eV as possible. Perhaps a better benchmark is to try to resolve some of the splitting between the Si^0 ($2p_{3/2}$ and $2p_{1/2}$) lines.[11] Those who are ambitious may wish to try for the $Al^0(2p)$ splitting.

The previous discussion is actually very incomplete because most analyses of high-resolution scans are not centered exclusively around a singular peak, but depend very much on the shoulders, spin-orbit companions, shake-up satellites, and other loss lines. Further, it is often extremely useful to generate high-resolution data from more than one element in the same scan. For these reasons the author generally prefers 20 eV rather than 10 eV scans and also often supplements his analyses with wider scans to incorporate all of the (aforementioned) useful features. Generally, however, the resolution, and related features, fall off significantly with scan width so that "high-resolution" scans of greater than 50 eV are not recommended. (Do not be fooled by the apparent peak features produced by computer spreading of 1000-eV scans. These results are useful indications, but should not be used for binding energies in the tenth of eV range or specific quantifications.)

H. OPERATIONS UNDER LESS THAN OPTIMAL CONDITIONS

No matter how much care is employed in maintaining an ESCA system, sooner or later the unit will be operated for periods of time under less than optimal conditions of vacuum and/or spectrometer performance. The situation is similar to one's favorite car — some of the slippage of performance is so gradual and subtle we forget that it "used to run better". In addition, most operations face some type of time pressure. The results must be generated now — in a week — this month. All of these pressures tend to make us put off those difficult-to-accomplish, time-consuming calibrations, bake-outs, etc. Fortunately, many of the desired results can be adequately produced under less than optimal conditions, but this realization can be somewhat dangerous for the inexperienced user. As noted above, ESCA almost always generates impressive results. Whether those results are indicative of the sought-after problem or, as indicated above, are beguiling "impersonators" created by a fickle nature depends very much on the problem at hand, and may require as much experience and know-how to adjudge properly as a complex methodology. No road maps can be constructed here to substitute for that experience, except to note that the situation may be dictated by the nature of the problem being studied. Thus, if the materials being examined include chemisorption on single crystal surfaces, and the latter is important to the results, then a compromise in vacuum integrity may be totally unacceptable. On the other hand, if the problem centers on generating chemical information about very sparsely doped metals on a ceramic support, then any reduction in signal to background may be devastating. All things considered this is a difficult area that mandates: (1) experience and knowledge, (2) operating under the best conditions *possible,* and (3) above all, "know your spectrometer".

I. WHAT ARE THE LIMITS?

Perhaps the most important asset of ESCA can also be its greatest deficit. The spectroscopy almost always yields results, often very diverse results (a tantalizing protrusion of peaks) — even when very little effort is applied to achieve those results. Unfortunately, the genesis of this profusion may be the surface that nature has provided for the often unsuspecting analyst, rather than the sought after surface, and further if that analyst is ill schooled it is very easy to make illicit interpretations of these results. (This fact was disturbingly demonstrated in the hundreds of incorrect ESCA analyses provided in the first 2 years of investigation of the new high Tc superconductors — see Section VIII in Chapter 8.[22]) It is vital, therefore, that the analyst and her/his ESCA become a team capable of the ready distribution of knowledge with recognition of the pluses and minuses of what is being analyzed. The author of this book has found very few analyses that are useless, even when mistakes or initial misjudgments are made. *ESCA is an area of analysis that often permits one to make "lemonade from lemons"*. Therefore, rather than discard mistakes, put them aside and return to them after the problem has been corrected. You will be surprised how often interesting bits of additional information are hiding in these "mistakes".

J. PROBLEMS
1. Charging

An explanation of charging as both a problem and potential asset is presented elsewhere in this book. Very little, however, is said in those discussions of the practical aspects of the problem. From a practical point of view, charging (as exemplified by a variably sized nonchemically induced shift upfield in binding energy) may arise whenever the sample is not Fermi edge coupled to the spectrometer (probe). This means that charging may arise for: (1) a sample that is a relatively poor conductor, i.e., one that cannot "drain" (inside of the timeframe of an ESCA measurement) the macroscopic positive charges induced in the sample by photoelectron ejection or (2) for a sample that is an acceptable conductor, but one that is electronically isolated from any conductive source, e.g., the spectrometer probe, by an effective sea of nonconductors. In either of these cases, positive charge will build on, or about, the surface of the sample inducing the previously described charging shifts.

Thus, from a practical point of view, charging (and Fermi edge decoupling) may result from the inherent nature of the (insulating) sample or from the method in which the ESCA experiment is designed. In some cases the latter may be a vital part of the experiment (see the section (VII.F) on small cluster shifts in Chapter 9). However, it is also relatively easy to design an experiment featuring improper coupling. The user should become adroit at including proper checks and flow diagrams designed, in part, to avoid this problem.

2. X-ray Satellites

As described above, when conventional ESCA is utilized with direct focus X-ray units (usually $Al_{K\alpha}$ or $Mg_{K\alpha}$) the XPS spectra that result will produce and register photoelectrons from all aspects of the X-ray line source. In the case of the two major sources employed, this means not only photoelectrons ejected by the principal lines (centered at 1486.6 eV for $Al_{K\alpha}$ and 1253.6 eV for $Mg_{K\alpha}$), but also those produced by any satellite lines associated with these X-ray sources. A typical example is shown in Figure 3-13, where the $Mg_{K\alpha}$ satellites at 9.6 and 11.5 eV (with respect to the position of the principal line) have produced two photoelectron peaks that have ~5 and ~3.5% of the intensity of the principal photoelectron line. It should be noted that all common X-ray line sources have these satellites and, in addition to the two prominent satellites mentioned above, the Mg anode will also register about one half dozen additional, smaller, satellite lines. In general, the relative intensities of the latter are so weak they are difficult to "extract" from the background.

Often these satellite lines do not create any problems, and they may even be viewed as an additional useful feature for calibration purposes; however, one must always be cognizant of their presence. A number of published cases exist in which mysterious identifications have proven to be forgotten satellites. A more difficult problem arises when the satellites cover critical peak structures of some less-pronounced ingredient of the total system. In order to circumvent the latter problem, nearly all spectral treatment subroutines provided by the manufacturers include some form of satellite subtraction routine. All of the latter work to some degree, but unfortunately none of these work completely. For example, one should examine the case that results in trying to read the near Fermi edge spectra of metals with deep d bands, such as the Zn° in Figure 3-8. Despite these problems, numerous cases have been recorded demonstrating reasonable success with these subtraction methods. In any case, one should always question results that demand for "subtracted regions" the same level of precision for quantification and chemical analysis as for other regions of the spectrum.

This problem area is, of course, one of those for which use of a monochromatic X-ray source completely eliminates the difficulty. All XPS monochromators are designed to employ just a segment of the total X-ray lines (usually $Al_{K\alpha}$), and that segment never includes the X-ray satellites.

3. Sample Damage

It has been common since its beginning to refer to ESCA as a method for *nondestructive analysis*. In part, this claim owes its origin to the fact that the ESCA process only requires the ejection of electrons, and not the removal or movement of nuclei — a necessity of the so-called *destructive methods of analysis*. Based on this more general

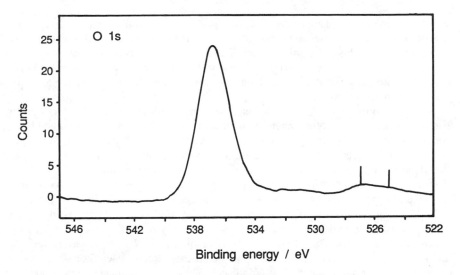

Figure 3-13 ESCA O(1s) spectrum achieved with conventional $Al_{k\alpha}$ anode for A1PO_4-8 molecular sieve (charge shift not removed). Note X-ray satellite peaks located 9.6 and 11.5 down field with respect to the principal O(1s) line.

designation, whether or not the material was, in fact, altered by the measurement process is only pertinent if the analysis is statistical, and thus requires repeated surveys in a common region of the material being analyzed. The latter is, of course, a necessity in ESCA. For this reason, lack of damage is a goal, effectively reached for most materials, but definitely not reached for many others.

In ESCA analysis several types of materials have produced near-classic histories in the area of damage. Thus, for example, those practitioners examining catalysts were inclined to agree absolutely with the no-damage claim and often left materials in the X-ray beam for days without any concern, whereas ESCA had gained such a bad reputation among polymer scientists that it was considered unusable by many polymer labs in the early to mid 1970s. Of course, neither absolute contention was valid.

Historically, the first dramatic example of sample damage in ESCA was the well-documented beam reduction of numerous cupric (CuII) compounds to the Cu(I) state, e.g., $2Cu(II)O \rightarrow Cu(I)2O + \frac{1}{2}O_2 \uparrow$. A number of incorrect identifications were rendered in these early studies. Experience soon demonstrated that the principal culprit was the extremely intense direct beam of the older ESCA systems. However, more detailed study showed that it was not the bombarding X-rays themselves, but rather the thermal effects (caused by the X-rays source and beam) that were inducing the copper reduction. Thus, Frost et al.[23] were able to achieve a variety of Cu(II) ESCA analyses by employing a cryoscopic sample holder and reducing the temperature of the sample to approximately 100°K.

In the course of such studies it was discovered that the indirect beam of a monochromatic X-ray source was often sufficient to prevent these beam damage effects. This panacea proved not to be total, however, as cases have been reported of monochromator-caused beam damage, even in catalysis. Thus, the author has discovered that long-time beam exposure of an oxidized system consisting of Mo(VI), Te(IV), Ce(IV), and Ni(II) on SiO_2 will produce a distinct violet-blue "stain" in the region of X-ray exposure after several hours, see Chapter 8. This is apparently the result of the beam reduction of some of the Mo(VI) oxide to $MoO_x(2 < x < 3)$ oxide, but the result is a most peculiar effect because the original oxidized catalyst contains, in fact, a complex quatenary oxide,[24] and it is this unit that is apparently reduced in the beam. Thus, similar exposure of simple $Mo(VI)O_3$ on SiO_2 does not create the effect. This example is presented to demonstrate the complexity of this potential problem. Thus, the previous copper examples are not properly specified, since beam and (its companion) thermal-induced reductions tend to be specific to particular compounds, and may be substantially different for complex compounds compared to simple oxides.

Another example of interest also demonstrates the complexity of this effect. Following examination of certain zeolite systems in a monochromatic (HP) ESCA for several hours the author has noted that many of these systems exhibited a readily detectable dark stain in the region exposed to the X-ray beam. The stain was most dramatic in the region directly exposed to the beam (approximate 1×5 mm line) but also was found to "spread" over an area far outside this region. Follow-up, comparative analyses were performed using the "damaged" samples and fresh ones examined under cryoscopic conditions, etc., and it was possible both to control and identify the origin of the effect as due to the damage of partially hydrated, select, zeolites, apparently producing sites of mixed hydrated cations and trapped electrons.[25] Of greatest importance, however, was the fact that ESCA analysis of the materials pre- and post-beam alteration revealed little or no change in composition, or bonding chemistry, of the complex silica-aluminate system. Thus, this would appear to be one case where the eye is actually more sensitive than the ESCA. Perhaps it suggests that these effects were induced deep within the zeolite by some sort of cascade process, but this has not been confirmed.

As stated above, polymeric ESCA studies have had a far more difficult history. Briggs[26] has pointed out that one major concern has generally not materialized. Thus, polymers do not result in excessive outgassing. In fact, they are generally less of a problem in this regard than typical heterogeneous catalysts. On the other hand, the author has noted that certain tenuous polymeric oxides, e.g., polyethylene and graphite oxides, are somewhat unstable, even in a monochromator beam, producing a continuous reduction in [O/C] and a slight increase in the background pressure. Briggs[26] has also noted that, with care, most polymeric systems can be successfully examined in a conventional ESCA and with use of a monochromator the time of successful beam exposure can be dramatically increased. He notes that this is even true for systems with additives. In this regard, polymers containing some of the halides are the least stable. Concerns about the important fluorocarbon systems have been substantially circumvented, but beam damage of the chlorine-containing materials (e.g., polyvinyl chloride) seems most dramatic.[26] Once again, adroit use of cryoscopic techniques has mitigated many of the problems.

Studies of biomedical surfaces have been extremely limited (see Section VII of Chapter 8). Most of the cases studied have involved replacement materials and their biocompatibility, and most of these cases are polymeric and subject to the above problems. There is little doubt that certain tissue samples will be adversely affected at room temperature by the ESCA beams; however, analyses planned for the near future are probably going to involve *in vitro* frozen sections, under cryoscopic conditions, that should curb much of the potential beam damage.

It should be noted that, in general, the X-ray beams employed in ESCA are less damaging than the electron beams of Auger or the ion beams of sputtering and the ion spectroscopies.[1] This is particularly true when the latter are focused with high energy and current density in a particular spot. This damage problem is, of course, not observed in small spot ESCA when the latter is achieved by selective extraction of a segment of the photoelectron output signal.[11] It will, however, be a growing problem for those cases where the X-ray beam is focused by some lens device, although use of a monochromator should mitigate this difficulty.

4. Poor Vacuum

The concept of a "poor" vacuum is for many ESCA users very similar to an amateur evaluating "good" art — one knows what *it* is only after one experiences it. Thus, it is possible to make meaningful ESCA analyses of some samples under pressures that are as much as 10^{+3} torr poorer than needed for many other materials. The process is therefore one of balancing compromise with integrity. Every user has discovered, however, that certain high surface area ceramic materials, such as alumina or zeolite powders, are such good adsorbers of O_2 and $H_2O_{(V)}$ that they will "bleed" for hours at $\sim 10^{-7}$ torr against the best pumping system. Therefore, if rapid analysis of these materials is necessary, it may have to be done at a relatively elevated pressure. Generally, despite this "compromise" the latter analysis will yield the same results as those obtained if one waits the required half day and then performs the analyses at $\sim 10^{-9}$ torr.

The same results are not true, of course, for most chemisorption studies. Older analysts will recall the early years of commerical ESCA when vacuums of 1×10^{-7} torr were the best achieved. In these cases, nearly every metal foil or single crystal was so rapidly covered with oxide, and related products, that few studies bore any useful fruit, and a number of incorrect interpretations were published. It is important for the reader to recall that a gas with a sticking coefficient of 1 will adsorb a monolayer in 1 s at 10^{-6} torr! One of the most important and dramatic examples of this problem is outlined in Figure 3-14 where a simple sputter cleaning-ESCA analysis of aluminum in a Hewlett-Packard (HP) ESCA was attempted with the pressure in the analysis chamber held at $\sim 2 \times 10^{-9}$ torr. The sputtering was achieved with Ar^+ at 1×10^{-5} torr. The entire (prep) chamber was filled with this pressure of Ar during sputtering, and then the pressure of 2×10^{-9} torr was reestablished, and the Al foil was reinserted into the analysis chamber. The total length of time between sputtering steps and ESCA examination was always less than 30 min. Based on the presputtering examination of Al(2p) one can determine that before sputtering the well-maintained, high purity, $(4 \times 9$'s) Al foil is covered by *the expected* mix of oxide and hydroxide (STP), air-induced oxidation products, e.g., Al_2O_3, AlOOH, and $Al(OH)_3$, to an apparent thickness of ~ 20Å. Based upon the calibrated performance of the sputtering gun, it is anticipated that ~ 4 to 5 min of sputtering will remove all of this "natural passivation" growth. However, repeated sputtering from 2 to 200 min produced essentially the same Al(2p) results — the ESCA detected presence of a moderate increase in Al°, and a substantial presence of oxidized Al. All things considered (repeated checking of gun performance, etc.), it is apparent that the sputter etching is succeeding in removing all (or at least nearly all) of the oxidized Al, but during the return of the foil from prep to analysis chamber at 2×10^{-9} torr the extremely reactive Al° is essentially "gettering" oxygen out of the vacuum chamber making it impossible, under these conditions, to realize an oxide-free surface of Al°. In fact, detailed studies, by others,[27] of aluminum oxidation have established that pressures in the range of 1×10^{-10} torr are needed. This experience is not atypical for many reactive surfaces, and it merely reinforces the belief that if one wishes to investigate pristine surface processes, e.g., submonolayer chemisorptions on single crystals, it is advisable to develop a system in which background pressures in the 10^{-11} torr range are achievable!

5. Ghost Peaks and Mystery Shifts

It should be noted that almost all ESCA systems are subject on rare occasions to the generation of mysterious peaks that appear at some unusual, but regularly spaced, intervals from the "expected" line positions. These peaks are often intriguing, and occasionally may be beguiling enough to fool the unwary analyst into a lengthy and very disheartening investigation. Generally, these mystery peaks come under the collective heading of "ghost peaks or mystery shifts" and, although they may be caused by a variety of problems, two culprits deserve particular mention.

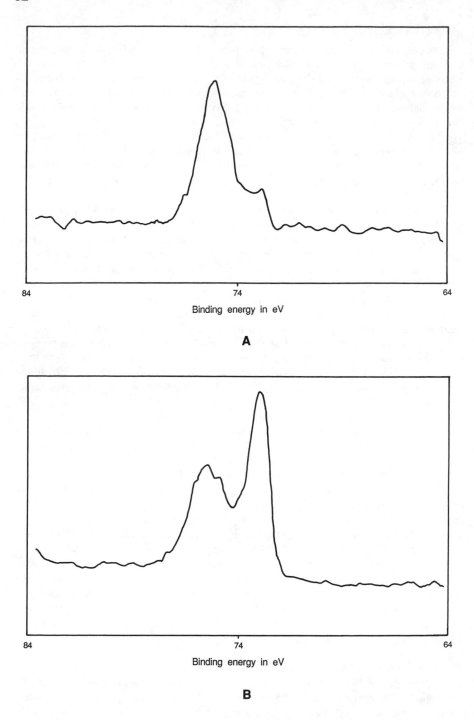

Figure 3-14 XPS study of Al_2O_3/Al interface demonstrating that a vacuum in the 10^{-9} torr range is not sufficient to clean and maintain an oxide-free surface of Al. Note the Al_2O_3 structure in the pre- and post-sputtered systems.

All ESCA systems are subject to the regular interjections of very precise offset voltages and fields that are designed to do such things as establish the binding energy scale and convert kinetic energy to binding energy, etc. These processes require precise performance of various power supplies and intricate interconnection between spectrometer components and the operating computer. At times, some of these components may malfunction. Usually when they do so, part of the total unit completely fails and the scaling processes are shifted by the amount produced by the component when functioning properly. Well-schooled analysts and repair technicians, therefore, can often use the size of the improper shift to assist in locating the malfunctioning component.

Perhaps the most common source of ghost peaks is the presence of a significant impurity on the X-ray anode (or the isolation window in the case of a monochromator). Thus, during dual anode production it is relatively easy to get a little Mg in the Al, or vice versa. The ready alloying of these elements makes it difficult for common plators and depositors to insure the integrity of these targets, even when not generating a dual anode. Almost all anode systems are produced on copper blocks that are employed as heat sinks and through which water is cycled to cool the anode. The amount of Al, and/or Mg, employed is generally quite small, and these thin layers are constantly being eroded during X-ray generation. As a result, it is not uncommon to occasionally expose some Cu to the cathode (electron) bombardment. This means that $Cu_{L\alpha}$ X-rays may be imparted to the sample.

In both of the aforementioned cases the ghost peaks should occur at regular spacings, offset from the "standard" peaks by amounts that correspond to the differences in energies of these various X-ray lines. (Remember that your ESCA system is computer scaled to subtract the photon energy of the prescribed X-ray line.) It should be noted that, although these ghost peaks are "valid" photoelectron peaks (generated from inappropriate anode sources), they are generally not exploitable results except in trouble shooting. The sporatic sites and/or morphological irregularities involved in producing these "ghosts" make them poor candidates for high-resolution analysis.

It is also true that anodes may be subject to deposition and/or reaction with many of the other common contaminants evolved in the ESCA system. Thus, micro-X-ray ejections from oxides, sulfides, carbon species, and many other illicit sources may occur. These are generally not major problems, but may, over time, make contributions to any eroding performance. This is particularly the case for direct sources, conventional anodes that are immersed in the analysis chamber (although these anodes are now generally also shielded with a metal foil!). The very sensitive monochromator anodes are usually isolated in their own chamber away from the contaminants of the analysis region.

6. Other Problems

Several problems are less common, but when they occur they may be difficult to identify. These include:

- Saturation — It is possible with some ESCA systems, when examining clean specimens with very high photoelectron cross sections, to produce a substantial amount of photoelectrons at a single spot in energy space that are dumped at an extremely rapid rate into a very small range in the electron detecting system. One can visualize this as occurring when a particular register in a computer is being filled and refilled faster than it is able to drain back to zero response. When this occurs certain types of detectors may become overloaded (saturation). As a result they may stop counting. Since this effect is most prone at the centroid of any photoelectron peak, the net results will be the fictitious appearance of two pseudo peaks representing the weaker walls of the Gaussian distribution. The resulting double peak structure looks very intriguing, but is, of course, a fictitious construct. The author has produced this effect while examining a clean copper foil with a HP-590A ESCA employing its original phosphor plus Vidicon T.V. camera. Later implementation of a solid-state ceramic two-dimensional detector in the same ESCA system with the same Cu foil did not produce any saturation. Part of the problem with the first system was the tendency of the phosphor to amplify the situation by generating a large number of electrons from each electron impact. It is anticipated that this saturation problem should generally not occur for the newer vintage ESCAs.

• Anode Plating or Deposition Problems — In many instances, either economics or suspension of manufacture makes it necessary for the users laboratory to enter the anode refurbishing business. Often this process is the first of its kind to occur under the user's direction. Because fresh anodes are seldom investigated in detail, many users are unaware of the best procedures of production, and the care that must be employed in this process. We have already alluded to the important contaminant problem and this should be coupled with the necessity for very tight controls over the integrity of the plating bath and/ or target species. It is also vital that the rate and energetics of the depositing film be thoroughly monitored and understood. The interfacial integrity, bulk mechanical properties, and surface morphology may all conspire to "distinguish" a successful replacement anode from a failure. Unfortunately, unless one has the services of an experienced expert, only trial and error will provide the proper guidance.

REFERENCES

1. **Briggs, D. and Seah, M. P., Eds.,** *Practical Surface Analysis,* 2nd ed., John Wiley & Sons, Chichester, England, 1990.
2. **Gelius, U. and Siegbahn, K.,** *Faraday Discuss. Chem. Soc.,* 54, 257, 1972.
3. **Frost, D. C., Ishitani, A., and McDowell, C. A.,** *Mol. Phys.,* 24, 861, 1972.
4. **Wheeler, D. R. and Pepper, S. V.,** *J. Vac. Sci. Technol.,* 20, 226, 1982; **Chaney, R. and Barth, G.,** *Fresenius Z. Anal. Chem.,* 329, 143, 1987.
5. **Gelius, U., Basilier, E., Svensson, S., Bergmann, T., and Siegbahn, K.,** *J. Electron Spectrosc. Rel. Phenomena,* 2, 405, 1974.
6. **Kelly, M. A. and Tyler, C. E.,** *Hewlett-Packard J.,* 24, 2, 1972.
7. **Kelly, M. A.,** More-powerful ESCA makes solving surface problems a lot easier, *Ind. Res. Dev.,* 26, 80, 1984.
8. **Tonner, B.,** private communication.
9. **Barr, T. L.,** Applications of electron spectroscopy to heterogeneous catalysis, in *Practical Surface Analysis,* 2nd ed., Briggs, D. and Seah, M. P., Eds., John Wiley & Sons, Chichester, England, 1990, chap. 8.
10. **Gelius, U., Asplund, L., Basilier, E., Hedman, S., Helenelund, K., and Siegbahn, K.,** *Nucl. Instrum. Methods Phys. Res. B,* 1, 85, 1984.
11. **Gelius, U., Wannberg, B., Baltzer, P., Fellner-Feldegg, H., Carlsson, G., Johansson, C.-G., Larsson, J., Münger, P., and Vegerfors, G.,** *J. Electron Spectr. Rel. Phenomena,* 52, 747, 1990.
12. **Powell, C. J., Erickson, N. E., and Madey, T. E.,** *J. Electron Spectr. Rel. Phenomena,* 17, 361, 1979.
13. **Siegbahn, K.,** *Royal Soc. London,* 1985.
14. **Barr, T. L., Yin, M. P., and Varma, S.,** *J. Vac. Sci. Technol. A,* 10, 2383, 1992.
15. **Barr, T. L.,** Recent advances of ESCA in corrosion research: natural passivation and its consequences, *Corrosion 90,* NACE, Houston, 296, 1, 1990.
16. **Barr, T. L.,** *Appl. Surface Sci.,* 15, 1, 1983.

17. **Kowalczyk, S. P., Ley, L., McFeeley, F. R., and Shirley, D. A.,** *J. Chem. Phys.,* 61, 2850, 1974.
18. **Ley, L.,** *Photoemissions in Solids,* Cardoma, M. and Ley, L., Eds., Springer-Verlag, New York, 1978, chap. 1.
19. **McFeeley, F. R., Kowalczyk, S. P., Ley, L., Cavell, R. G., Pollack, R. A., and Shirley, D. A.,** *Phys. Rev. B,* 9, 5268, 1974.
20. **Barr, T. L. and Yin, M. P.,** *J. Vac. Sci. Technol. A,* 10, 2788, 1992.
21. **Wagner, C. D., Riggs, W. N., Davis, L. E., Moulder, J. G., and Mullenberg, G. E.,** *Handbook of X-ray Photoelectron Spectroscopy,* Perkin-Elmer Corp., Eden Prairie, MN, 1979.
22. **Fowler, D. E., Brundle, C. R., Larczak, L., and Holtzberg, F.,** *J. Electron Spectroc. Rel. Phenomena,* 52, 323, 1990.
23. **Frost, D. C., Ishitani, A., and McDowell, C. A.,** *Mol. Phys.,* 24, 861, 1972.
24. **Barr, T. L., Fries, C. G., Cariati, F., Bart, J. C. J., and Giordano, N.,** *J. Chem. Soc. Dalton Trans.,* 1825, 1983.
25. **Barr, T. L. and Lishka, M. A.,** *J. Am. Chem. Soc.,* 108, 3178, 1986.
26. **Briggs, D.,** *Practical Surface Analysis,* 2nd ed., Briggs, D. and Seah, M. P., Eds., John Wiley & Sons, Chichester, England, 1990, chap. 9.
27. **Batra, I. and Kleinman, L.,** *J. Electron Spectrosc. Rel. Phenomena,* 33, 175, 1984.

Chapter 4

Quantum Mechanical Arguments

I. INTRODUCTION TO THE TOTAL (PRIMARILY ELASTIC) PROCESS

The quantum mechanics of the photoelectron process has been considered previously in detail by various researchers, particularly Feibelman and Eastman,[1] Cederbaum,[2] Mahan,[3] Pendry,[4] and others.[5] Because of our multifacited goal, we present herein only an outline of those arguments, paying particular attention only to those parts that deal specifically with several relatively new processes that are featured in this book (Chapters 6 through 9).

A number of years ago, Berglund and Spicer[6] developed a useful model of the photoelectron phenomenon as a three-staged scattering process. In this book we follow the lead of Sunjic,[7] who has presented a particularly succinct version of this model. The three steps include:

1. In the first step we consider the electronic state of the solid material of interest to be described by a many-electron state function, $|\Psi_I\rangle$, which we hope to adequately model as a properly symmetrized set of product functions of one-electron orbitals. Thus, this system is often described as an *initial state* in which correlation is being ignored.[8] The incoming (X-ray) photon that is nearing the surface of the solid may be described by a vector potential $\vec{A}(\vec{r}, t)$. As pointed out by Sunjic, during stage 1 this potential is already being altered from its free form as it begins to interact with the solid system.

2. As the photon field enters the solid system, it interacts with the current density, $\vec{J}$, of the material, evolving a linear coupling and scattering a particular electron, i, of momentum $\vec{P}_i$. In the linear coupling form, the photon-material scattering potential assumes a symmetric form:

$$V = -\frac{e}{2mc}\left(\vec{A}\cdot\vec{P} + \vec{P}\cdot\vec{A}\right) \qquad (4\text{-}1)$$

where, as pointed out by Martin and Shirley[8] and Saguiron et al.,[9] if we consider the field of the vector potential only through the dipole approximation, then we need consider only the polarization part of $\vec{A}$, labeled here as $\vec{\varepsilon}$. In this regard, Feynman[10] and others[11] have shown that to the level of the dipole approximation

$$\frac{\vec{P}}{m} \propto v\vec{r} \qquad (4\text{-}2)$$

When considering the radiation as a plane wave,

$$\vec{A}(\vec{r}, t) = \left(\frac{8\hbar c^2}{v}\right)^{1/2} \vec{\varepsilon}\cos\left(vt - \vec{k}\cdot\vec{r}\right) \qquad (4\text{-}3)$$

where v is the frequency of radiation spreading into the material with wave number $\vec{k}$ at $\vec{r}$. Then, in the dipole approximation

$$V \propto \vec{A} \cdot \vec{P} \qquad (4\text{-}4)$$

and

$$\vec{A} \cdot \vec{P} \propto \vec{\varepsilon} \cdot \vec{r} \qquad (4\text{-}5)$$

Thus, the quantum development can be represented at this point in terms of either a time-dependent or time-independent, many-body scattering relation. This has been described by Liebsch[12] and reiterated by Saguiron et al.,[9] i.e.,

$$\mid \Psi_F > \,= \int d^3 r_1 \ldots d^3 r_N G(\vec{R}, \vec{r})(V) \mid \Psi_I > \qquad (4\text{-}6)$$

where G is the exact many-body Green's function that motivates the "evolution" of $\mid \Psi_I >$, the many-body, initial state, to $\mid \Psi_F >$, the corresponding many-body, final state, as a result of the involvement of the potential V.

This problem is obviously beyond exact solution, but it may be substantially simplified by evoking the previously mentioned dipole approximation and the restriction mentioned in step 1 above to approximate the various exact $\mid \Psi >$ with properly symmetrized products of one-body *orbitals*, realized, at most, up to the Hartree-Fock (HF) limit.

3. With these restrictions the completion of the photoelectron process may be expressed as a V-perturbed one-electron process evolving against a *possible* background of final-state interacting (excited) electron-hole pairs[7] (i.e., for the scattering cross section, σ_{dipole}),

$$\text{I}$$

$$\sigma_{dipole} \propto \frac{1}{h\nu} \mid < \psi_f(\vec{r}, \vec{k}) \mid \vec{\varepsilon} \cdot \vec{r} \mid \phi_i(\vec{k}) > \mid^2$$

$$\text{II}$$

$$\mid < \Psi_F(N-1) \mid \Psi_I(N-1) > \mid^2 \qquad (4\text{-}7)$$

$$\text{III}$$

$$\delta(\varepsilon - \varepsilon_{h\nu} - \varepsilon_{fs})$$

where ϕ_i is a valid one-electron (perhaps HF) orbital from which the electron is to be scattered (i.e., in this model ϕ_i is an orbital part of $\Psi_I(N)$ and $\psi_f(r,k)$ is the outgoing photoelectron state of the scattered electrons). The term I is, thus, the rendition of the photon-induced electron scattering into the photoelectron state, described to the level of the dipole approximation. The outgoing nature of this situation depends upon the appropriate choice of the one electron state, ψ_f. If the photoelectron proceeds to the detector without further interaction, it is reasonable to approximate ψ_f as a simple plane wave; however, if some form of intermediate (even elastic) scattering occurs between emission and detection, a more complex form may be necessary. Thus, as shown in detail by Saguiron et al.,[9] if one considers the (ready prospect) of the photoelectron emitted by atom 1 being diffracted by its neighbors, it may be necessary to consider ψ_f as, at least, a spherical wave. In addition, as we shall describe in the section on photoelectron diffraction, the question is still being considered as to whether this scattering may be adequately treated inside a single-scattering cluster (SSC) model[9,13] or whether the arguments of multiple scattering are needed.[14]

Outside of these elastic events the most important secondary situations that can occur as part of stage 3 of the photoelectron process are the secondary inelastic scatterings that result in relaxations and satellite formation. The prospects for these events are carried in the scalar product in part II of Equation 4-7 and their energetic development (or lack thereof) will be reflected in the number of terms that appear in the delta function of part III of Equation 4-7.

II. INELASTIC SCATTERING PROCESSES

A. RELAXATIONS

In this part of the process, we are trying to connect the *final (deep hole) state* [characterized by $E_F(N-1)$ in Equation 2-2] back to the *initial state* [characterized by $E_I(N)$ in Equation 2-2]. In order to attack this problem, we must introduce some additional quantum mechanical considerations. As suggested above, the appropriate many-body states may be reasonably expressed in terms of properly symmetrized products of one-electron orbitals. In these cases it is constructive to consider the behavior of the $N-1$ balance of occupied orbitals during and after the photoelectron ejection. As noted above in Chapter 2, Section II, it may be reasonable to consider that these $N-1$ electron orbital states are unaffected by the photoelectron ejection. When the Ψ states are evolved to the HF level, this lack of change is referred to as the frozen orbit approximation.[15] (Although more appropriate for molecular than for solid-state systems, the loss arguments realized in the following discussions generally apply for either type of system.)

1. Frozen Orbit Approximation

We first consider the solution to the initial state, N-electron Schrödinger equation:

$$H(N)\Psi_o^A(N) = E_o^A(N)\Psi_o^A(N) \tag{4-8}$$

As an approximation we will consider the following approximate Hamiltonian:

$$H^o(N) = \sum_{i=1}^{N} F_i \tag{4-9}$$

where F_i is the ith Fock Hamiltonian, i.e.,

$$\langle \phi_m | F_i | \phi_m \rangle = \langle \phi_m | h_o(i) | \phi_m \rangle$$
$$+ \sum_p \left\{ \langle \phi_m \phi_p | v_{ij} | \phi_m \phi_p \rangle - \frac{1}{2} \langle \phi_p \phi_m | v_{ij} | \phi_m \phi_p \rangle \right\} \tag{4-10}$$

and where

$$h_o(i) = -\left(\frac{1}{2}\nabla_i^2 + \frac{Ze^2}{r_i}\right)$$

(4-11)

and

$$v_{ij} = \frac{e^2}{r_{ij}}$$

(4-12)

This implies (1) that a possible (approximate) wave function, Φ_o, for Equation 4-8 may be constructed from Slater determinants of the orbitals in a single particle (one-electron, independent particle approximation) set, $\{\phi_n\}$ and (2) that each of these ϕ_n's are the ground state, HF orbitals for our A system of N electrons, i.e., that they are the "best" basis set of one electron orbitals in terms of their ability to describe (minimize) the energetics of the Ath system. Any improvement upon this Φ_o solution for the A system would require the introduction of the previously mentioned correlation (i.e., the explicit introduction of electron-electron interaction) in place of the average self-consistent field generated in the HF approximation.[15] The restriction to the HF limit is thus one that is often imposed in ESCA to avoid these complications and, in turn, allows the retention of the ability to say that an ejected photoelectron comes from a specific orbital, e.g., $Cu(2p_{3/2})$.

Therefore, in this approximation,

$$\Psi_o \rightarrow A\{\phi_1(1)\ldots\phi_n(N)\} \equiv \Phi_o$$

(4-13)

(i.e., a single Slater determinant for a closed shell system) where ϕ is an HF (spin) orbital and A is the appropriate (anti) symmetrizer and normalization function. Note that the subfix o assumes a ground state system.

Returning now to Equation 2-2, we see that we have described (up to the HF limit) a solution to the second energy term on the right side of the equation, i.e.,

$$E_{I_A}(N) = \sum_{i-1}^{N} \varepsilon_i$$

(4-14)

where the ε_i's are obtained from

$$F_i\phi_i = \varepsilon_i\phi_i$$

(4-15)

Thus, ε_i is the HF energy for the ith state described by the HF orbital ϕ_i. Working through the details of this *independent particle approximation* one finds that, when the N-body wave function is a properly symmetrized *product* of the N single-electron orbitals, the N-body energy, E_I, is a *sum* of these individual orbital energies (i.e., Equation 4-14).

If we further suppose that the only thing accomplished by the interjection of the (X-ray) photon is the removal of the electron from, for example, the kth orbital (i.e., nothing else happens to the system A), then

$$H^\circ(N-1)\Phi_o^{-k}(N-1) = E_F^{-k}(N-1)\Phi_o^{-k}(N-1) \qquad (4\text{-}16)$$

where, $\Phi_o^{-k}(N-1)$ differs from $\Phi_o(N)$ only through the "disappearance" of orbital k. Therefore, if this "restriction" is logical, then

$$\Phi_o^{-k}(N-1) = A\{\phi_1(1)\ldots\phi_{k-1}(k-1)\phi_{k+1}(k+1)\ldots\phi_N(N)\} \qquad (4\text{-}17)$$

and

$$E_F^{KT}(N-1) = \sum_{i=1}^{N}\varepsilon_i - \varepsilon_k \qquad (4\text{-}18)$$

and therefore from Equations 2-3 and 4-14 to this level of solution

$$E_A^m = -\varepsilon_k \qquad (4\text{-}19)$$

Thus, inside the bounds of this approximation, *the measured (m) photoelectron energy is equatable directly to the original (pre-photoelectron experiment) orbital energy and its (kth electron) removal by the XPS experiment merely provides the means to measure that (initial state) property.* Note that this "approximation" presupposes that the remaining $N-1$ orbitals will (1) either not change at all during the XPS experiment or (2) that any change they do experience will be "well after the fact" and thus not "perceived by" the photoelectron. This supposition is often called the *frozen orbital approximation.* It was originally derived (theoretically) by Koopmans and is known to theoreticians as Koopmans' theorem: thus, the superfix "KT" in Equation 4-18.[16]

If this approximation is reasonably valid, we may then consider a second molecular unit (B) that also contains the same element and easily describe the (Siegbahn) chemical shift, ΔE^m, as the difference in orbital energy for these two chemical states (Equation 2-4). Thus, inside this approximation the measured difference between binding energies for two (or more) units is a *true chemical shift,* i.e., inherently a measure of *only* the (initial state) chemical change between these two systems.

Unfortunately, the Koopmans' theorem, or frozen orbital approximation is just that — an approximation — *and* it only applies (with quantitative precision) to a few systems. Generally, there are substantial differences between the results realized on the left side of Equations 2-3 and 2-4 and those calculated employing Koopmans' theorem. Also, there are numerous, often significant, secondary peaks or relaxations realized in the experimental spectra. None of these facts can be described inside the limits of Koopmans' theorem. This necessitates the documentation of final state effects that may lead to relaxations and secondary peaks, and thus remove some of the "chemical" from the (Siegbahn) chemical shift. It has been noted that the relaxation effects to be described are those for a particular orbital of a particular atom in two different chemical states. Therefore, some have suggested that these relaxation effects may often cancel when one considers the chemical shift subtraction (Equation 2-4). In fact, sometimes substantial parts of this relaxation do cancel. However, as demonstrated by Davis and Shirley, the relaxation to be described has both an intraatomic and extraatomic part.[17] The latter, which depends primarily upon the interaction of a particular atom with its neighbors, will often vary substantially when those neighbors are changed, making the relaxation effect quite different for states 1 and 2 in a $1 \rightarrow 2$ chemical change.

2. Modification of Koopmans' Theorem for Slow Electrons

If the electrons that remain in the $N - 1$ electron system are "allowed" to relax following ejection of a photoelectron (from ϕ_k), then they may all evolve in time into a new set of orbitals, such that:

$$\Phi_F(N-1) = A\left\{\phi_1'(1)...\phi_{m-1}'(m-1)\phi_{m+1}'(m+1)...\phi_N'(N)\right\} \qquad (4\text{-}20)$$

where the orbitals $\{\phi_i'\}$ differ from the initial state set $\{\phi_i\}$ in that the former have "relaxed" (following photoelectron ejection) in an attempt to compensate for the presence of the core hole. This problem may be attacked directly, by trying to solve $H_F(N - 1)\Psi_F = E_F\Psi_F$. However, even SCF calculations are difficult for large hole-ion systems and, therefore, other forms of approximate treatments are often employed. One way to model this process[18] is by assuming a polarization of each (retained) ϕ_i', i.e.,

$$\phi_i' = \phi + \sigma x \qquad (4\text{-}21)$$

where X is a polarization correction to ϕ as a result of the long-time retention of a hole in the vicinity of state i and δ is the appropriate weighting. Now, it should also be noted that, since our SCF H $(\sum_i F_i)$ contains the orbitals explicitly, the (eigen) Hamiltonian for the best, one-body solution to the final state also must be changed to recognize this alteration, i.e.,

$$H_F^O(N-1) \cong \sum_i^{N-1} F_i' \qquad (4\text{-}22)$$

where

$$\langle \phi | F_i' | \phi \rangle = \langle \phi | h_o(i) | \phi \rangle$$
$$+ \sum \left\{ \langle \phi\phi_p' | v_{ij} | \phi\phi_p' \rangle - 1/2 \langle \phi\phi_p' | v_{ij} | \phi_p'\phi \rangle \right\} \qquad (4\text{-}23)$$

and

$$F_i'\phi_1' = \varepsilon_i'\phi_i' \qquad (4\text{-}24)$$

This approximation is often referred to as "adiabatic"[15,18] to designate that the $N - 1$ electron balance of the system will, following photoelectron ejection, slowly (in a relative sense) drift into the previously defined relaxed states without any discontinuous change in the Hamiltonian. Thus, comparing any point in time j with its immediate predecessor $j - 1$, we find that $E(j) = E(j - 1)$.

In this case it has been shown by Hedin and Johansson[18] that the new final states $\{\phi_i'\}$ will shift the actual measured binding energy of the outgoing photoelectron by an amount that often may be significantly different from that predicted by the frozen orbital approximation, i.e.,

$$E_{A(j)}^m(1) \cong +\varepsilon_{A(j)}^{KT} - E_{A(j)}^R \qquad (4\text{-}25)$$

where $E_{A(j)}^R$ is defined as the *relaxation energy* experienced by the jth orbital of the Ath atomic system.

Imposing the predefined HF limitation, and further truncating the resulting solution at a convenient point, Hedin and Johansson[18] have shown that one may express E^R as

$$E^R_{A(j)} = \frac{1}{2} < \phi^A_j \mid V_p \mid \phi^A_j >$$

(4-26)

where ϕ^A_j is the (pre-photoelectron) HF solution for the jth orbital in the A system and V_p is defined as a polarization potential, given by

$$V_p = \sum_i \left(J_{ij} - \frac{1}{2} K_{ij} \right)$$

(4-27)

where J_{ij} is the coulombic term, i.e.,

$$J_{ij}\phi_j(t) = \int \frac{\phi'_i(s)\phi'_i(s)\phi_j(t)}{r_{ts}} d\tau_s$$

(4-28)

and K_{ij} is the corresponding exchange integral, i.e.,

$$K_{ij}\phi_j(t) = \int \frac{\phi'_i(s)\phi_j(s)\phi'_i(t)}{r_{ts}} d\tau_s$$

(4-29)

where t and s label any two electrons retained in the system.

As mentioned above, this description provides a reasonably accurate rendition for the final state behavior of a system that suffers a (relatively) slow ejection of a photoelectron.[18] Many observed photoelectron peaks fall into this classification. They are generally seen as singular (principle only) discrete lines in the resulting spectrum. Thus, a reasonable theoretical description of their realized E^m_b may be achieved with Equations 4-25 and 4-26.

Unfortunately, however, many photoelectrons do not exit the material system in this slow fashion. Rather they often rapidly "jump" out to the detector (in a relative sense) before this adiabatic relaxation has a chance to occur. Thus, these "fast electrons" will experience a dramatic or sudden upheaval as they evolve into the quantum world of the N − 1 electron system. This type of phenomenon is best represented in the "sudden approximation."[15,19]

B. SATELLITES — THE LOSS PROCESS
1. The Sudden Approximation

Any type of sudden (fast) electron ejection may cause a discontinuous change in the N − 1 electron Hamiltonian, i.e.,

$$H(N) \rightarrow \left[H_F(N-1) + h(1) \right]$$

(4-30)

Note that, as required, the "total" N electron Hamiltonian is still continuous! However, the change expressed in the N − 1 part of the Hamiltonian [from $H_I(N-1)$ to $H_F(N-1)$]

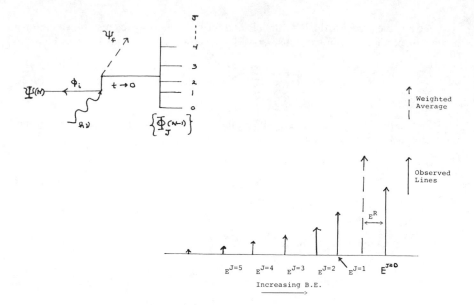

Figure 4-1 Schematic representation of the (intrinsic) shake up process.

is discontinuous.[19] At the same time, of course, the wave function for the N–1 electron set must experience a continuous change from the eigenspace of $H_I(N-1)$, $\{\Phi_I(N-1)\}$, to the eigenspace of $H_F(N-1)$, i.e., $\{\Phi^J_F(N-1)\}^\infty_{j=0}$, where the infinity symbol indicates both discrete and continuous states, and where now not only does the J=0 state contribute (as in the adiabatic transition), but the entire manifold of eigenstates of $H_F(N-1)$ contribute.[20] Therefore (assuming, for convenience, only discrete states), continuity of the wave function demands the transformation (photoelectron ejection) at time, t^α, with $\alpha \to 0$, such that

$$\Psi_I(N-1) = \sum_{J=0}^{\infty} C_J \Phi^J_F(N-1) \tag{4-31}$$

where

$$H_F(N-1)\Phi^J_F(N-1) = E^J_F(N-1)\Phi^J_F(N-1) \tag{4-32}$$

As noted, all of the N – 1 electron hole ion states contribute to Equation 4-31 and 4-32, and

$$C_J = <\Psi_I(N-1) \mid \Phi^J_F(N-1)> \tag{4-33}$$

Thus, the system, beginning in state $\Psi_o(N)$, before completion of the interaction with $\vec{A}$, is extensively and rapidly disturbed by the "speedy" ejection of electron k (from orbital ϕ_k) in a manner that "shakes" the system[21] into the manifold of eigenstates of $H_F(N-1)$ as depicted in Figure 4-1.

Further manipulations lead to a version of the Manne-Aberg sum rules:[21]

$$E_{F_A}^{KT}(N-1) = \sum_{j=0}^{\infty} |C_J|^2 E_F^J(N-1) \tag{4-34}$$

and

$$\varepsilon_{k_A}^{KT} = \sum_{J=0}^{\infty} |C_J|^2 \varepsilon'_{J_A} \tag{4-35}$$

where C^2_J is the probability weight for each "shake up state" and ε'_J is the orbital energy in the *relaxed*, final state that "announces" the "transition" into the $N - 1$ electron manifold[21] (i.e., for a hypothetical spectrum with loss lines as shown in Figure 4-1.)

Thus, this shake up-type process produces a multiple (weighted) transition to the relaxed final states, the eigenstates of $H_F(N - 1)$.[21] The weightings are such that the manifold sum realizes the singular, unrelaxed, initial state solution $E^{KT}_{F_A}(N - 1)$. These transitions are *intrinsically (hole)* induced.[22] Spectroscopically, we may prefer to describe all of this in terms of a properly weighted (hole ion) spectral function, $N_+(\omega)$, where ω is the transition frequency:[23]

$$N_+(\omega) = \sum_{I=0}^{\infty} |< \Phi_o^I(N-1) | \Phi_J^F(N-1) >|^2 \delta(\Delta E) \tag{4-36}$$

where

$$\delta(\Delta E) = \delta(E - E_J) \tag{4-37}$$

As mentioned above and described below, two "extremes" of this physical problem have previously been examined in detail both theoretically and experimentally.[23]

2. The Two Extreme Cases
a. *Molecular (Chemical) Solids*

In these cases, the emphasis is on the word "molecular" with any solid-state involvement considered to contain regions of interest describable by molecular (or atomic) orbitals. The emphasis is on local bond effects rather than delocalized symmetry and k space. These cases are thus characterized by well-described orbital to orbital-type transitions that collectively may evolve toward band to band. The resulting discrete peaks* are sometimes referred to as Manne-Aberg shake-up effects[23] to differentiate them from those described below. In some cases these transitions are of the charge transfer type, as apparently occurs for CuO^{24} (see, for example, Figure 4-2). We will expand upon select aspects of these arguments in Chapter 6 on loss spectroscopy.

---

* Note that, in addition to the discrete (shake-up) peaks, a continuum of shake-off peaks may result (see above).

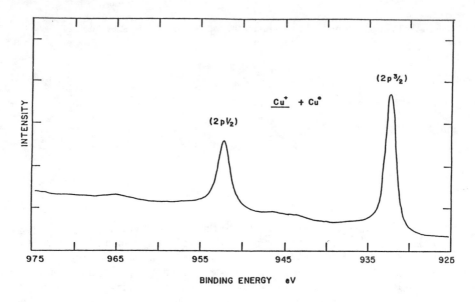

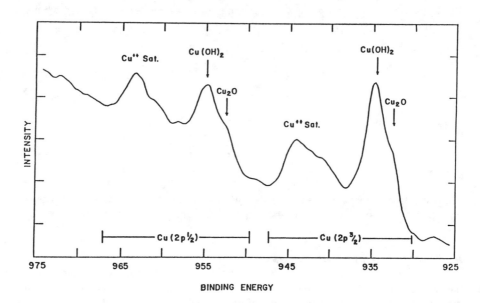

Figure 4-2 Representative (Manne-Aberg) shake-up demonstrating its use in chemical analysis: Cu°, Cu(I), and Cu(II). Only Cu(II) produces distinct shake-up.

b. Metals, with Zero Band Gap

As also described above, if the sudden ejection of the photoelectron occurs for certain metals (band gap energy $E_g = 0$), it has been shown that a form of collective or plasmon-type excitations will dominate the satellite loss scene. The featured properties of this process will be a series of regularly spaced discrete peaks that fall off in intensity in an exponentially decreasing, Poisson array. Aluminum produces an excellent example of this effect, as was first published by the Shirley group[25] (see Figure 4-3). Because of the

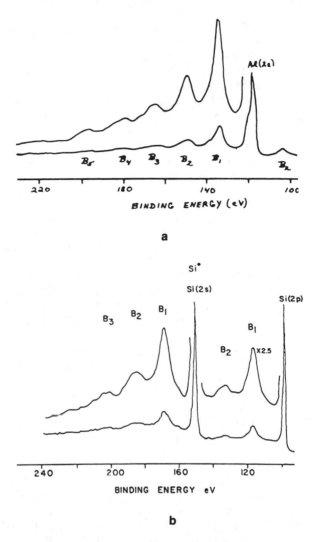

Figure 4-3 Representative (Mahan) plasmon dominated loss cases: (a) Al° and (b) Si°.

historical development of the explanations for these effects, Gadzuk[23] has referred to these collective phenomena as Mahan shake-up features.[26]

Some uncertainty still exists as to the degree of extrinsic (electron-induced) vs. intrinsic (hole-induced) character in these loss spectra (see below). Our principle concern, however, will be to consider what happens to these loss features when the ingredients for the Manne-Aberg and Mahan forms are mixed. This occurs for many semiconductors, and even some insulators. This feature is discussed in more detail in Chapter 6 on loss spectroscopy.

REFERENCES

1. **Feibelman, P. J. and Eastman, D. E.,** *Phys. Rev. B,* 10, 4932, 1974.
2. **Cederbaum, L. S.,** Non-single-particle excitations in finite Fermi systems, *J. Chem. Phys.,* 62, 2160, 1975.
3. **Mahan, G. D.,** Theory of photoemission in simple metals, *Phys. Rev. B,* 2, 4334, 1970.

4. **Pendry, J. B.,** Theory of photoemission, *Surface Sci.,* 57, 679, 1976.
5. **Sunjic, M. and Sokcevic, D.,** Inelastic effects in X-ray photoelectron spectroscopy, *J. Electron Spectrosc. Rel. Phenomena,* 5, 963, 1974; **Gadzuk, J. W. and Sunjic, M.,** Excitation energy dependence of core-level X-ray photoemission spectra line shapes in metals, *Phys. Rev. B,* 12, 524, 1975.
6. **Berglund, C. N. and Spicer, W. E.,** *Phys. Rev. A,* 136, 1040, 1964.
7. **Sunjic, M.,** Theoretical problems in solid state photoemission, *Studies Surface Sci. Catal.,* 9, 112, 1982.
8. **Martin, R. L. and Shirley, D. A.,** *Phys. Rev. A,* 13, 1476, 1976.
9. **Saguiron, M., Bullock, E. L., and Fadley, C. S.,** The analysis of photoelectron diffraction data obtained with fixed geometry and scanned photon energy, *Surface Sci.,* 182, 287, 1987.
10. **Feynman, R. P.,** *Quantum Electrodynamics,* W. A. Benjamin, New York, 1961, chap. 1.
11. **Bjorken, J. D. and Drell, S. D.,** *Relativistic Quantum Mechanics,* McGraw-Hill, New York, 1964, chap. 7.
12. **Liebsch, A.,** *Phys. Rev. Lett.,* 32, 1203, 1974; **Liebsch, A.,** *Phys. Rev. B,* 13, 544, 1976.
13. **Fadley, C. S.,** The study of surface structures by photoelectron diffraction and Auger electron diffraction, in *Synchrotron Radiation Research: Advances in Surface and Interface Science,* Bachrach, R. Z., Ed., Plenum Press, New York, 1993, in press.
14. **Poon, H. C. and Tong, S. Y.,** *Phys. Rev. B,* 30, 6211, 1984; **Tong, S. Y., Poon, H. C., and Snider, D. R.,** *Phys. Rev. B,* 32, 2096, 1985.
15. **Szabo, A. and Ostland, N. S.,** *Modern Quantum Chemistry: Introduction to Advanced Electronic Structure Theory,* Macmillan, New York, 1982.
16. **Koopmans, T. A.,** *Physica,* 1, 104, 1933.
17. **Davis, D. W. and Shirley, D. A.,** *J. Electron Spectrosc. Rel. Phenomena,* 3, 137, 1974.
18. **Hedin, L. and Johansson, A.,** Polarization corrections to core levels, *J. Phys. B,* 2, 1336, 1969. For examples of more direct calculations of the relaxed states realized in Equation 38 and their necessity, one should consult **Basch, H. and Snyder, L. C.,** *Chem. Phys. Lett.,* 3, 333, 1969; **Bagus, P. S.,** *Phys. Rev. A,* 139, 619, 1965; **Schwartz, M. E.,** Direct calculation of binding emergies of inner-shell electrons in molecules, *Chem. Phys. Lett.,* 5, 50, 1970; **Schwartz, M. E., Coulson, C. A., and Allen, L. C.,** Dipole moments, atomic charges and carbon inner-shell binding energies of the fluorinated molecules, *J. Am. Chem. Soc.,* 92, 447, 1970; **Brundle, C. R., Robin, M., and Basch, H.** *J. Chem. Phys.* 53, 2196, 1970.
19. **Schiff, L. I.,** *Quantum Mechanics,* 2nd ed., McGraw-Hill, New York, 1955, chap. 31.
20. **Aberg, T.,** *Phys. Rev.,* 156, 35, 1967.
21. **Manne, R. and Aberg, T.,** Koopmans theorem for inner-shell ionization, *Chem. Phys. Lett.,* 7, 282, 1970.
22. **Fadley, C. S.,** Peak intensities and photoelectric cross sections, *Chem. Phys. Lett.,* 25, 225, 1974.
23. **Gadzuk, J. W.,** Many-body effects in photoemission, in *Photoemission and the Electronic Properties of Surfaces,* Feuerbacker, Filton, and Willis, R., Eds., John Wiley & Sons, New York, 1978, chap. 5.
24. **Kim, K. S. and Davis, R. E.,** *J. Electron Spectrosc. Rel. Phenomena,* 1, 251, 1972; **Kim, K. S. and Winograd, N.,** Charge transfer shakeup satellites in X-ray photoelectron spectra of cations and anion of $SrTiO_3$, TiO_2 and Sc_2O_3, *Chem. Phys. Lett.,* 31, 312, 1975.

25. **Kowalczyk, S. P., Ley, L., McFeeley, F. R., Pollack, R. A., and Shirley, D. A.,** *Phys. Rev. B,* 8, 3583, 1973; **Pollack, R. A., Ley, L., McFeeley, F. R., Kowalczyk, S. P., and Shirley, D. A.,** *J. Electron Spectrosc.,* 3, 381, 1974.
26. **Mahan, G. D.,** Electron energy loss during photoemission, *Phys. Stat. Solid B,* 55, 703, 1973.

Chapter 5

Established ESCA Procedures

I. QUANTIFICATION

Quantification is one of the most important and, in many cases, *the* most important feature of ESCA. Surprisingly, although this capability was recognized in the pioneering work in Uppsala,[1] the development and explanation of the degree and methodology of realizing quantitative information in ESCA is still a very active, necessary area of research.[2] Part of the reason for this is the evolution of the spectroscopy itself. Thus, as ESCA expands to encompass new and often more challenging techniques, the methods and scale of precision for its quantification need to be reexamined and often revised. Another evolving area of concern arises due to the somewhat uncertain nature of the materials being analyzed in ESCA. Thus, the fact that ESCA seems to detect material to a depth, η, that in one sense is considered a surface analysis and in another a subsurface examination makes the exact definition of its quantification very difficult. In particular, we must keep in mind that the predisposition of a factor labeled as an *escape depth*, λ, can be a very beguiling fact. Thus, if it is suggested that the *average* ESCA observation is rendered to a depth of (for example) 30 Å, this generally tends to immediately produce the type of image presented in Figure 5-1a, where we have blocked out a 30-Å cylindrical section of a material with an atomically flat surface and the predetermined diameter of the X-ray beam analysis area (in this case assumed to be circular and of a rather typical 1- to 2-mm diameter). Many casual practitioners are inclined to employ this image immaterial of the exact conditions imposed in the measurement. Unfortunately, this type of image is often far from accurate, and may not even be suggestive of anything with even qualitative merit. A major problem is, of course, that the actual depth of detection, η, of an element, labeled as A, may be expressed as

$$\eta = \lambda_M (E_{AX}) \cos \theta \qquad (5\text{-}1)$$

where M defines the particular matrix containing A, and the appropriate photoelectron is emitted from the X core level with energy E at angle θ with respect to the surface normal. Thus, as mentioned earlier, even for an ideal sample the actual escape depth will vary (sometimes substantially) with the energy of the source, the photoelectron binding energy, and the angle of emission. This means that the depth of photoelectron detection, for a given ESCA system, may vary dramatically from measurement to measurement, and even vary inside the same measurement.

The ready assumption of an "ideal" regularity for the materials in question, such as that depicted in Figure 5-1, may prove to be an even larger problem. Thus, although simple to model, and often suggested by visual, even light microscope analysis, most materials are not nearly as surface regular as depicted in Figure 5-1a. In fact, many have cracks or other displacements as suggested in Figure 5-1b, whereas others are prone to steps and kinks, and still other surfaces are in the form of microscopic boulders and indentures that may assume a variety of shapes, clusterings, and stackings. Some renditions of the these types are displayed in Figure 5-1c. Obviously, the presence, or lack of presence, of these nonuniformities may have a dramatic effect on the ESCA-detected quantification, since

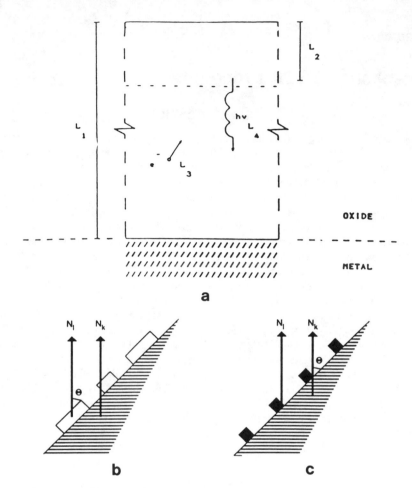

Figure 5-1 (a) Representative renditions of surface situations expected during ESCA analysis of absorbent (e.g., oxide) of thickness L_1 on a metal substrate. X-ray photons to depth L_4 can create photoelectrons as deek as L_3, but only those to depth L_2 are detected. (b) Surface with patch deposition or segregation. (c) Useful rendition of regular overlayers or substrate.

the previously mentioned escape depth often places the major extent of the ESCA analysis on the same scale as the average size of many of these "deformities".

As suggested above, quantification in electron spectroscopy has been the active research area of a number of the best practitioners in the field. Originally, the efforts in ESCA were championed by Wagner,[3] who, among other things, has contributed several painstakingly researched, empirically derived, atomic, sensitivity factor scales.[4] Suggestions as to the proper usage and the extent of applicability of these tables also have been provided by Wagner et al.[5] and Nefedov et al.[6] Subsequent developments in quantification have been led by two senior electron spectroscopists, Seah (primarily for Auger spectroscopy)[7] and Powell[8] (primarily for ESCA). (Both are mentioned here because of the near total interchangeability of many of the results in these two fields.) Because of the existence of several excellent, recent treatments of this subject by both Seah[9] and Mroczkowski[10]), we only outline the details herein, providing more than cursory comments only when there is a direct connection to our other highlighted areas.

The feature measured in an ESCA experiment and directly related to quantification is expressable, as expected, in terms of an electron current, or intensity factor, generally labeled herein as I_{AX}, where both A and X are defined as above, and [9-11]

$$I_{AX} = \sigma_{AX}(h\upsilon)D(E_{AX}) \int_{x}^{y} {}_{\gamma=0}^{\pi} \int_{x}^{y} {}_{\phi=0}^{2\pi}$$

$$\times L_{AX}(\gamma) \int_{x}^{y} {}_{y=-\infty}^{\infty} \int_{x}^{y} {}_{w=-\infty}^{\infty} J_0(wy)T(wy\gamma\phi E_{AX}) \qquad (5\text{-}2)$$

$$\times \int_{x}^{y} {}_{z=0}^{\infty} N_{ax}(wyz)e^{-Z/\lambda_M(E_{AX})\cos\theta}dzdwdyd\phi d\gamma$$

where $\sigma_{AX}(h\upsilon)$ is the quantum mechanical photoelectron cross section for the emission of an electron from the X core level of atom A as a result of the "consumption" of a photon of energy, $h\upsilon$, D is the detection efficiency of the photoelectrons, $L_{AX}(\gamma)$ is the atomic angular asymmetry of the photoelectron intensity, J defines the properties of the X-ray line in the detection plane, T is the transmission function for the analyzer of the spectrometer, N_A is the atomic density, z the detected distance into the sample in question, and λ is defined as before. As pointed out by Seah,[11] Equation 5-2 and/or its corresponding form for AES are far too complex to employ in their entirety. As unwieldy as Equation 5-2 may seem, however, it is still specifically constructed for the so-called homogeneous, semi-infinite slab model and therefore must be extensively modified if the sample is nonuniform.

Calculations of absolute intensities are generally not attempted. Although many of the characteristics of the outgoing photoelectrons may (and generally should) be determined, it is also necessary to accurately and simultaneously gauge the properties of the incoming X-ray flux and the detection efficiency of the spectrometer in order to make these calculations. All of this generally is cumbersome, if not impossible. Shortcutting schemes are readily available, and these have often been employed with reasonable success, but it is in this area that the often-used arbitrary gauge of precision of $\pm10\%$ appears, and even this rather poor capability may be stretching the facts. Generally, however, absolute quantification is not needed. In almost all cases, particularly in practical analysis, the analyst is interested in *relative* knowledge. Thus, questions such as, "What is the silicon to aluminum ratio in that zeolite?" or "How much more potassium is there on the surface of thin film A than B?" are indicative of typical requests. Fortunately, in this form of analysis, one may generally simplify the complex equation relating intensity and concentration, and the relative precision may be substantially improved. In fact, if one is careful in the selection of the materials and peaks involved, it is may be possible to employ the following:

$$\frac{I_{AX}}{I_{BT}} \simeq \frac{\sigma_{AX}N_{AX}}{\sigma_{BT}N_{BT}} \qquad (5\text{-}3)$$

For this relationship to apply, it is necessary that the different photoelectrons AX and BT be emitted under nearly identical circumstances, i.e., from the same place in the same (or, at least, very similar) materials and with kinetic energies such that

$$E_{AX} \sim E_{BT} \qquad (5\text{-}4)$$

When these criteria are not observed, it may be necessary to include in Equation 5-3 all

Table 5-1 **ESCA Detected relative quantification for select zeolites: [Si/Al]**

Material	Representative materials examined [Si/Al]		%Na⁺
	Bulk (reputed)	Surface (determined)	
Na A	1.0	1.1	100
Na Y	2.4	2.6	100
Ca X	1.25	1.3	<5
Ca Y	2.4	2.7	10
NH_4 Y	2.4	2.8	10
Kaolinite	1.0	1.1	0
Bentonite	2.0	2.2	0
α-SiO_2	—	—	0
Silica Gel	—	—	0
α-Al_2O_3	—	—	0
α-Al_2O_3	—	—	0

Note: Each example listed is one of several commercial systems examined. All are reputed to be of excellent purity. Our ESCA survey scans suggest that claim to be correct.

terms that are energy dependent, particularly λ_M. An example that we feel demonstrates a reasonable use of Equation 5-3 was accomplished when we generated the Si/Al ratio for a series of zeolites employing the Si(2p) and Al(2p) peaks, as displayed in Table 5-1. Even in this case, however, a slight compensation was included for the fact that the more energetic Al(2p) are sampled at a slightly greater depth than the Si(2p).

It is possible, of course, to employ the terms determined by Wagner,[4,5] labeled as sensitivity factors, K. Often the latter are determined from measurements of purportedly standard reference materials, perhaps coupled with the σ calculations of Schofield[12] and perhaps the $\lambda_M(E_A)$ values calculated by Penn.[13] The resulting values, which have been presented in a variety of sources,[11] have generally been pruned of many problems and may be considered a useful alternative for semiquantitative results, but one must be very cautious, as various assumptions regarding homogeneity and depth of analysis are being arbitrarily employed.

In many cases, particularly in applied studies, the relative analysis is actually what we refer to as relative/relative. Thus, for example, the A/B ratio for sample 1 may be compared to the A/B for sample 2. Now, if all features are ideal,

$$\frac{\left(I_A/I_B\right)_1}{\left(I_A/I_B\right)_2} \simeq \frac{\left(N_A/N_B\right)_1}{\left(N_A/N_B\right)_2} \tag{5-5}$$

Interestingly, the use of Equation 5-5 may actually relax the requirements of homogeneity and the concept of an infinite slab. If the samples are subject, for example, to inhomogeneities, then it is only important that they be equally so. Consider the results for several zeolites reported in Table 5-2. These results are for relatively smooth, similarly prepared wafers pressed from uniformly sized powders.

In general, however, lack of surface uniformity and various types of inhomogeneities have played a major role in the interpretation of surface analysis. In ESCA, researchers who are interested in catalysis, particularly dopant dispersion with its generally variable sized particulate formation, have extensively studied these questions and proposed a number of models.[14–16] Fung, in particular, has introduced specific factors to take into

Table 5-2 **ESCA Detected relative/relative quantification for zeolites: [Si/Al]ₐ/[Si/Al]ᵦ**

	[NaX]/[NaY] compositional formulae	[NaX]/[NaY] surface (XPS)[a]
Na	1.57	1.47
Si	0.77	0.77
Al	1.57	1.53
O	1.01	1.08

[a] Based upon several measurements of same and different systems, ± 0.5.

account the various shapes that may occur in any undispersed phase.[17] More recently, Fadley et al.[18] have provided fairly detailed descriptions of the modifications expected for Equation 5-2 when overlayers, patches, noncontinuous overlayers, and variable concentration gradients result. Figure 5-1 displays some of these model systems. One must be cautious when applying these models, as they depend upon a certain regularity of variable morphology that may be difficult to produce or even verify. For the correct cases, however, the methods should be very valuable.

Angular resolution studies constitute one of the principal areas for quantification in ESCA. This has been a matter of concern for several research groups, in particular the outstanding work of Fadley et al.[18] The basic angular resolution experiment, such as presented in Figure 5-2, seems to strongly suggest the detection of regular thin film overlayers. Cursory analysis indicates, however, that these experiments are readily susceptible to surface roughness, shading, and even the instrument response function. All of these features may play a major role in the detected N_A when plotted vs. $\lambda_M \cos(\theta)$. We will expand further on this aspect of quantification in the following section on angular resolution ESCA.

II. ANGULAR RESOLUTION-PHOTOELECTRON SPECTROSCOPY (AR-PES)

A. BACKGROUND

It is certainly true that, depending upon one's point of reference, the limited depth generally detected in an ESCA experiment may be either a benefit or a hindrance. In reality there is no single standard depth that results from conventional ESCA analysis, although it is common for ESCA practitioners to refer to depth limits of ~40 ± 15 Å. Even though this range of values is just a reference point, it is not difficult to see why conventional ESCA may be viewed as *too* surface sensitive for some potential uses, whereas it is *not surface sensitive enough* for others.

The X-ray photons generally employed to eject the photoelectrons (typically $Al_{k\alpha}$ or $Mg_{k\alpha}$) are able to penetrate for thousands of angstroms into most of the solids to be analyzed. These photons will create photoelectrons at all accessible depths, but due to the relatively short mean free path of electrons (see Figure 2-4) only those electrons ejected at or near the surface are able to escape materials capture and make their way to the detector.

Close scrutiny of the mechanism of the ESCA process and Figure 5-2 indicates that photoelectrons (and other types of electrons) ejected at substantially different energies may be emanating from quite different depths. This is one reason why one should generally pick photoelectron peaks that are fairly close in binding energy for relative quantitative analysis (see the separate discussion on quantification in Section I). As an example, we chose to ratio the Ge(3d) and O(2s) peaks in a study of possible

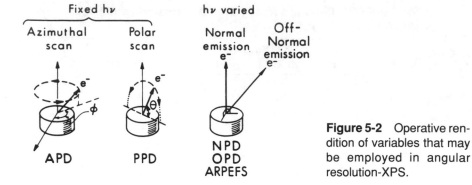

Figure 5-2 Operative rendition of variables that may be employed in angular resolution-XPS.

nonstoichiometries in germania systems, rather than the much more intense (but very widely split) Ge(2p) and O(1s) (see Table 5-3).[19]

B. AR DEVELOPMENT AND PROCEDURE

As with so many aspects of ESCA, angular resolved photoemission (AR-XPS) had its origins in the pioneering studies in the research group of Kai Siegbahn.[1,20] These studies were initiated not long after it was determined that ESCA was a tool that examines the surface and near-surface regions of solids. Simple arguments in electron optics and knowledge of their escape depth suggested that, by rotating the sample, X-ray source, and collection optics in order to obtain near grazing angles, one may achieve a substantially enhanced surface sensitivity. Although simple to argue, the actual experimental procedures needed to achieve this feature without substantial reduction in resolution and sensitivity require adroit manipulation.

A variety of scientists, in particular Fadley,[18,21] diligently examined and expanded the AR technique to include variable angle studies (both polar and azimuthal) at fixed energy and variable energy at fixed angle situations. The procedure proved to yield not only elemental qualitative and chemical analysis with depth, but also to provide reasonable quantification, surface geometry, and valence region information.[18,21] The general geometric description provided by Fadley, and reproduced herein (Figure 5-2), is a useful feature to study. Although the variable energy approach suggests the use of a synchrotron source, one may achieve some useful AR-XPS data with a conventional system employing two or more X-ray anodes, or even by monitoring photoelectron peaks with distinctly different E_b. All of these efforts have succeeded so well that all commercial manufacturers regularly provide AR options on their XPS systems, operable even in the small spot mode.

The angular resolved procedure may, of course, also be exploited in the UPS mode, and numerous useful studies of the presence of alterations in band character, surface states, preferential surface orientations, etc. have been provided in AR-UPS.[22]

The extension of the procedure to exploit the channeling and diffraction of surface emitted photoelectrons from crystals (PED) has also developed into a separate procedure.[21] The latter is described later in some detail in Chapter 9.

C. THE USE OF ANGULAR RESOLUTION

Because of the extreme importance of surfaces and interfaces in the behavior of thin films and in cases of adsorption or surface segregation, the ESCA literature of the last decade is full of descriptions of the use of the conventional AR technique. As a rather basic example, consider Figure 5-3 in which Fadley describes the use of AR in the vicinity of the interface between SiO_2 and $Si°$.[18]

In a somewhat more involved example, our own research group employed AR to examine the O(1s) and Pt(4p) and several of their satellites during the thermal induction

Table 5-3 **Select [O/Ge] values determined employing O(2s) and Ge(3d) peaks for variously treated germanium oxides**

State	Process	[O/Ge][a]
a	Fresh thin film deposition	2.0
b	Following ion alteration	1.4
c	Air exposure partial oxygen reinjection	1.65
d	Thermal (total) oxygen reinvestment	2.0

[a] ± 0.2.

of a modest oxygen adsorption and slight oxidation of platinum. Examination of the changes in relative sizes of the peak structures in both Figures 5-4a and b are readily indicative of the potential depth of field and sophistication of the angular resolution procedure. In Figure 5-4a, for example, we find a partial answer to one of the major questions in surface science. Most noble metals (e.g., Pt) do not readily form surface oxides when exposed to air, and yet following sputtering to remove any "superficial" oxide a Pt(111) surface still reveals a noticeable peak at ~532 eV, long attributed to dissolved (and presumably reacted) oxygen. Examination of the angular resolution spectrum for this material (Figure 5-4a) suggests that the persistent (most prominent) part of the peak structure around 532 eV is, in fact, due to the Pt° ($4p_{3/2}$) loss line (see Chapter 6), rather than that due to oxygen.

Examination in Figure 5-4b of the angular resolved peak structure between 540 and 565 eV for two stages in the sputter "cleaning" provides further evidence in this case. The resulting peaks all appear to be due to loss lines (Chapter 6). Grazing incidence (13°) of the outer surface demonstrates that those peaks around 555 to 560 eV are most prominent. These are due primarily to O(1s) losses from adsorbed oxygen and carbon oxides. The possible peak near 550 eV may be Pt $4p_{3/2}$ losses due to the modest formation of PtO. The peak structure around 546, on the other hand, is relatively weak on the outer surface and much more prominent in the bulk of the sputtered Pt(111). This suggests therefore that it is due to the second $4p_{3/2}$ plasmon loss line of Pt° (Chapter 6). All of this suggests a surface presence for the detected oxygen.

The AR method has been the pivotal feature in even more detailed analyses. Consider, for example, a film of high-purity, catalytically prepared (but purportedly unreactive) polypropylene, subjected to a variety of oxidative treatments.[23,24] For our purposes, we examine herein the results following exposure to ambient air at standard temperature and pressure (STP) (labeled by our group as natural passivation or NP) and simulated sea water (3% NaCl). The systems to be studied were examined in the form of small coupons with AR-XPS, Figure 5-5. This procedure was employed rather than sputter etching to permit a detailed chemical analysis from the outermost surface layers to a depth of ~60 Å. Sputter etching would permit analysis to an even deeper level, but we have discovered that even moderate energy Ar⁺ will substantially sputter reduce the organic by-products generated in the treatments under consideration. The resulting lack of depth realized by AR-XPS did not turn out to be a major problem, however, because most of the critical changes induced by the various treatments are realized in the outer 60 Å, and are thus discerned in detail by AR-XPS.

After exposure to STP air,[23,24] polypropylene (despite its visual indication of inertness) is subjected to modest oxidation, producing a variety of products. Instead of just an indiscernible mixture, high-resolution analysis (employing as binding energy standards the detailed XPS analyses achieved on pure organic systems with different functional groups by Clark[25] and Briggs[26]) is able to demonstrate that the NP process produces a variety of discernible oxidized by-products. Repeated angular resolution analyses suggest

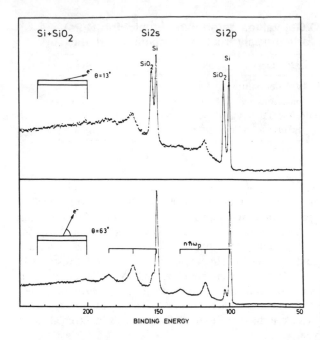

Figure 5-3 Typical angular resolution result: SiO$_2$ thin film on Si°.

that these products are realized (produced and deposited) in repetitive layers.[23,24] The products formed and their semidiscrete interfaces are depicted in Figure 5-5. One should note that the resulting oxidized carbon products do not entirely cover the surface, but do seem to form a product distribution that is still present more than 35 to 40 Å below the outer surface.

The general lack of oxidized products formed (during NP), and the fact that they always occur in mixtures, greatly complicates detailed analysis. Fortunately, the degree of oxidation is dramatically enhanced following exposure of the polymer to various simulated liquid environments, particularly sea water. In all of these fluid exposure cases

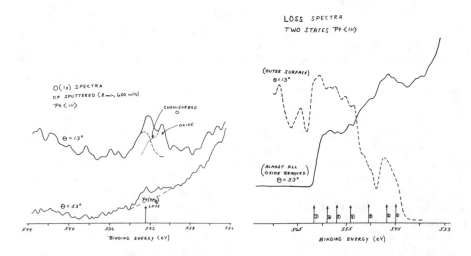

Figure 5-4 More involved angular result: oxidation of Pt° (note satellite changes with depth). States (1) and (2) due to Pt° losses, states (4) through (7) due to chemisorbed oxide.

the large preponderance of retained C–H- and C–C-type carbons still ensures that the C(1s) spectra will be dominated by these species making positive analysis of the C(1s)-oxygen species very difficult. Therefore, most of the identifications were rendered using AR versions of the resulting O(1s) spectra. Several typical examples[23,24] of these spectra are presented in Figure 5-6.

Despite the aforementioned interesting and perhaps even dramatic results, they are not as significant as those realized during NP of the polypropylene. As mentioned above, STP air seems to oxidize only about one third of the polypropylene units.[23,24] (It should be noted that total oxidation should produce one carbon-oxygen unit for every three carbons, i.e., one for each propylene monomer unit.) The AR discerned course of the NP of polypropylene (see Figure 5-5) suggests that there may be some type of relatively ionic unit formed near the interface between the polymer and its oxide products. The latter seem to occur ~35 Å into the evolving layers of oxidized material. The species found outside these ions seem to form the thickest layer detected, and also one that exhibits fairly good integrity. This layer is primarily composed of single bonded (ether-type or epoxide) species. As the layer extends outward it begins to exhibit C=O (carbonyl-type) species.

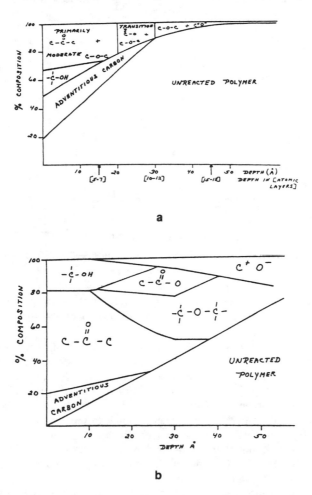

Figure 5-5 Qualitative rendition of oxide layer structure produced on polypropylene detected by AR-XPS: (a) natural passivation; (b) simulated sea water.

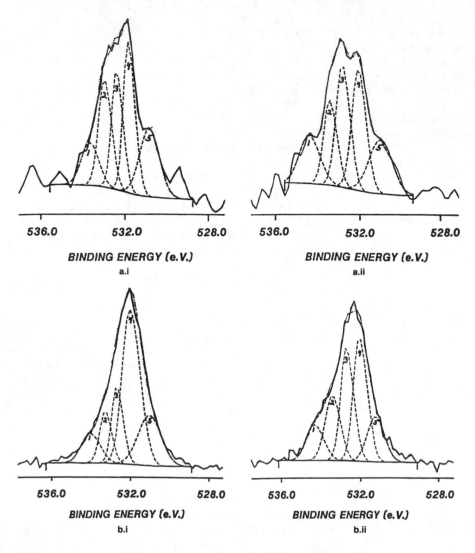

Figure 5-6 Select angular resolution ESCA study of O(1s) on polypropylene: (a) natural passivation (air oxidation) (i) 8°; (b) simulated sea water (i) 18°, (ii) 38°.

In these initial carbonyl layers the ether-type species are still detected, suggesting an interfacial region with mixed ether-carbonyls (esters). As the AR-XPS analyzes closer to the surface of the oxidized products, a region is detected which is apparently composed of primarily carbonyl-type species. Finally, a very thin layer composed primarily of hydroxide (i.e., alcohol-type, C–OH units) is found on the outermost surface. It should be noted that there is also a fairly extensive deposition in this region of the species dubbed (by surface analysts) as adventitious carbon (see Chapter 7).[23] The latter is deposited on the surface of all air-exposed materials, and produces XPS spectra that so closely resemble those for the polypropylene that its initial presence complicated the present analysis, particularly its semiquantitation.[23]

Many other cases of AR may be described. In the latter parts of this book we present a short description of substantially more involved techniques realized, starting with the basic technology so far developed.

III. ESCA AND ION SPUTTER ETCHING[27]

Few scientific marriages are as shaky as that between the various surface spectroscopies and ion sputter etching. It is not so much that ion etching does not deliver what it promises, but rather that the union may be too demanding to succeed with any grace.

A. IDEAL CASE

Ideally, ion beam sputter etching is accomplished in either the principal analysis chamber or in a side chamber that may be accessed without any serious compromise of the surface integrity of the material being investigated. This means that the chambers may be separated by some type of high integrity valve, thus permitting the inherently "dirty" sputtering process to be accomplished in a region that does not risk compromise of the various analysis tools. Before one opts for this type of setup, however, one should be aware that time and distance may not be your friend. For example, an $Al°$ foil sputter cleaned of oxide in a side chamber at 10^{-9} torr will reform significant amounts of oxide before one may return the foil to the analysis chamber. The gas employed for ion etching may be chosen from one of several possibilities, but $Ar°$ is by far the most common choice. These gases must be very pure (at least 99.9%) for this type of use. The gun may be a simple (nondifferentially pumped) "ionization" system, with little or no capability to alter the size or raster the crudely generated ion beam. Generally, however, the best results and the best control of the potential problems are achieved by employing a differentially pumped gun* with superior control over both beam size and movement. (Filling the analysis chamber with 5×10^{-5} torr of Ar is often not conducive to optimal analysis.)

Generally, the beam energy employed for sputter etching may be varied, and having a gun that can be employed over a range from several hundred to ~10 keV is useful. For a variety of reasons it is common for Auger spectroscopies (performing "profiles") to use a somewhat higher energy (commonly ~5 keV), whereas ESCA scientists, doing something more generally labeled as "sputter cleaning", often employ a lower energy, typically 1 to 5 keV. Practitioners must realize that the "depth of field", uniformity, sputtering yield, and chemical damage all depend very much on the type of gas employed, size and orientation of the beam, and its energetics. The yield, Y, is, as expected, an inverse function of the surface bond energy,[27] and also directly proportional to the reduced mass of the combined target and ion unit. The rate of sputtering is directly proportional to the ion flux and yield, and inversely proportional to the density of the target material. Generally, one needs to employ heavy ions (e.g., krypton and/or xenon) to sputter effectively heavy atoms. One should avoid the temptation to sputter etch either very slowly or very fast. The former may result in the extensive development of any preferential sputtering, whereas the latter will almost certainly contribute to implantation and other "knock on" events.

In general, the ideal sputter cleaning situation may be represented by the use of 5×10^{-5} Ar^+ ions at 1 keV to remove adventitious carbon and oxide from the surface of a zinc foil. In this case, the relative stability of the oxide (ZnO), and its singular nature (ZnO is the only common oxide of zinc), both contribute to allow the process to succeed in a nearly optimal fashion, see, for example, Figure 8-5. It should be noted that the various XPS spectral features indicative of ZnO shift quite smoothly to those for $Zn°$, and such key "signposts" as line width suggest that, even without any annealing step, the final sputtered surface exhibits reasonable structural, as well as chemical, integrity.[28] Another, perhaps more intricate, example involves the removal of the oxides produced on a film of GaSb as exemplified in Figure 5-7.

---

* Generally, one that employs a tightly controlled leak valve to introduce select amounts of the sputtering gas into the back of the gun. The chamber in front of the gun is thus only affected by the sputtered gas.

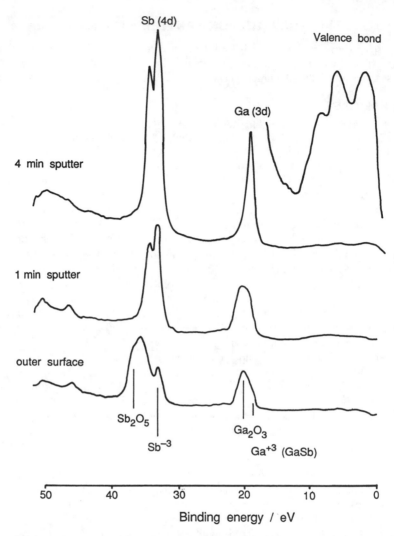

Figure 5-7 An example of the successful use of sputter cleaning: GaSb.

It is, of course, a common feature that the science of sputter etching for analysis is divided by electron spectroscopists into two areas: (1) depth profiling and (2) sputter cleaning, and that (1) is used as a label when sputter etching is employed during AES and (2) is commonly reserved for sputtering during XPS studies. In fact, these distinctions are often fictitious and their use may be incorrect. Thus, it is relatively common to employ sputter etching in conjunction with XPS to profile a material in depth, and it is also common to employ the technique for cleaning in conjunction with AES. It is true, however, that certain fundamental differences of practice are central to these overemployed distinctions. Thus, whereas it is a relatively straightforward process to operate a sequentially controlled sputter etching step with an AES step, i.e., the electron gun employed for the AES need not be turned off during ion etching, this is not true for XPS. Simultaneous operation of an ion gun and X-ray units in the same chamber is generally not acceptable practice. This means that either the XPS spectroscopist ion etches in a side chamber or the two processes are conducted as totally separate, somewhat time-consuming steps, with ion gun on-X-ray off, and then the reverse, etc. As a result of the latter, although XPS profiling can often be accomplished, it is generally laborious, and often fraught with problems induced by delay and start-up uncertainties. For this reason, whereas depth profiling is a standard part of the AES operation retinue, it is often a rather specialized side venture for XPS.

Sputter etching, of course, may also be effectively applied to many nonmetallic materials as well as metals. Thus, it has been used by the author to investigate the classic SiO_2 on $Si°$ system, see Figure 2-1b, but unfortunately, in that same study, we find that it is at the sensitive (and most interesting) interface where the process tends to alter, rather than just expose, sublayers.[29]

It also is important to note that the process of sputter etching is, to a great extent, the mirror image of sputter deposition. Thus, within the scope provided by the previously mentioned criteria of gas type, process dimensions, and beam energy, one should expect the targets (and to some extent the evolving substrates) in sputter deposition to experience the same alterations that result during sputter etching. This fact is mentioned because the author is aware of several cases of deposition scientists who would never permit sputter etching to be employed during the analysis of their materials.

B. PREFERENTIAL SPUTTERING[27,29]

Unfortunately, the zinc example mentioned above proves to be the exception rather than the rule. Thus, all compounds and alloys, and even many metals, generally suffer from some form of *retained* ESCA-detectable surface degradation as a result of the use of sputter cleaning. Fortunately, the amount and/or the type of this degradation is often not so severe as to negate the use of this process. It should be noted by the novices in the field that this is often a key point of disparity between some of the purists in surface science and those on the more practical side. In fact, there are some surface scientists who will not consider any problem in which sputter etching may be involved. To this author, this stance is too extreme, but it is very important for those who employ sputter etching to realize that it is a destructive technique and as such *always leaves some damage*. The only question is whether the analyst can cope with this damage. ESCA spectroscopists in particular should not fool themselves into believing that, because they do not detect any alteration, no damage has occurred. The relatively large beamed, statistical, chemically dominated nature of the ESCA experiment is *often* insensitive to the localized, structural-type damage that generally occurs. If the analyst feels that the microtopography of the analyzed surface is of importance, then various additional steps up to and including annealing must be included.

Besides structural alteration the principal problems induced by sputter etching are *preferential sputtering* and its commonly (XPS) detected adjunct — *sputter reduction*.

Perhaps the most obvious point where preferential sputtering occurs is during sputter-ing of alloys. Because of the nature of the thermodynamics involved in the formation (and destruction) of alloys, it is common for the sputtering process to "eject" preferentially one of the two (or more) constituents in these systems. More often than not, it is the lightest of the components that is most favorably extracted, although a variety of other features besides relative mass play a role in this process. Many of these aspects have been considered in depth by the combined discussions of Hoffmann[27] and Seah[30] and the interested reader is referred to those excellent sources for details. As indicated in those chapters, this is an area that has been dominated by AES analysis, but there have been a number of interesting and useful contributions by XPS. It is important to realize that the formation of a surface and the production of compounds are also events that are domi-nated by *selective* thermodynamics.[30] This means that, if someone has produced an alloy of two constituents, e.g., $A_{30}B_{70}$, then the surface tension-induced segregations (described in detail by Miedema[31] and others[32]) will generally significantly alter the surface compo-sition away from 30 to 70%. At the same time, the two components A and B will often experience dramatically different reactivities toward the ambient fluids encountered (particularly the O_2, H_2O, and CO_2 in the air) and as a result preferential surface (or interfacial) "stackings" of oxidized "by-products" will occur, further altering the relative concentration in the surface-near surface region, away from 30 to 70. On top of this, of course, is then placed the added proviso of preferential sputtering, during attempts to analyze the resulting situation in depth. It should be readily apparent that this was one area where a significant percentage of the early studies were fallacious. This does not mean that one cannot employ a combination of electron spectroscopy and sputter etching to analyze the surface of alloys. A reasonably successful example employing XPS to study the oxidation of brass is described in Section IV of Chapter 8.[23,33]

Because of the intimate interplay in these areas between the layers formed and their chemistry, many of the "negative" issues raised are natural points for analysis with XPS, particularly *sputter reduction*. This was, in general, a simultaneous observation made by several groups, but probably the most significant early determination was that of Kim et al.[34] In their analysis, they documented the fact that for oxides there is a propensity for the generally lighter oxygen atoms to be preferentially sputtered. Kim et al.[34] pointed out that the selectivity of this feature has an obvious connection with the relative thermodynamic stability of these oxides. Subsequent studies carried out in our laboratories and in others have demonstrated that, in addition to the thermodynamic consideration in the sputter reduction of oxides, there are several general rules that may be stated:

1. If there is any propensity for an oxide system to form a lower valent form, it will do so to some extent during sputter etching.

2. Thus, all of the common higher valent oxides, e.g., Fe_2O_3, SnO_2, MnO_3, Sb_2O_5, PdO_2, etc. (when sputter etched in Ar^+ under conventional conditions) will readily suffer sputter reduction to their "common" lower valent forms.

3. Generally, significant amounts of all of the (common) lower valent oxides will be created in this process, e.g., sputtered TiO_2 will result in mixtures (in the newly created surface) of TiO_2, Ti_2O_3, and TiO (see below).

4. As indicated in the statement above, all of the processes of sputter etching evolve to a steady-state finale. Thus, there is no such thing as *total* sputter reduction. Some presence of the "reactant", e.g., TiO_2, always remains no matter how long the process is run.

5. Generally, only the relatively common suboxides (e.g., PbO or FeO) are produced in substantial amounts by the sputter reduction process.[34] However, in almost all cases at least a moderate amount of all possible suboxides may be created if the energetics, gas type, and dimensions are properly organized. In some intermediate cases we have discovered that significant amounts of uncommon suboxides may be readily produced — see the discussion of GeO and AlO in Section II.D of Chapter 7.

6. It is also uncommon for most oxides to be sputter reduced all the way to the elemental metal (or metalloid).[34] Thus, the XPS detection of some M° (e.g., Cu°, Fe°, Si°, etc.) generally means that these species were present in the presputtered subsurfaces. However, this is not always the case — some metal oxides, e.g., anatase, may be completely sputter reduced. Also, one must be aware of knock on and other migratory effects and not be too quick to assign geometrical origins to the realized results.

7. These arguments must be extensively modified to describe the effects of sputtering on mixed or complex oxides (e.g., $SrTiO_3$ and NaY zeolite[29]). It is a general truism that no mixed oxide may be sputtered without some degree of degradation as well as surface removal. The degradation induced in these complex oxides is often quite different from that experienced by simple A_xO_y-type oxides, but may end up (at steady state) resembling the latter. This occurs because one of the first processes apparently often induced into complex oxides during conventional sputter etching is the conversion of the complex oxide into its simple oxide "projenitors", i.e.,

$$A_z B_s O_w \rightarrow A_x O_y + B_m O_n$$

The latter species are then subjected to their expected types of alteration (e.g., sputter reduction). Representative examples of this behavior are presented for zeolites in Section II.D of Chapter 7.

8. It should be apparent that, although we have developed these arguments around inorganic oxides (and these are by far the most investigated systems), the hypotheses apply generally to almost all inorganic systems and indeed some organic materials. Several general factors are playing a role in these results. Thus, the fact that most common metals are heavier than the conventional anions and many of the metals involved readily exhibit several oxidation states tends to generalize the arguments.

9. When considering these effects it is important to keep in mind the statement made in the preface of this book. *None of these concepts are absolute. At best, they represent generalities or signposts.*

Reprise — It is very important to note that most of the effects mentioned, or alluded to, above may, in fact, be halted, and often reversed, by a proper choice of conditions. Thus, one of the most effective ways of doing this for oxides is to sputter etch in a plasma formed from a mixture of a certain amount of O_2 and the inert (carrier) gas (e.g., Ar). The relative amount of oxygen in this process can be varied substantially, but should not be 100%. (The sputter etching scientist should look for guidance here from the sputter deposition scientists skilled in depositing oxides.) The art form being employed here is obviously the regrowth of oxides to balance their destruction. It all sounds much easier to do than is actually the case, but presumably one should be able to selectively halt the aforementioned quasi-stepwise alterations at any stage. Very little effort has yet gone into the "analysis" of this type of process.

At this point we will describe several, select cases of differential, or destructive, sputter etching, but before we do so we should reiterate one of our central themes. *All results, no matter how negative in appearance, should be viewed as useful information, if they can be controlled, analyzed, and understood.* This is as true for ion sputter etching as it is for any feature in surface studies. Thus, it is certainly true that many, if not most, surfaces suffer degradations due to such features as differential sputtering during etching, but it is also true that most of this degradation is selective, depending very much on the chemistry and physics of the surface, and if the results of sputtering can be ESCA analyzed valuable information about the original surface may result.

C. SPECIFIC CASES OF SPUTTER ALTERATION

The first case to be mentioned is an attempt to describe the status of the air-induced oxides (natural passivation, NP) formed by STP air on a high-purity Fe° foil. In this case, it is tempting to analyze the results ex post facto, and try to describe the oxidation products, layer by layer, formed to a steady-state thickness of less than 50 Å by combining high-resolution XPS with moderate energy (1-KeV) Ar^+ sputter etching. In this regard, initial (presputtered) XPS shows an outer layer dominated by iron (III) products suggesting a predominance of outer layer hydroxides, i.e., $Fe(OH)_3$, on top of Fe_2O_3, with Fe_3O_4, and the mixed oxides-hydroxide, slightly lower in depth.[23] Initial sputtering (Figure 5-8) rapidly reveals, however, a dramatic change with little retention of Fe (III) species, even in the Fe_3O_4 form. The dominant product detected within 10 Å of the outer surface is Fe(II)O. This species remains predominant all the way to the metallic interface where a relatively flat surface of Fe° is detected. It should be noted that the XPS also detects significant amounts of carbonaceous by-products, apparently deposited throughout the oxide "lattice", first carbonates and adventitious carbon, and then other species.

The problem with this analysis can be seen if one tries to reverse the process and XPS monitor the chemistry of the layers formed (over a period of several days) when clean Fe° is exposed to STP air. In this case, the initial iron oxide layer is again found to be FeO, but, unlike that detected during sputter etching, the initial FeO layer is found to be relatively thin and is instead rapidly converted to Fe_3O_4. Also, the outer Fe_2O_3, and particularly the $Fe(OH)_3$ layers are found to be much thicker. All of this, with its suggested "hysteresis" loop, are displayed in Figure 5-8.

The rationale for the apparent discrepancy can be found in the beguiling nature of the effect of the ion beam on the air-deposited iron oxides. It turns out that both Fe_2O_3 and Fe_3O_4 (and perhaps FeOOH) are subject to sputter reduction. As a result, the thicknesses of the different oxide layers formed are all dramatically displaced.

This suggests that the growth study is the more accurate reflection of the NP layers formed on Fe°, but even that interpretation may be in error. It turns out that to create the surface of Fe° (free from oxides) we sputtered the oxide layer away. This unfortunately also created in the newly exposed Fe° layer "sites" that can only be described as hyperactive. As a result, the initial exposure to air induces reactions that generally will not occur on an "inactive" surface. As a result of this activation we have found that, at the interface between Fe° and the reforming oxides, significant deposits of Fe_3C are found — but these carbides are not present when we approached this interface while sputtering inward. They are apparently *produced* by interaction of the adventitious carbon with the "activated" metal surface.[23]

We have also discovered that the latter type of sputter-induced (carbide) by-products have been evolved at the metal (metalloid) air interface of a number of sputter-cleaned surfaces (e.g., Zr, W, Si, etc.).[23]

Another interesting example of the analysis of sputter reduction concerns a study performed by John Hackenberg, a former member of my research group, of the sputter "evolved" chemistry of titanium oxides.[35] In those studies, very pure samples of TiO_2 in both the anatase and rutile forms were examined. There are a number of well-accepted publications of the principal ESCA spectra for both of these in the published literature.[36] Both materials were submitted by Mr. Hackenberg to (XPS) identical sputtering histories (1 keV Ar^+), until they reached a steady state in their evolution. Both systems suffered significant sputter reduction, *but the degree to which they were reduced was dramatically different*, with anatase exhibiting near total reduction to suboxide species, whereas the rutile was only moderately reduced. Further, only the anatase exhibited significant reduction all the way to Ti°. As mentioned above, this total reduction is unusual under these conditions.

The aforementioned examples were presented because they demonstrate that given the correct circumstances it is possible to take a negative result — in this case sputter reduction — and turn it into a useful auxiliary analysis tool. Thus, the oxidized iron results provide information about the NP of iron, and the stability of its oxides, whereas the titanium oxide study suggests useful facts about the relative structural and chemical stability of the two most prevalent structural forms of its common oxide.

IV. ULTRAVIOLET PHOTOELECTRON SPECTROSCOPY (UPS) AND GAS PHASE ANALYSIS

Although this book is primarily intended as a presentation of the merits and demerits of present day XPS as it is applied to solids, we would be remiss if we did not mention some of the attributes and special features of ultraviolet photoelectron spectroscopy (UPS) and gas-phase analysis achieved by either method.[37]

The standard approach employed for UPS is to couple a resonance light source to a conventional electron spectrometer. Most of the major manufacturers produce lamp

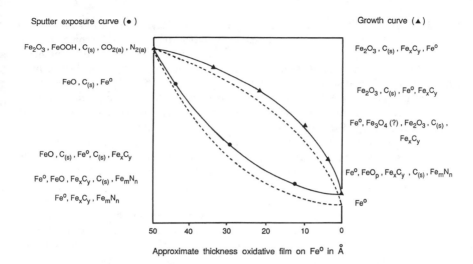

Sputter exposure curve (●)

Fe_2O_3 , FeOOH , $C_{(s)}$, $CO_{2(a)}$, $N_{2(a)}$

FeO , $C_{(s)}$, Fe^0

FeO , $C_{(s)}$, Fe^0 , $C_{(s)}$, Fe_xC_y

Fe^0 , FeO , Fe_xC_y , $C_{(s)}$, Fe_mN_n

Fe^0 , Fe_xC_y , Fe_mN_n

Growth curve (▲)

Fe_2O_3 , $C_{(s)}$, Fe_xC_y , Fe^0

Fe_2O_3 , $C_{(s)}$, Fe^0 , Fe_xC_y

Fe^0 , Fe_3O_4 (?) , Fe_2O_3 , $C_{(s)}$, Fe_xC_y

Fe^0 , FeO_p , Fe_xC_y , $C_{(s)}$, Fe_mN_n

Fe^0

50   40   30   20   10   0

Approximate thickness oxidative film on Fe^0 in Å

Figure 5-8 Example of a type of "sputter hysteresis", i.e., upper curve is oxidized layered chemistry produced on Fe during growth, whereas lower curve is chemistry "revealed" following sputter profile.

systems that may be coupled directly to their (most exploited) XPS systems. A number of different gases may be used in these resonance lamps, but most often the choice is helium, with the He(I) line at ~21 eV and the He(II) line at ~41 eV, the favorite excitations. The older lamps operated using a quartz capillary discharge, and required moderately high pressures and currents. A newer design has permitted significant reductions in both current and pressure. This has enhanced the capability to generate the well resolved He(II) line with its more substantial energy range. Very low operating pressures may be employed if cyclotron resonance confinement is realized in a microwave discharge.[37]

In general, UPS exhibits superior resolution compared to XPS, thus permitting the realization of substantial vibrational structure during gas phase analysis.[37] A disadvantage of UPS is its limited energy range and thus its inability to realize the key core lines often needed for careful qualitative and quantitative analysis of systems of uncertain composition.

It should be noted that the cross sections realized by UPS are often the reverse of those realized using XPS. This has to do with the differences of the accessible final state functions, Ψ_F, and the nature of the dipole vector activated by the photon field. At low photon energy (UPS), atomic and molecular orbitals with extensive eccentricity, i.e., p or π-type orbitals, will have larger cross sections than those with substantial density around the center, i.e., s or σ-type orbitals; the reverse is true for XPS. Thus, the general features displayed in 5-9A are usually realized when both UPS and XPS are used to examine the same gases.

This type of intensity ratio arises for both gas-phase and solid-state analyses. Thus, XPS analyses of graphite exhibit almost no density for the π bands, whereas the resulting σ bands are quite pronounced, see Figure 5-10. To some extent, the reverse is true for UPS investigations of graphite.[38]

Studies of gas-phase phenomena are special challenges for both UPS and XPS. In both cases, it is necessary to make use of an acceptable gas cell. The permissible vapor pressures, temperature control, and the capabilities of differential pumping all play key roles in the realized results. It is, of course, important to confine the gas to be analyzed to the region of photon bombardment, while maintaining as much vacuum as possible in the analyzer. Acceptable gas cells are not difficult to construct, introduce, or maintain, but they are not readily compatible with solid-state detectors, so that generally some time and adjustment (or expense) must be employed to convert from one procedure to another.[37]

The most important differences between gas and solid photoelectron analysis centers around the lack of band structure and a Fermi edge in the former case. In the gas phase, valence electron states may be reasonably approximated as atomic or molecular orbitals, whereas the related states are generally broadened into bands in the case of solids. The zero of binding energy for a gaseous system is the least bound valence orbital, and this is generally considered the ground-state vacuum level. For the case of solid substrates, one may always realize a zero in energy classified as a Fermi level, although this state often has nothing to do with the situation formally defined in the free electron model, and one can imagine hypothetical situations in which no Fermi level actually exists, e.g., an insulator at absolute 0 K. Having found, or at least defined the Fermi level, E_F, one may also realize for solids a vacuum level, as that state separated from E_F by the work function ϕ_S. However, it is perhaps formally more correct to assume the reverse, i.e., that there is for all solids a vacuum level, and generally a Fermi level can be defined, rather than the other way around.

The correct coupling of these various energy zeros when, for example, a gas is adsorbed onto a solid is not a trivial concept, creating a situation that cannot be achieved without making some assumptions with respect to work functions and many-electron vs. one-electron models.

In general, gas-phase photoelectron analyses, whether employing XPS or UPS, are accomplished without trying to establish a rigorous zero of binding energy. Typical examples of simple gases that have been examined by both XPS and UPS, Figure 5-9A, demonstrate the superior resolution, but limited range, of UPS.[37] (Note, for example, the vibrational features realized by UPS, Figure 5-9B.) One should also note that XPS also reveals a readily discerned set of secondary (loss) peaks. These peaks are shake-up lines, indicative of secondary transitions from one molecular orbital state to another. They are, of course, closely associated to the plasmon transitions described in Chapter 6. The latter require the presence of valance bands that are needed to provide the plasma collective,

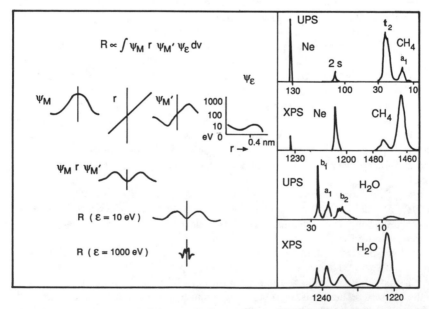

Photoelectron kinetic energy eV   **A**

whereas shakeup effects only require readily available excited orbitals that may receive final state transitions from valence orbitals that are properly (symmetry) coupled.

Photoemission cross sections of molecules are subject to an asymmetry parameter, β, that may be employed to describe the angular distribution of molecular orbitals.[37,39] The angular-dependent, differential cross section of photoionization may be expressed as

$$I(\alpha) = \frac{\sigma_{tot}}{4\alpha}\left[1 - \frac{\beta}{4\pi}(3\cos 2\alpha - 1)\right] \qquad \textbf{(5-6)}$$

In order to realize the angular distribution of the orbitals one needs to employ polarized

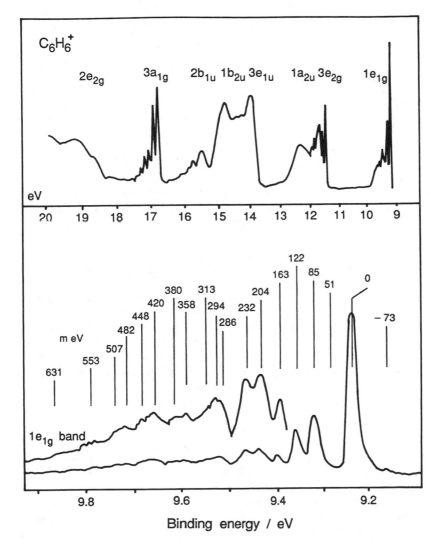

Figure 5-9 (A) Contribution of various aspects of wave function [core, valence (s or p), etc.] to the photoelectron transition moment, as a function of the energy of the photoelectron. (B) He I photoelectron spectrum of benzene at various levels of resolution.

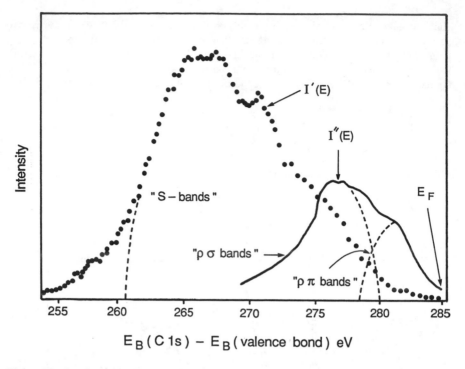

Figure 5-10 Analysis of the XPS valence band structure of graphite. (After McFeeley et al.[38]) Note the specific subsectioning of the contributions due to particular valence orbitals, particularly the π density and its relative lack of intensity.

radiation. This can be achieved in UPS by using a monochromator, a tool that enhances the general resolution in UPS.[39,40] Elliptical polarization is naturally achieved using synchrotron radiation.

Even with a superior monochromator the process of valence orbital mapping is a complicated one. One may try to determine the orbital distribution from experimental β values or reverse the process. The correlation may require substantial data at a variety of photon energies. Such energetic tunability is natural for a synchrotron source, but is more difficult in conventional UPS. Siegbahn has described several experimental setups employing polarizable laser beams to produce photoelectrons in vaporous samples or molecular beams.[40] Detailed analyses of the many features inherent to analysis of the asymmetry parameter, β, are described in detail by Ghosh.[37]

REFERENCES

1. **Siegbahn, K., Nordling, C., Fahlman, A., Nordberg, R., Hamrin, K., Hedman, J., Johanson, G., Bergmark, T., Karlsson, S.-E., Lindgren, I., and Lindberg, B.,** ESCA — atomic molecular and solid state structure studied by means of electron spectroscopy, *Nova Acta Regiae Soc. Sci. Ups.,* Ser. IV, Vol. 20, 1967.
2. **Tougaard, S.,** Formalism for quantitative surface analysis by electron spectroscopy, *J. Vac. Sci. Tech. A,* 8, 2197, 1990.
3. **Wagner, C. D.,** How quantitative is ESCA? — an evaluation of the significant factors, in *Quantitative Surface Analysis of Materials,* McIntyre, N. S., Ed., ASTM STP 643, American Society for Testing and Materials, Philadelphia, 1978, chap. 2.
4. **Wagner, C. D.,** *Anal. Chem.,* 44, 1050, 1972; **Wagner, C. D.,** *Anal. Chem.,* 49, 1282, 1977.

5. **Wagner, C. D., Davis, L. E., Zeller, M. V., Taylor, J. A., Raymond, R. H., and Gale, L. H.,** *Surface Interface Anal.,* 3, 211, 1981.
6. **Nefedov, V. I., Serguskin, N. P., Band, I. N., and Trzhakovskaya, M. B.,** *J. Electron Spectrosc. Rel. Phenomena,* 2, 383, 1973.
7. **Seah, M. P.,** *Surface Interface Anal.,* 2, 222, 1980.
8. **Powell, C. J.,** The physical basis for quantitative surface analysis by Auger electron spectroscopy and X-ray photoelectron spectroscopy, in *Quantitative Surface Analysis,* McIntyre, N. S., Ed., ASTM STP 643, American Society of Testing and Materials, Philadelphia, 1978, chap. 1.
9. **Seah, M. P.,** *Surface Interface Anal.,* 9, 85, 1986.
10. **Mroczkowski, S. J.,** The effect of electron transmission functions on calculated Auger sensitivity factors, *J. Vac. Sci. Tech. A,* 7, 1529, 1989.
11. **Seah, M. P.,** Quantification of AES and XPS, in *Practical Surface Analysis,* Briggs, D. and Seah, M. P., Eds., John Wiley & Sons, Chichester, England, 1983, chap. 5.
12. **Schofield, J. H.,** *J. Electron Spectrosc. Rel. Phenomena,* 8, 129, 1976.
13. **Penn, D. R.,** *J. Electron Spectrosc. Rel. Phenomena,* 9, 29, 1976.
14. **Angevine, P. J., Vortuli, J. C., and Delgass, W. N.,** *Proc. Int. Congress Catal.,* 6, 611, 1977.
15. **Kerkhof, F. P. J. and Moulijn, J. A.,** *J. Phys. Chem.,* 83, 1612, 1979.
16. **Windawi, H. and Wagner, C. D.,** Use of ESCA in the characterization of heterogeneous materials, in *Applied ESCA,* Windawi, H. and Ho, F. F.-L., Eds., (Chemical Analysis Ser., Vol. 63), Wiley-Interscience, New York, 1982, chap. 9.
17. **Fung, S. C.,** *J. Catal.,* 56, 454, 1979.
18. **Fadley, C. J., Baird, R. J., Siekhaus, W., Novakov, T., and Bergstrom, S. A. L.,** *J. Elec. Spectrosc. Rel. Phenomena,* 4, 93, 1974; **Fadley, S. C.,** Instrumentation for surface studies: XPS angular distributions, *J. Electron Spectrosc. Rel. Phenomena,* 5, 725, 1974.
19. **Barr, T. L., Mohensian, M., and Chen, L. M.,** *Appl. Surface Sci.,* 50, 71, 1991.
20. **Siegbahn, K., Gelius, U., Siegbahn, H., and Olsen, E.,** *Phys. Lett. A,* 32, 221, 1970. This is also one of the introductory expositions of photoelectron diffraction.
21. **Fadley, C. S.,** Angle-resolved X-ray photoelectron spectroscopy, *Prog. Surface Sci.,* 16, 275, 1984.
22. **Feuerbacher, B., Fitton, B., and Willis, R. F., Eds.,** *Photoemission and the Electronic Properties of Surfaces,* John Wiley & Sons, New York, 1978; **Cordona, M. and Ley, L., Eds.,** *Photoemission in Solids,* Vol. 1 and 2, Springer-Verlag, Berlin, 1978; **Plummer, E. W. and Eberhardt, W.,** *Review in Advances in Chemical Physics,* Vol. 49, Prigogine, I. and Rice, S. A., Eds., John Wiley & Sons, New York, 1982, 533.
23. **Barr, T. L.,** Recent Advances of ESCA in Natural Passivation, presented NACE National Convention, Las Vegas, April 1990, to be published.
24. **Barr, T. L. and Yin, M. P.,** ESCA studies of the oxidation of polymer surfaces: polypropylene, *J. Vac. Sci. Technol.,* in preparation.
25. **Clark, D. T. and Harrison, A.,** *J. Polym. Sci., Polym. Chem. Ed.,* 19, 1945, 1981.
26. **Briggs, D.,** Applications of XPS in Polymer Technology, in *Practical Surface Analysis,* Briggs, D. and Seah, M. P., Eds., John Wiley & Sons, Chichester, England, 1983, chap. 9.
27. **Hoffmann, S.,** in *Practical Surface Analysis,* 2nd ed., Briggs, D. and Seah, M. P., Eds., John Wiley & Sons, Inc., Chichester, England, 1990, chap. 4.
28. **Barr, T. L., Yin, M. P., and Varma, S.,** *J. Vac. Sci. Technol. A,* 10, 2838, 1992.
29. **Barr, T. L.,** *Appl. Surface Sci.,* 15, 1, 1983.
30. **Seah, M. P.,** in *Practical Surface Analysis,* 2nd ed., Briggs, D. and Seah, M. P., Eds., John Wiley & Sons, Chichester, England, 1990, chap. 7.

31. **Miedema, A. R.,** *Zeit. Metallkunde,* 69, 455, 1978.
32. **McLean, D.,** *Grain Boundaries in Metals,* Oxford University Press, London, 1957; **Williams, F. L. and Nason, D.,** *Surface Sci.,* 45, 377, 1974; **Wynblatt, P. and Ku, R. C.,** *Proc. of ASM Materials Science Seminar on Interfacial Segregation,* Johnson, W. C. and Blakeley, J. M., Eds., ASM, Metals Park, OH, 1979, 115.
33. **Barr, T. L. and Hackenberg, J. J.,** *Appl. Surface Sci.,* 10, 523, 1982.
34. **Kim, K. S., Baitinger, W. S., Amy, J. W., and Winograd, N. W.,** *J. Electron Spectrosc. Rel. Phenomena,* 5, 351, 1974.
35. **Hackenberg, J. J. and Barr, T. L.,** unpublished.
36. **Briggs, D. and Seah, M. P., Eds.,** *Practical Surface Analysis,* 2nd ed., John Wiley & Sons, Chichester, England, 1990.
37. **Ghosh, P. K.,** *Introduction to Photoelectron Spectroscopy,* John Wiley & Sons, New York, 1983.
38. **McFeeley, F. R., Kowalczyk, S. P., Ley, L., Cavell, R. G., Pollack, R. A., and Shirley, D. A.,** *Phys. Rev. B,* 9, 5268, 1974.
39. **Werme, L. O., Grenberg, B., Nordgren, J., Nordling, C., and Siegbahn, K.,** *Nature,* 242, 453, 1974; **Wannberg, B., Veenhuizen, H., Mattsson, L., Norell, K.-E., Karlsson, L., and Siegbahn, K.,** *J. Phys. B,* 17, L529, 1984.
40. **Siegbahn, K.,** *J. Electron Spectrosc. Rel. Phenomena,* 52, 111, 1990.

Chapter 6

Loss Spectroscopy in ESCA: Its Causes and Utilization

I. INTRODUCTION

As noted above, the XPS or ESCA process is made complicated in part because of the ready presence of a variety of secondary interactions that often shift the principal peak and may, in some cases, also provide one or more discrete, satellite peaks. Although much of the inherent simplicity of the origin explanation of ESCA vanishes with these *loss* effects, a wealth of additional, often unique, analytical information is also realized. Certain aspects of this capability are now being employed on a regular basis by XPS scientists. These features, therefore, will only be outlined herein, since they are interjected in part throughout the text. We will, however, in this chapter describe in some detail an approach that emphasizes a relatively unexploited feature of the collective electron nature of some of these losses that seem to arise in many ESCA results.

The loss peaks that have been employed in the past by ESCA scientists generally fall under the heading of shake-up phenomena.[1] In general, these may be considered to be the loss features that correspond directly to those that occur through rapid transitions, except that, in the case of the most-used shake-up transitions, any collective behavior in the system may have largely collapsed.

The presentation to follow is designed to fill in some of the gaps that exist in XPS loss spectroscopy which we designate herein as ESCALOSS. In view of the numerous detailed discussions in the literature for several of the select cases, this book only expands upon and ties together the rather diffuse, general outline presented above in the theoretical section in several key areas. For those interested in a more general treatment, they should also consult the most pertinent parts of the literature in this field. Particularly, if one is interested in the XPS loss spectra involving our predefined crystalline, conductive solid, we recommend the reports of Langreth[2] and Gadzuk.[3] On the other hand, if one is concerned with the materials we are classifying as molecular solids, papers by Manne and Aberg,[4] Fadley,[5] and the Shirley group[6] should be very informative. In those cases of our principal concern (i.e., the extensive range of materials that are not described by either of these extreme models) there are only a few considerations. Interestingly, the most detailed discussion of the effects that may occur when these two cases are mixed seems to have been presented in the often-forgotten pioneering reports of Pines and Bohm[7] and of Nozières and Pines.[8] Thus, these authors not only established the existence of the plasmon as a fundamental quantum particle, but also considered in some detail what occurs with plasmons as the solid under consideration evolves from one adequately described by a free electron gas model [zero periodic potential, i.e., $V(r) = 0$], through the Bloch formulation [regularly varying $V(r)$], to cases that may only be approximated by our predefined, *molecular solid*. The missing feature in the Pines and Nozières studies (for those interested in ESCALOSS) is that they considered only the extrinsic (electron induced) loss spectra, produced in a system by outside electrons without the creation of deep holes.

Mahan,[9] Langreth,[2] Gadzuk,[3,10] Sunjic,[11] Doniach,[12] and others[13] have greatly expanded on these plasmon arguments for cases in which core holes are a necessary adjunct (e.g., during photoelectron and X-ray spectroscopies). Thus, these authors have tried to describe situations in which intrinsic (hole-caused) losses and hole-induced couplings and shifts accompany the extrinsic effects in the loss spectra. Unfortunately, all of this had been done by the latter group without consideration of the possibility of collective (e.g., plasmon) vs. noncollective losses in a nonfree electron gas-type environment. This oversight has led some practitioners who are only slightly acquainted with the field to incorrectly assume that plasmons only occur in systems involving conductive materials.[14] Thus, one of our principal goals herein will be to demonstrate that plasmon-like effects contribute to the loss spectra of semiconductors and (even) insulators.[7] Unlike Pines et al.,[7,8] we will do this for ESCALOSS, with its accompanying hole-induced effects.

A. RELATED PLASMON-TYPE TRANSITIONS

The bulk collective modes of interaction that are the principal concerns of this section of this book represent only a fraction of the possible types of collective effects. For example, it is quite possible for collective behavior to be restricted to particular regions of a material, such as at its interfaces or surface. The results of the latter (labeled as surface plasmons) are generally produced by any relatively pure system that also yields reasonable bulk plasmons. This suggests that surface plasmons are often also achieved in cases where bulk plasmons exhibit free electron character. In these cases, the surface plasmons have a frequency

$$W_s \cong \frac{W_p}{\sqrt{2}} \tag{6-1}$$

where W_p is the bulk plasmon frequency, as exemplified below in Equation 6-16. A representative example occurs for In° in Figure 6-1. Even in the case of very clean samples, at conventional XPS incidence, the intensity of the surface plasmons are generally substantially less than their equivalent bulk form (reflecting on the two-dimensional character of the former). If one resorts to grazing incidence in the manner described in Chapter 5 however, the ratio of I_s/I_p experiences the anticipated increase. These features have been exploited in a variety of forms of analyses.[15]

A number of other types of specialized plasmons have also been proposed. For instance, several groups have experimentally and theoretically described the special collective situations that result when adatoms are introduced into select matrices.[12] These situations are of special significance in cases related to catalysis and composite formation.[16] There are a number of interesting, unique features to the theory of these adatom plasmons that have only recently been totally described.[17] Experimentally, they often require superior resolution and sensitivity[16] to yield spectra of analytical quality.

Nozières and Pines[8] have also suggested that it should be possible to achieve low energy plasmon-type transitions due to collective behavior in modestly populated con-

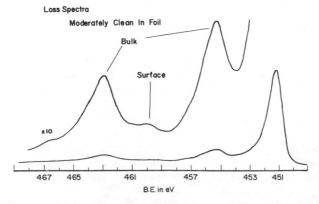

Figure 6-1 ESCALOSS for In° demonstrating E_L and moderate surface plasmon.

duction bands. As Pines has shown, these states, with their relatively small N values, should produce discrete transitions in the infrared region.[8] The quantum mechanics of the latter processes represents another interesting set of select situations, but their use in analysis awaits improvements in measurement techniques.

B. THE EARLY HISTORY OF LOSS SPECTROSCOPY AND PLASMONS

It is, of course, important to realize that an entire science of plasmon-oriented spectroscopy involving inelastic electron scattering predated its development as a part of ESCA and other forms of photoelectron spectroscopy. Pines[7] and Nozières and Pines[8] earlier described a variety of these studies in support of their theoretical arguments. Shortly thereafter, a number of scientists extended this extrinsic-only spectroscopy by introducing a UHV form. One of the most important of these early UHV studies was the work of Powell and Swan,[15] who described in some detail the loss spectra of readily oxidized metals, such as Al°. A number of other research groups have extended the latter studies to include nonmetallic systems, even nonconductive materials.[18]

Another form of UHV loss spectroscopy has been developed over the last 2 decades centered around the detection of the very low-energy, inelastic, loss transitions induced by the various vibrational modes produced by surface and near-surface bonds. This form of low-energy, loss spectroscopy has become so popular, it is often the only type suggested by the designation electron energy loss spectroscopy, or EELS.[19] The procedures, technology, and theory of EELS have been extensively developed, particularly by Ibach[20,21] and, although the method requires a very sensitive, stable spectrometer, its use may be worth the effort, as it is often the most informative tool for descriptions of the bonding in chemisorption and other surface studies.[19-21]

II. UTILIZATION OF XPS-INDUCED LOSS SPECTROSCOPY (ESCALOSS): ADVANTAGES AND DISADVANTAGES

It should be obvious that one of the principal reasons that some ESCALOSS results have appeared in the literature is the natural tendency of scientists in a particular area (including electron spectroscopists) to try to exploit every unique feature related to a particular technique, immaterial of its true utility. Therefore, since this is a book dedicated to "useful ESCA", we should in fact ask the following question: What is there about the ESCALOSS method that would recommend it as a *unique, useful*, supplemental procedure to the nonexpert user of electron spectroscopy? In order to properly explore this question we need to examine some of the possible shortcomings of "conventional" ESCA, and ask whether ESCALOSS provides valuable information in these areas.

As described above, the principal purpose of ESCA seems to be the definition of the chemistry and to some extent structure (both fixed and evolving) of a materials system in its surface and near-surface region. In order to accomplish this with "conventional" ESCA, it is necessary to be able to resolve the principal core level peaks with their proper sizes and shapes and have them fixed in their *proper* binding energy positions. It is, of course, also important that these positions exhibit discernible shifts from one position to another as the chemistry evolves. Although, in principal, "conventional" ESCA can often achieve all of these features, there are frequently cases in which shortcomings arise and, unfortunately, many of these cases seem to be exacerbated during studies of more practical systems. These shortcomings would seem to include the following.

A. CHARGING AND FERMI EDGE DECOUPLING

We have previously described the charging problem, noting how it may be removed in many of the examinations of applied materials.[22] As we have shown, however, if the goal is chemical analysis, then, due to the inherent difficulties in establishing the Fermi edge, distinct problems may still arise in accurately (and universally) determining the binding energy scale, even if charging can be removed. It would be advantageous, therefore, to have access to an ESCA-based procedure that produces peaks and shifts that are not adversely affected by charging, and particularly by the floating Fermi edge problem.[22] As we shall describe below, ESCALOSS, because of its reliance on energy differences, exhibits this attribute.

B. LARGE, RELATIVELY EASY TO DISCERN SHIFTS, WITH CHANGES IN CHEMISTRY

As mentioned above in a number of important cases, the conventional (Siegbahn) chemical shifts in ESCA are often very small. This occurs for a variety of reasons, some of which we have described elsewhere in more detail. Often these small shifts cannot be detected with commercial ESCA systems, e.g., $In° \rightarrow InSb$ using the $In(3d)$. In many of these cases it turns out that there are corresponding, much larger shifts in the ESCALOSS splittings that may be easily detected. Some examples of this attribute are presented below (see Table 6-1). The reasons behind these attributes for ESCALOSS are presented later in this chapter.

C. NONCHEMICAL BASIS FOR CHEMICAL SHIFTS

As described elsewhere in some detail, the positions of the binding energy peaks in ESCA result not only from initial state features, such as differences in chemistry, but also from final state effects, such as the previously described relaxations induced into the photoelectron peak position by the contraction of the valence shell inward (toward the nucleus) *as* the photoelectron is ejected.[23] Thus, the peak position of a photoelectron line has a multifaceted origin. Unfortunately, the relaxation contribution to a principal core line for a particular element in one chemical state (e.g., $Al°$) may differ significantly from the relaxation contribution for the same element in another chemical environment (e.g., Al_2O_3). This means that one cannot attribute the total detected shift, ΔE^m, in binding energy (e.g., for $Al° \rightarrow Al_2O_3$) to initial state chemical changes, and thus the separation of chemical from nonchemical causes may be difficult.[23] This tends,

Table 6-1 **Representative ESCALOS, ΔE_L, relative to common chemical shifts, ΔE_s (S $\equiv$ Siegbahn) for select compounds in eV $\pm$ 0.3 eV**

From–to	Element and state	Principal (Siegbahn) shift	XPS loss shift
$Sl°–SiO_2$	$Si(2p)$	3.4	5.2
$Al°–Al_2O_3$	$Al(2p)$	2.2	8.6
$In°–InSb$	$In(3d_{5/2})$	0.2	1.8
$In°–InN$	$In(3d_{5/2})$	0.4	4.0
$In°–In_2O_3$	$In(3d_{5/2})$	1.0	7.0
$GaSb–GaAs$	$Ga(2p_{3/2})$	0.4	0.9
$GaSb–Ga_2O_3$	$Ga(2p_{3/2})$	1.0	1.9
$Ga_2O_3–Ga(OH)_3$	$Ga(2p_{3/2})$	0.3	4.0
$Sb°–InSb$	$Sb(3d_{5/2})$	−0.2	−2.7
$Sb°–GaSb$	$Sb(3d_{5/2})$	−0.1	−0.7
$Sb°–Sb_2O_5$	$Sb(3d_{5/2})$	2.5	2.6
$Sn°–SnO$	$Sn(3d_{5/2})$	1.2	6.0

in some instances, also to be injurious to efforts to relate ESCA shifts to other parametric features that are purported to describe changes in chemistry.[23] *As we will describe below, the shifts detected with ESCALOSS, on the other hand, are generally either devoid (or nearly devoid) of relaxation and other noninitial state effects.*

On the other hand (as will become apparent hereafter), ESCALOSS is not always a panacea for these problems, as the required statellite peaks are generally weak and broad, often making establishment of loss shifts a different problem.

III. THE PLASMON DOMINATION ARGUMENT

As mentioned above, the process of electron loss spectroscopy involving plasmons was first described in some detail by Pines, Bohm, and Gross[7,24] and later by Pines and Bohm,[7] and particularly Nozières and Pines.[8] These descriptions involved situations in which the excitation particle (always an electron) employed to excite the plasmons is issued and detected outside of the materials involved (e.g., in a manner of the inelastic electron scattering that may be produced in electron microscopy). Thus, the additional fields and problems induced by the presence of the hole(s) which occur during X-ray and photoelectron spectroscopy were ignored. Also considered in the treatment by Pines et al.[7,8] were situations in which forms of excitation, other than plasmons, are excited by the excitation source. In particular, Pines et al.[7,8] considered cases in which the valence and/or the conduction band electrons in a solid may act individually, rather than collectively (i.e., plasmon-like), and thus experience band-to-band type transitions. Pines et al.[7,8] demonstrated in some detail that the apparent "competition" between these band-to-band transitions and the plasmon transitions depends both on the nature of the solid material involved and/or the conditions of the experiments.

The previous discussions of the loss spectroscopy that occurs during a photoelectron experiment, on the other hand, were considered almost exclusively from two entirely separate points of view: plasmons, for free electron metals,[2,3] and shake-up-type transitions for molecular solids.[1,2,4,5] Possible dispersions of the plasmons were considered as small perturbations of the plasmon field, but *still in the context of the free electron model.* Thus, the latter become slight displacements of a harmonic oscillator (the plasmons) in which the free electron gas was also shown to suffer an infrared catastrophe.[2,25,26] The latter can be solved *exactly* in the long time limit *(for a free electron gas)* producing a term that merely "disperses" the resulting plasmon dominated peaks up field by small amounts referred to as Nozières-de Dominicis shifts,[26] as shown in Figure 6-2.

The problem with these two extremes is that neither one addresses the many systems of interest between these cases (e.g., semiconductors, many alloys, composites, and even certain oxides and polymers). Our analysis, on the other hand, suggests that even the latter systems exhibit loss transitions that seem to be strongly influenced, perhaps even dominated, by plasmon-type (collective) effects, but transitions that are also significantly perturbed by the individual electron (or molecular) behavior of the system. The latter property is obviously not completely explicable in terms of the Nozières-de Dominicis effect. In fact, it is readily apparent that these systems cannot be properly modeled by the simple free electron model.[2] The presence of plasmon dominated loss peaks in cases of semiconductor and even insulating materials was described in some detail by Pines[7] and Nozières and Pines[8] in their pioneering studies of plasmons and, to some extent, by our group for cases of photoemission losses.[27-33] Because the details of this "mixed" case have never adequately been described for situations involving both electron and hole-loss excitations, we present here the rudiments of that argument.

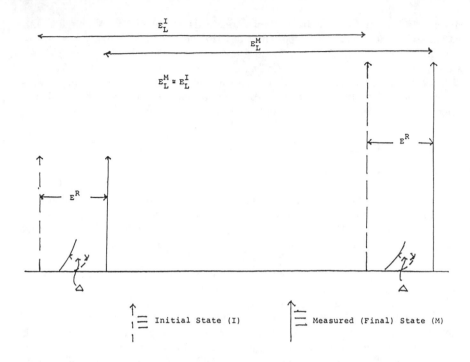

Figure 6-2 ESCALOSS results with moderate dispersion of both ΔE^R and Δ (where Δ corresponds to Nozières-de Dominicis shift).

IV. GENERAL QUANTUM MECHANICAL BASIS FOR ESCALOSS

A. THE PINES AND NOZIÈRES ARGUMENTS (EXTRINSIC ONLY)

There are a number of essentially equivalent ways of representing the electronic system of a standard solid about to be subjected to photoelectron ejection (with the accompaniment of any subsequent loss phenomena). One of the most expressive ways is to employ a perturbative partition of the system Hamiltonian, using the advantages of creation and annihilation operators in either position or wave vector space, i.e., as before, we begin with the Schrodinger equation,

$$H\Psi(N) = E\Psi(N) \tag{6-2}$$

where now we assume that

$$H(N) = H^\circ + \lambda H^1 \tag{6-3}$$

where N defines the number of system electrons being considered and, as in Chapter 4, $H^\circ \equiv$ an appropriate, approximate sum of *independent, one electron problems*, and $H^1 \equiv$ any resulting electron-electron interaction terms over all 3m individual electron terms, and all $3(N - m)$ collective coordinates (see below). In this manner, as shown by Pines et al.[7,8]

$$\Psi(3N) \rightarrow \int \Phi(3m)\chi[3(N-m)] \tag{6-4}$$

where $\int$ represents an appropriately symmetrized, product partitioning of the coordinate space of the system designed to reflect the possible presence of both a space of individually functioning electrons and "regions" where these electrons only "evolve" collectively.

1. H° — The Various Models

Depending upon the nature of the systems involved, any one of several possible H° may be chosen and constructed in terms of the previously mentioned creation and annihilation operators.

a. Valence Electrons Only

1. The free electron model

$$\sum_k \frac{k^2}{2m} C_k^+ C_k \qquad (6\text{-}5)$$

where C_k^+ is the appropriate electron creation operator in k space and C_k is the corresponding annihilation operator. Three cases arise, with two extremes:

  a. No collective (plasmon) terms (i.e., H^1 above will lead only to Nozières-Dominicis effects).[26]
  b. All valence electron interactions are collective (plasmon-type) (i.e., H^1 as in Lundquist,[13] Mahan,[9] and Langreth[2] models).
  c. Plasmon *dominated* interactions, dispersed by electron-electron interactions, as in the Langreth[2] and Gadzuk[3] models.

2. The Pines model[7]

$$\sum_i \left[\frac{P_i^2}{2m} + V(\vec{r}_i) \right] C_i^+ C_i \qquad (6\text{-}6)$$

where C_i^+ is the appropriate electron creation operator in position coordinates (i), etc., i.e., H° now includes a potential exhibiting the lattice periodicity, $V(\vec{r})$. The Φ^o that are eigenstates of H° now contain Bloch-type functions.

3. The Fock Hamiltonian, as earlier in Equation 4-10.

$$\sum_i F_i \qquad (6\text{-}7)$$

This is our previously defined "molecular solid", or Gadzuk's, Manne-Aberg shake-up[3,4] case.

b. Inclusion in H° of Certain Core Electron Effects

1. Inherently as a part of the Pines model.
2. These may also be added simply as polarization corrections to the Pines or the free electron model. Thus, the core inclusion can be shown to modify the plasmon dispersion (see below) from

$$1 = \frac{4\pi^2 e}{m} \sum_k \frac{f_k}{w_p^2 - w_k^2} \quad \text{to also include} \quad \frac{4\pi e^2}{m} \sum_i \frac{f_i}{w_p^2 - w_i^2} \qquad (6\text{-}8)$$

where f_k and f_i are, respectively, representative of valence and core electron induced, oscillator strengths, and w_p, w_k, and w_i are the appropriate frequencies for the free plasmon, intra (1) or inter (2) band, individual valence electron transitions, and appropriate core level transitions, respectively. All of this may in some instances be made more tenable by noting that

$$\varepsilon_c = 1 + \frac{4\pi e^2}{m} \sum_i \frac{f_i}{w_i^2} \tag{6-9}$$

and therefore, as we shall argue, the approximation

$$w^2 \cong \frac{w_p^2}{\varepsilon_c} \tag{6-10}$$

may be employed, where ε_c is a reasonable rendition for a static, core, dielectric constant.[7] One can employ certain empirical arguments for the latter in order to avoid quantum mechanical calculations.

3. The inclusion of core electrons in the Fock Hamiltonian above is generally a much more difficult proposition that may substantially enhance the scale of the Hartree-Fock calculations.

c. Explicit Inclusion in H° of Electron-Electron Correlation

It is also possible, of course, to expand on the initial treatment of the system, and explicitly include some degree of the electron-electron correlation in our H° for any of the cases in Section IV.1.a. As shown by Nozières and Pines,[8] such alterations will still permit a detailed perturbative treatment of the plasmon problem as part of H[1], but will naturally substantially increase the complications.

2. H¹ — The Various Types of Electron Interactions

Having described various possible H° cases, we can continue to follow Pines,[7] and Nozières and Pines[8] and describe the *extrinsic-only* H[1]:

$$H^1 = H_{SR} + H_{coll} + H_{int}$$
$$\quad\text{(i)}\quad\quad\text{(ii)}\quad\quad\text{(iii)} \tag{6-11}$$

where

1. H_{SR} — Short-range-electron pair (e.e) interactions. (Part of these would move into H° for case c above.) According to Pines,[7] these interactions may be expressed as

$$H_{SR} = \sum_{k > k_c} \frac{2\pi e^2}{k^2} \left(\rho_k^+ \rho_{-k} - N \right) \tag{6-12}$$

where ρ_k^+ and ρ_{-k} are the appropriate, respective, electron *pair* creation and annihilation operators, for example,

$$\rho_k = \sum_i e^{i\vec{k}\cdot\vec{r}_i} C_i^+ C_i \qquad (6\text{-}13)$$

In this case, k_c defines the range in k space over which the electrons behave as a collective, (plasma-like) entity. Thus, above k_c the plasmon is no longer a valid particle, and one must consider all electrons as exhibiting only individual behavior. (Note that the phrase "short range" occurs because position space is the reciprocal of k space!) In general, it is crucial to the present argument that k_c be relatively large for most of the *conductors* under consideration, i.e., that the short range (e.e) effects enter a plasmon-dominated free electron situation as reasonable (relatively small), perturbations. (This type of phenomenon was adroitly argued for the photoelectron case by Langreth,[2] and later by Gadzuk.[3])

2. H_{Coll} — long-range (H_{LR}) collective (plasmon) fields

$$H_{coll} = \sum_{k<k_c} \left[\frac{P_k P_k^*}{2} + \frac{w_p^2 Q_k^* Q_k}{2} \right] \qquad (6\text{-}14)$$

where P_k and Q_k are the appropriate momentum and coordinate operators, respectively, for the plasmons. For some purposes it is more convenient to rewrite this expression in terms of appropriate (boson) creation and annihilation operators:

$$H_{LR} = \sum_{k<k_c} w_k a_k^+ a_k \qquad (6\text{-}15)$$

where, without dispersion for the free electron model above

$$w_k \rightarrow w_p = \left(\frac{4\pi N e^2}{m} \right)^{1/2} \qquad (6\text{-}16)$$

where N is the total valence electron density, and the effective mass $m^* \rightarrow m$ for cases of maximum k_c, i.e., where there is no valence or core dispersion. As we shall see, reduction in k_c (introduction of dispersion) can be approximated as a deviation of m^* from m, and also by a reduction of the collective coordinate space from N to N′, where

$$N' = k_c^3 / 6\pi^2 \qquad (6\text{-}17)$$

and, where the difference 3 (N − N′) = 3m is that part of the general coordinate space that "retains" its individual electron character.[7] Note the previous discussion of the wave function in Equation 6-4.

3. $H_{int} \equiv$ dispersions of the collective field

a. Dispersions That Result from Interactions Between Collective and Individual Valence Electrons

As Pines[7] has shown, these dispersion effects are dominated by linear interaction terms

$$H'_{int} = \sum_{k<k_c} V_k Q_k \qquad (6\text{-}18)$$

where V_k is our rendition of a general interaction operator. There are also related nonlinear interaction terms that mix several plasmons with a particular electron. As we shall show later, the short-range individual electron (e.g., band to band) transitions also play a direct role in this dispersion.

b. Dispersions That Result from Core Interactions (See, for Example, Equations 6-8 and 6-10)

It should be noted that all such interactive effects depend crucially on two factors: (1) the relative size of the interactive oscillatory strength, f_j, specifically the strength of the interaction between a particular (jth) electron state and the appropriate plasmon, and (2) the relative similarity of the energies of any individual electron transitions, hw_i, e.g., valence-to-conduction band or near core state-to-conduction band, to the appropriate plasmon transition, hw_p. As we shall describe below, these features may be best expressed in the form of the development of a standard perturbation theory.[30]

B. LOSS SPECTRAL DEVELOPMENT FOLLOWING PHOTOELECTRON EJECTION (INCLUSION OF INTRINSIC PROCESSES)

As we have mentioned above, the problem of relating the developments of Pines and Bohm[7] to a photoelectron experiment is that the latter evolves in a hole ion environment, which was not considered by Pines.

In view of the results to be described, it is important to consider the character of the temporal development with the photoelectron ejection involved. A detailed study of this problem has been made by Meldner and Perez[34] and Müller-Hartmann et al.[35] In this regard, they noted that there is a substantial difference between the behavior of a photoelectron ejected from a deep core state [e.g., In(3d) in In°] compared to one from a substantially delocalized (e.g., valence band) state [e.g., In(5p) in In°]. The former may be considered to be *recoiless*[35] and thus to retain its *suddenly* evolving character[34] throughout the photoelectron process (initiation to detection), i.e., the presence of the hole may be expressed as

$$H_{DH} = \varepsilon b^+ b \qquad (6\text{-}19)$$

where DH designates a typical, *deep* hole of constant energy, (ε), that has been suddenly created by the Fermion operator, b^+.[35] In view of the persistence of this field during a photoelectron experiment (as well as the photoelectron itself), a much more complex Hamiltonian arises that compared to Equation 6-11 must include interactions with two "excitation" particles: (1) the photoelectron induced (extrinsic) processes and (2) the deep hole-induced (intrinsic) processes. A reasonable rendition of this Hamiltonian, H_T, may be expressed as[2]

$$H_T = \varepsilon b^+ b + H^\circ + H^{(1)} + H^{(2)}_{int} \qquad (6\text{-}20)$$

where H° and H^1 have been previously defined and

$$H^2_{int} = \sum_{k,k'} V_{kk'} D_k^+ D_k bb^+ \qquad (6\text{-}21)$$

The latter is a generalized relation depicting the interaction of the deep hole with the two types of electron fields (whose creations are symbolized by D^+) previously attributed to the solid system, i.e., interactions with the individual electrons:

$$H_{int}^{(2)SR} = \sum_{k,k'>k_c} V'_{k,k'} C_k^+ C_{k'} bb^+ \tag{6-22}$$

and the collective (plasmon) type field

$$H_{int}^{(2)LR} = \sum_{k,k'<k_c} V'' \left(a_k + a_k^+ \right) bb^+ \tag{6-23}$$

where the arrow points to Q_k above the term.

It should be noted that these last two interactions are, in fact, the solid-state physics equivalent of those parts of the final state, hole ion Hamiltonian that provide for the sudden evolution into the eigenstates, $\{\Psi_f(N-1)\}$, of that system: all of this was outlined above in Chapter 4 where the present argument corresponds to the case referred to by Gadzuk as "Mahan shake-up".[3,4] (One must be very careful here with the counting procedures employed as the electrons are fermions and the plasmons are bosons and their respective wave functions will reflect this in their statistics.)

As Langreth[2] was the first to demonstrate, and Gadzuk[3] has clarified, the shifts and states that result from H_T may be determined exactly, if one assumes that the system is adequately described by a free electron solution with no, or only modest, dispersions (i.e., starting with the H^o in the free electron model). In this regard, one obtains a photoelectron probability that contains several terms, including an electron (extrinsic) induced term

$$P(n,z) = e^{-z/\lambda} (z/\lambda)^n / n! \tag{6-24}$$

where z is the distance within the solid where the photoelectron originates and creates n plasmons and λ is the mean free path of those plasmons. In the total, sudden manifold, *where plasmons predominate for a free electron system*, the term in Equation 6-24 must be convoluted with a contribution due to the sudden creation of the deep hole (intrinsic) effects, i.e.,

$$P(w) = \sum_{n=0}^{\infty} |<0|n>|^2 \, \delta \left(hw - E^0 + \Delta \varepsilon^R - nhw_p \right) \tag{6-25}$$

where we are considering the so-called principal, no loss peak ($n = 0$), and only losses due to discrete, collective (plasmon-type) transitions ($n = 1, 2$, etc.).[2,3,34]

As mentioned above, it is possible to obtain an exact solution for the term ΔE^R (the relaxation correction), if the system is assumed to behave only as a manifold of slightly perturbed, free electrons.[2] In this case, ΔE^R may be divided into two distinct effects:

1. *Long range*

$$\Delta E_{LR}^R \rightarrow -iaw_p = \sum_{k<k_c} \frac{g_k^2}{w_k^2} w_p \tag{6-26}$$

where, due to the nature of the problem, a is a constant shift

2. *Short range*

$$\Delta E_{SR}^{R} \rightarrow i\Delta t + \text{ additional smaller corrections} \qquad (6\text{-}27)$$

where t is the time and Δ is also a constant rendition of the previously mentioned, catastrophically induced, long wavelength (infrared) manifold of transitions. The latter were originally realized by Nozières and de Dominicis[26] and, where finite, should shift the centroid of each peak (both nonloss and loss) upfield (to higher binding energy). As demonstrated by Anderson,[25] this shift is in effect an infrared catastrophe that results from the fact that the resulting hole-particle situations are in fact bosons. An excellent discussion of the nature of these shifts with figurative renditions was provided by Langreth.[2] Our corresponding rendition of these shifts appears in Figure 6-2.

It should be noted that Müller-Hartmann et al.[35] have demonstrated that, if the aforementioned hole is not sufficiently deep (e.g., for a valence band photoemission), then the previously described relaxation shifts should be muted. The intrinsic transitions, however, may still occur.[35]

C. FURTHER ARGUMENTS ON THE NATURE OF THE ELECTRON-INDUCED DISPERSIONS IN LOSS SPECTROSCOPY

In the case where the free electron model still applies, i.e., where E_g (the band gap) < 1.2 eV, following Pines,[7] it is easy to see that dispersions of plasmons may be described in terms of those features often referred to in solid-state physics as the $\vec{k} \cdot \vec{P}$ approximation.[36] In this method (sometimes also labeled as the near free electron model[37]), the electron traversing the lattice is assumed to experience small, constant, perturbative accelerations that reflect the lattice periodicity. As a result, energy shift terms related to

$$\Delta E_{Dis}^{(s)} \propto \sum_{r} \sum_{i} \frac{| \vec{k} \cdot \vec{p}_i |_{sr}^2}{m^2 \left(\Delta E_{sr}^0 \right)} \qquad (6\text{-}28)$$

must be added to hw_p, where i labels the particular valence electrons involved, and the $\Delta E_{Dis}^{(s)}$ modulates the effects of the free electrons in state s by introducing the modest "impact" of the lattice, represented by all other states [r]. ΔE_{sr}^0 is the energy separation of states s and r at $k = 0$. This correction seems to act to produce a nonlinearity in k space in that the free electron behavior, originally expressed by Equation 6-5, no longer applies. In fact, energetic changes occur as if the electron is forced to accelerate (or, alternatively, decelerate) by the factor ΔE_{Dis}, as it (effectively) crosses the lattice. As mentioned above, it is common to express this acceleration as if the rest mass of the electron's involved in plasmon formation have been altered from their "free" value, m, to an effective value, m*.[36] In this case, the dispersion of the free electron, plasmon frequency may be reasonably expressed as in Equation 6-8.

In addition to this effect, when E_g > 1.2 eV we note that dispersion implies that the short-range interactions of the electrons also must be finite, and we must include higher order correction terms that should be proportional to k^2. If the system is described by the H^0 given in Equation 6-6 and H^1 in Equations 6-11 and 6-14, then we must also allow for the effect of band-to-band transitions (from, say, state 0 to all q possible electron transition states) on this process. As Pines[7,8] has shown, this leads to the relationship:

$$\left[\frac{\partial^2 \rho_k}{\partial t^2} + w_p^2 \rho_k\right]_{qo} = -w_{qo}^2 (\rho_k)_{qo} + \sum_{k' \neq k} \frac{4\pi e^2}{m(k')^2} \vec{k} \cdot \vec{k}'(\rho_k' \rho_k)_{qo} \qquad \textbf{(6-29)}$$

where the first term on the right is the linear electron-plasmon ($\vec{k} \cdot \vec{P}$) correction (as described above), and the second is a corresponding nonlinear electron (k) – plasmon (P) – electron (k') correction.

In this case, we may then write

$$w_{qo}(\rho_k)_{qo} = \left|\sum_i \left(\frac{\vec{k} \cdot \vec{p}_i}{m} - \frac{\hbar k^2}{2m}\right) e^{i\vec{k} \cdot \vec{x}_i}\right|_{qo} \qquad \textbf{(6-30)}$$

where $\vec{x}_i$ locates the point of the linear interaction. Following Pines,[7,8] we can expand these "corrections" in the form of a perturbative expansion where

$$w^2 = w_p^2 + \frac{4\pi e^2}{\hbar k^2} \sum_q \frac{2 |w_{qo}(\rho_k)_{qo}|^2}{w^2 - w_{qo}^2} w_{qo} + \text{higher order corrections} \qquad \textbf{(6-31)}$$

Dispersion terms, such as those on the right in Equation 6-29, may be generally expressed in the form of Equation 6-18. It is important to note that all of the terms in this relationship may be obtained from independent measurements of the spectroscopic properties of the particular electronic system under consideration.

Perhaps it is useful at this point to stop and quickly review those features of the ESCA process that have been described so far in this section.

We have noted that

1. Many materials with zero, or relatively narrow, band gaps have collective loss features, described as plasmons, that often provide a significant presence in the ESCA spectrum.
2. *All* of the peaks detected in the ESCA spectrum exhibited by materials of this type are shifted by relaxation effects that can be reasonably modeled by Equations 6-26 and 6-27.
3. In addition, we have noted that only the "most pristine" plasmon sources will produce loss lines split from the no-loss peak by the hw_p of Equation 6-16, i.e., most systems will experience some additional dispersion (perturbation) shift, and a reasonable approximation for many of these shifts is provided by Equation 6-31.

Viewed from another perspective, the completely undressed, free electron plasmon expressed in the intrinsic relation of Equation 6-25 must generally be modified to include hw in place of hw_p, where w reflects the dressing (dispersions) expressed in Equation 6-31. (It should be noted that this modification of the frequency of the collective effect also influences the extent of the relaxation; see, for example, Equation 6-26.)

All of these possible effects are hypothetically rendered in Figure 6-2.

D. ON THE SEPARATION OF THE HOLE AND ELECTRON-INDUCED LOSSES

We have thus suggested that loss spectra may be created by both the influence of the photoelectron itself (Equation 6-24) and what we have described as *the hole spectral function* (Equation 6-25). These terms are presented in the manner of Gadzuk[3,10] as two separable functions. This has been shown to be a reasonable concept when the photoelectron escape velocity is sufficiently fast to warrant the application of the sudden approximation (i.e., photoelectrons of several hundred electronvolts, or more). As the photoelectron slows, and remains for a longer period in the vicinity of the hole, an interference term also results, indicative of an adiabatically induced interaction of both excitation sources (recall the discussion in Section 4.II.A.2). This leads to modifications of the extent of relaxation, e.g., the *a* factor in Equation 6-26 is damped to

$$\left[a' = a - \left(\frac{e^2}{hv} \right) F \right] \tag{6-32}$$

where v is the velocity of the electron and F is a function that is proportional to the charge and velocity and inversely proportional to the plasmon frequency. The term a' thus modifies the plasmon relaxation, and the Δt factor in Equation 6-27[2,3] is also damped. These interference terms are generally considered to be inherent parts of the intrinsic (hole induced) excitation.[11]

As pointed out by several researchers,[2,3,11] the photoelectron experiment often does not yield situations that permit ready separation of the extrinsic and intrinsic contributions. This occurs, in part, because all photoelectron ejections must correspondingly create a hole, and thus the intrinsic excitations must always be present. However, as we have described above, the determining factor in *reducing* the relative contribution of intrinsic effect depends upon the energetic *location* of that hole.[29,34,35] In this regard, it is important to note that, whereas a localized deep hole may maximize the intrinsic contribution inside a particular layer an experiment can be constructed in which the photoelectron created in one layer *may* be forced to pass through another, chemically detached layer, before being detected. This *"displaced" photoelectron may produce loss transitions in the latter layer that are only of extrinsic origin.*[29,38] Examples that suggest the possible registration of this situation are described later in this section.

E. ESCALOSS
1. E_L and ΔE_L for Materials with Nonzero E_g

It is useful now to consider the expressions that we feel will most appropriately describe the loss splittings, $E^A_L(1)$, for atom A in chemical state (1) (as defined in Figure 6-3), and the change that results in these loss splittings when the state is changed from 1to2, $\Delta E_L^A(1 \rightarrow 2)$. It should be noted that Tables 6-2 and 6-3 contain comparisons of the experimental and select theoretical values of these types of terms for a variety of materials systems. In all cases the approximate dispersions and values are loosely related to the size of the band gaps, E_g. As we shall describe below this correlation is fairly general, but not universal.

Because of the relative ease in detecting and reproducing those generally distinctive features, we have labeled the registration of E_L and ΔE_L for analytical purposes as *ESCALOSS*.

2. Relaxations and Loss Peaks Inside the Free Electron Model

If one scrutinizes Equation 6-25, one will notice that the plasmon states defined by the quantum numbers n = 0, 1, 2, etc. form a series of poles as first displayed by Langreth.[2] In the latter description, each of these plasmon states form poles that are displaced upfield (to lower binding energy) by the ΔE_{LR}^R in Equation 6-25 from the positions established before considering hole-induced relaxations, but it should be noted that as expressed these

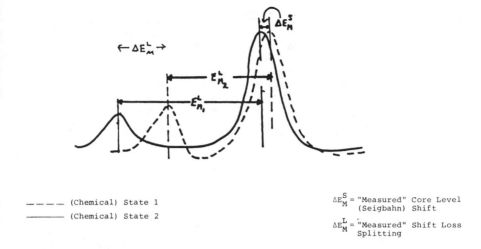

Figure 6-3 ESCALOSS showing figurative E_L and ΔE_L values.

Table 6-2 **Elemental, E_L values compared to other experimental ESCALOSS, non-ESCALOSS, E_L values, and calculated E_p (all results in eV[a])**

	Band[b] gap	hWp free electron	Barr XPS	Pollack[99b] loss	Pines[100b] optical	P & E (HEELS)[c]
In	0	12.5	11.5	11.8	12	
Sn	0.08	14.1	14.0	14.2	12	
Sb	0.1	15.1	N.A.	16.0	15	
Al	0	15.8	15.8	15.6	15	
Cd	0	10.5	9.4	9.3	20	
Si	1.17	16.6	17.8		17	16.9
Ge	0.74	15.6	16.5		17	16.4
C	5.2[d]	31.1	21.6[e]		22	

[a] Loss results from this lab ± 0.2 eV. Consult original references for other precisions.
[b] Intrinsic.
[c] P & E ≡ Philipp, H. R. and Ehrenreich, H., *Phys. Rev.*, 129, 1550, 1963.
[d] Diamond.
[e] Absorbed hydrocarbon.

ΔE_{LR}^R values are constants, aw_p, added onto each nw_p for each state, 1, of each element, A. In addition, if one considers the short-range relaxations, ΔE_{SR}^R, the resulting, weak, downfield shifts (generally in the infrared range) are also constant.[2] Thus, as demonstrated in Figure 6-2, even though the principal (no loss) peak and all of the plasmon-dominated loss peaks suffer significant relaxation effects, for dispersions out of the free electron model, the amount of the resulting peak shift is a constant for each peak; *therefore, as indicated in Figure 6-2, $E_L^A(l)$, the loss splitting for material A in state 1, and $\Delta E^A(l \rightarrow 2)$, the changes in loss splitting going from state 1 to 2 are independent of this particular relaxation.*[27,28,30,31]

Table 6-3 **Molecular E_L values**

	B.G.	**hWp**	**Barr XPS loss**	**This lab HEELS**	
GaAs	1.5	16.3	16.2		
InSb	0.25	13.9	13.5	14.0	
GaSb	0.8	14.8	15.3		
InN	1.7	19.0	15.5	16.0	
				Pines[100b]	**P & S[a]**
In_2O_3	3.6	22.5	18.5	19	
Al_2O_3	7.2[b]	N.A.	24.4[c]	23	22.2
SiO_2	9.0[b]	N.A.	23.0[c]	25	

[a] P & S ≡ Powell, C. J. and Swan, J. B., *Phys. Rev.*, 118, 640, 1960.
[b] Indirect.
[c] Average 2p-Variable.

3. Summary of Photoelectron Loss Splittings for Different Models
a. $E_g \leq 1.2$ eV

In this case, if we start with no dispersion we obtain

$$E_L^A(1) \cong h\left(\frac{4\pi N}{m}\right)^{1/2} = K\sqrt{N^A(1)} \qquad (6\text{-}33)$$

where K is employed to symbolize all constant terms and $N^A(1)$ is the "atomic" density of element A in "compound" 1, and

$$\Delta E_L^A(1 \to 2) \cong K\left[\sqrt{N^A(2)} - \sqrt{N^A(1)}\right] \qquad (6\text{-}34)$$

When we take into consideration the modest dispersion that may be realized through Equation 6-28 above, we find that we should modify Equation 6-33 to read:

$$E_L^{A'}(1) = K\sqrt{N^A(1)} + \frac{\hbar^2 k^2}{2m}\left(\frac{m}{m^*}\right) + \text{HOT} \qquad (6\text{-}35)$$

where HOT designates higher order terms. A somewhat different form of these corrections may be expressed by substituting $m^*_{1(A)}$ for m in Equation 6-16. In the case of Equation 6-34, this leads to

$$\Delta E_L^{A'}(1 \to 2) = k\left[\sqrt{\frac{N^A(2)}{m_2^*}} - \sqrt{\frac{N^A(1)}{m_1^*}}\right] \qquad (6\text{-}36)$$

Effective mass values, m^*, may be found in the literature for some of the appropriate systems. Based on Equation 6-36, therefore, the differences recorded in Table 6-2 between E_L and E_p for such systems may be reflected in these m^* values. These E_L results should, at least, suggest whether m^* is larger or smaller than m. One should be careful at this point, however, since, as we have outlined above in Equation 6-4, the processes previously described as short range may also produce a corresponding, effective reduction in the number of electrons experiencing collective effects from N to N′. As we shall show below, this effect also should be factored into Equation 6-36.

b. $1.2 < E_g \leq 3.5$ eV

By including the electron-plasmon dispersion arguments[7,8] suggested in Equations 6-29 and 6-31, we find that we may argue that

$$\Delta E_L^A(1 \rightarrow 2) = K\left[\sqrt{N^A(2)} - \sqrt{N^A(1)}\right]$$

$$-K'' \sum_{k<k_c} \left\{ \frac{1}{N^A(2)}[A(1)] - \frac{1}{N^A(1)}[B(2)] \right\} \qquad (6\text{-}37)$$

where, for example,

$$[A(1)] = \sum_q {}^{(1)} \, f_{oq}(k) \frac{w_{qo}(1)w_p - \left(w(1)_{qo}\right)^2}{w_k^2 - w_{qo}(1)^2} \qquad (6\text{-}38)$$

where $f_{oq}(k)$ is the previously defined oscillatory strength for a 0 to q transition.

In this case, therefore, we are assuming that the dispersion experienced by the free plasmon is primarily due to short-range effects conveniently expressed as quasi-perturbation corrections induced by noncollective band-to-band transitions, from state 0 to q.

At this point we should reconsider a key feature expressed about in both Equations 6-36 and 6-37. In view of the fact that our plasmons (and for that matter our individual electrons) are now (of necessity) "dressed" with dispersions, we must reexamine whether E_L and ΔE_L may still be considered free from relaxation effects. This feature is considered below in Section 6.IV.F.

c. $E_g > 3.5$

It should be noted that we have not proven that the loss spectra for materials with band gaps greater than ~1.5 eV are dominated by collective (plasmon) effects. In fact, it seems apparent that the relative importance of plasmons should generally decrease as the band gaps of the involved materials grow. Despite this, *it is our contention that the principal loss lines detected for most materials, even those with comparatively broad band gaps (including, for example, Al$_2$O$_3$[31] and even SiO$_2$[28]) are still primarily of plasmon type.* In this sense, we are following the lead of Pines[7] and Nozières and Pines,[8] and arguing against the implied thrust of the theoretical work of the 1970s. It should be obvious, however, that the following points must be considered as valid, whichever source one accepts as the principal cause for the loss transitions in question:

1. The size and influence of the plasmon contribution to any loss peak structure is directly proportional to the electron density at, or very near, the Fermi edge. Thus, as the valence band density of states shrinks, and/or the leading edge pulls back away from the Fermi edge (as happens, for example, for the sequence of materials In^o → InSb → InN → In$_2$O$_3$[27]), then the relative sizes of the loss peaks decrease, and the exclusive nature of any plasmon contribution must decrease.
2. Band-to-band and/or near core level-to-conduction band discrete loss transitions may alternate in intensity with the corresponding plasmon transitions. Thus, as the "effects" of the latter weaken, the relative contributions of the former generally grow.[7,27,29]

3. As various types of transitions of the right symmetry (e.g., plasmon and band-to-band) approach one another in size, and energy (see, for example, Equation 6-31), *these transitions will mix and create combined features.*[30] Thus, as we have argued, no matter what effect one accepts as the dominant influence of the resulting loss transitions, the experimental E_L is generally subject to extensive quantum mixing with any near degenerate transitions.[30]

The principal reasons for *our* supposition of plasmon domination for most E_L are the following:

1. The arrangement and position of the loss peaks — In numerous cases, even those with E_g not 0, multiple loss peaks have been detected (e.g., SiO_2,[28] InSb,[27] GaAs,[30] etc.), and most of these multiple features follow the Poisson distribution pattern required for harmonic oscillator-driven plasmons, rather than a pattern indicative of band-to-band-type transitions.

2. The lack of energetically appropriate band-to-band transitions — Based upon the ESCA determination of the positions of the valence band density, and any realized near core states, there are *often* no ready choices for possible band-to-band or near core-to-band candidates to explain many observed loss transitions.

3. Improper sequencing of loss transitions with chemical changes — It should be noted that the sequences observed in the loss splittings for select elements, ΔE_L (such as that reported in Table 6-4 for indium in going from $In^o \rightarrow InSb \rightarrow InN \rightarrow In_2O_3$), cannot be explained by invoking band-to-band arguments using the corresponding energy differences between the In(4d) and the valence and conduction bands.[27,30]

4. Further aspects of inappropriate band-to-band features — In the case of Al_2O_3[29] and SiO_2[28] (and many related oxides[30]), if one argues that the E_L are due to band-to-band transitions, then one must explain why, with a substantial difference in the corresponding valence band features and gaps (see Chapter 7), there is so little difference in the E_L values for these oxides and also why the small E_L differences detected seem to change in the wrong direction.

It may appear to the reader that the author is being somewhat equivocal at this point in referring to the types of material systems whose loss transitions may be described as "of plasmon origin". This equivocation is purposeful, as there is really no adequate, simple model that encompasses all experimental cases.

As indicated, we have found that the size of the band gap is one useful signpost of the degree of plasmon character, but a far from inclusive one. Thus, while there are certain periodic sequences of material for which as the band gap, E_g, shrinks their loss spectra become more and more plasmon dominated, there are other groups of elemental or compound species that exhibit this type of loss behavior long before E_g reaches zero, while still other materials with zero E_g may produce little or no significant plasmon dominated losses.

Thus, an alternative method of designation is obviously needed. Although, still incomplete, the following composite statements are proposed:

1. If a material has a zero band gap that results from the Fermi edge coupling (electron occupancy) of only s or s + p bands that are conductively configured (delocalized) at E_F, then readily discernible, plasmon-dominated loss transitions will result.

2. If the sole presence of delocalized s or s + p states at E_F is substantially mitigated by the ready, near E_F, presence of highly directed, localized orbitals (e.g., those of d and/or f type), then materials with this type of electron distribution will generally exhibit significant Nozières-de Dominicis (up field) peak distortions, at the cost of plasmon losses. Many of the elemental transition metals are ready examples of this type of system.

Table 6-4 **In° and In compound ESCALOSS results**

	Direct band gap	XPS this lab ± 0.2	XPS other lab ± 0.2	Optical ± 0.5	Heels this lab ± 0.5	Heels other labs ± 0.5	hWp free electron calculation
In°	0.0	11.5	11.8[a]	12.0	—	—	11.0[b]
							11.5[c]
InSb	0.25	13.3	—	—	13.5	13.0[d]	13.5
InN	1.7	15.5	—	—	16.0	—	19.0
In$_2$O$_3$	3.6	18.5	—	—	19.0	—	22.5

[a] Taken from Reference 61 and Reference 6.
[b] Taken from Reference 7.
[c] This lab.
[d] From Powell, C. J. and Swan, J. B., *Phys. Rev.* 118, 640,1960. With permission.

3. If there are no near-valent d or f states, then a system may still produce plasmon-dominated losses, even if its s + p states are "organized" and pulled back from E_F into a semiconductor format (e.g., Si°).

4. If the materials in question are wide band-gapped semiconductors or insulators, but do not have any near-valent d or f states, then the loss peaks of these materials are substantially dispersed, but are still most accurately described as "of plasmon origin" (e.g., Al$_2$O$_3$ and SiO$_2$).

5. If, on the other hand, the material in question is endowed with a modest E_g, and its valence band states are substantially populated by partially filled (or conversely partially unfilled), localized, directed d or f states, then arguments that initiate with band theory are generally not adequate to describe its loss spectra. These materials are some of the previously mentioned "molecular solid" systems, and their quantum character is best described using orbital models that lead to Manne-Aberg-type loss transitions (see Chapter 4). Examples of these materials include CuO, CeO$_2$, NiO, etc.

F. THE INFLUENCE OF DISPERSIONS ON RELAXATION EFFECTS IN E$_L$

A feature that was not considered in any previous discussion of the plasmon effect is that for semiconductors and insulators the relaxation shifts of both the principal (no-loss) line and the corresponding plasmon-dominated loss peaks should not be considered as a coupling of a (deep) hole state with a free electron plasmon (as in Equation 6-25), but rather as a coupling between that hole and a fully dressed (dispersed) plasmon,[7] such as that described in Equation 6-31. If these effects are properly considered, then one will find expressions such as

$$\Delta E_{LR}^{R'} = a'(q)w_q \qquad (6\text{-}39)$$

where $a'(q)$ will be needed to replace the simpler relation for (a) in Equations 6-26 and 6-32. The complications represented in Equation 6-39 are severalfold. First, one must begin by recognizing the electron-plasmon, dispersion shifted frequency, w_q, in place of w_p. Second, it must be acknowledged that the collective-hole coupling term, $a'(q)$, is now affected through this dispersed collective, rather than the free electron plasmon. This means that a' in Equation 6-39 may differ substantially in size from the a in Equation 6-26. Further, a' is now a complex function that will, to higher orders in perturbation theory, include the dispersive forces acting on the plasmon and thus introduce Fermion terms that will, in turn, couple with the Fermion hole. The result will not be exactly soluble into a constant term (such as a), and higher-order terms may depend upon the final state index, n, which will sum on the mixed collective-individual states (i.e., states that no longer index just the plasmons). If the latter is true, terms similar to the following will occur in the energetic delta function in Equation 6-25[8]

$$\delta(\Delta E) = \delta\left\{hw - \varepsilon + a'(q)w_q + \Delta t - n\left[w_q - a''(q)\right]\right\} \qquad \textbf{(6-40)}$$

where the factor $a''(q)$ is employed to represent any higher order effects in the dispersion that affect each of the n states differently. *These effects will therefore make the loss splitting E_L (and corresponding ΔE_L) relaxation dependent.* Based upon the nature of this argument, however, it is felt that any effect of this type will generally be quite small; therefore, it should be reasonable to ignore relaxation shifts in applying Equation 6-25 to most semiconductor (and even some insulator) loss shifts.

G. FINAL ARGUMENT FOR EFFECTS OF DISPERSION

In view of the previous arguments, if one wishes to determine the nature of the loss spectra in photoemission (and perhaps use it in analysis), one needs to estimate the relative contributions of the collective and noncollective behavior. In some cases we will find the latter to be similar in size to the former; however, as we have argued above, generally *for solids of major interest*, the collective (plasmon) features dominate. (Recall that despite the previous discussion there is no absolute proof for this statement other than the fact that it seems to support the resulting data.) On the other hand, Pines has demonstrated,[7,8] and we have pointed out,[27-33] that, for most nonconductive materials, the noncollective effects should be large enough to influence the plasmon feature and therefore cannot be ignored. Generally, however, despite the large size that might be predicted for some of these noncollective effects, one can generally construct reasonable arguments for their inclusion, in terms analogous to conventional perturbation theory.[30] For example (as we have seen), we may assume that a reasonable solution may be achieved through second order, i.e.,

$$E_L(i) \cong E_0^0(i) + E_0^1(i) + E_0^2(i) \qquad \textbf{(6-41)}$$

where, by analogy, $E_0^0(i) = E_P(i)$, the free electron, plasmon value, and i is employed in this section to collectively index both the element, e.g., (A) and its state, e.g., (1). Now, based upon previous arguments,

$$E_0^1(i) = 0 \qquad \textbf{(6-42)}$$

and

$$E_0^2(i) \cong \sum_{j=1} \frac{|<\phi_j \,|\, H_{int} \,|\, \phi_0>|^2}{\left(E_0^0 - E_j^0\right)} \qquad \textbf{(6-43)}$$

where $E_0^2(i)$ is thus the approximate change in the loss splitting [$E_L(i)$], created by the dispersion of the collective motion by the individual electron behavior. Thus, using Equation 6-31, it is reasonable to consider the energies E_j^0 to be appropriate band-to-band or near core-to-conduction band transition energies, and the numerator of Equation 6-43 corresponds to the interactive matrix elements in Equations 6-30 and 6-31. The latter tend to grow as the band gap broadens. In this regard, we may obtain a qualitative feel for the relative size and direction of the shift provided by $E_0^2(i)$ by noting that

$$E_0^2 \text{ increases in size as } |E_0^0 - E_j^0| \mapsto 0 \qquad \text{(6-44)}$$

that is, as the appropriate individual electron transitions to the conduction band become similar in size to that for the appropriate free electron plasmons. More specifically when

$$\left[E_p(i) - E_j^0\right] > 0, \text{ then } E_L > E_p \qquad \text{(6-45)}$$

and when

$$\left[E_p(i) - E_j^0\right] < 0, \text{ then } E_L < E_p \qquad \text{(6-46)}$$

A further concept that should be obvious is that when relationship 6-44 does not hold (i.e., when there is no band-to-band or core-to-conduction band transition of energy close to E_p), then $E_L \sim E_p$. Although these arguments are worthy of consideration, they should not be over-subscribed. One should note in the tables in this book and also the papers of Pines[7,8] that there are numerous exceptions.

It should also be noted that a lack of dispersion does not necessarily signify that the loss peak will be large in relative intensity,[6] nor does it necessarily indicate that systems containing a particular element in a particular state (i) with large dispersions will produce loss peaks with less intensity than those systems containing (i) with modest dispersion.

V. EXAMPLES OF THE USE OF ESCALOSS IN CHEMICAL ANALYSIS

A. ELEMENTAL SYSTEMS

The relative size of the previously described loss peaks seems to depend upon several factors — the primary one being the number and character of the density of states at or near the Fermi edge of the system.[7] Close study of loss size and type suggests that the best source of a system that provides a plasmon-dominated loss spectrum of relatively large size is a spherically symmetric (i.e., s state) or, at most, p-type valence density at or near the Fermi edge that is relatively free from any localized, directed, valence density (e.g., d-type) coming from the low binding energy parts of the valence region.[6] Therefore, *most* transition metals with large near-Fermi edge, directed, d-state densities should exhibit poor loss intensities, with little or no plasmon character.[6,30] Thus, for example, Al^0 ($3s^23p$) forms a well-recognized loss pattern, Figure 6-4, that is dominated by strong, plasmon peaks with little or no dispersions,[6,30,38] whereas Cu^0 ($4s^13d^{10}$), on the other hand, exhibits only a weak loss spectrum which may not contain any plasmon character.[39] As expected, In^0 also exhibits a strong, plasmon-dominated, loss pattern, similar to Al^0.[27,30] Interestingly, Cd^0, which has a near valent d band, still produces a significant plasmon-dominated loss spectrum,[6] apparently because its d band is pushed to relatively high binding energy, ~11 eV, separating it from the symmetric, plasmon-producing, s band state.

B. INDIUM COMPOUNDS

The use of ESCALOSS as a tool in chemical analysis requires, of course, that it be directly expressive of changes in chemistry. One of the first detailed demonstrations of this capability was provided in a detailed study of loss spectra produced by a variety of solid indium systems, including In°, InSb, InN, and In_2O_3, in both bulk and thin film form.[27] The critical aspects of the results are given in Table 6-4 and Figure 6-5, showing that the use of conventional, principal peak ESCA chemical shifts to differentiate between these systems is not feasible for all of these compounds, due to the vanishingly small size of the shifts.[27,40] In addition, the InN, and particularly the In_2O_3, are nonconductive systems that exhibit noticeable charging shifts, making the registration of conventional binding energies difficult. The loss splittings, E_L, of all these materials, however, as mentioned above, are independent of charging. The large size of these loss splittings and their progressive shifts make them intriguing chemical signposts.[27] It should be noted, however, that the size of the change in ΔE_L and the position of their E_L values are more involved than just a registration of the differences in valence state electron density per unit cell, N_i (i.e., the differences in their free electron plasmon shifts). The size of the difference between E_L and E_p (i.e., the dispersion), is found to increase with the band gap (Table 6-4), and thus also influence ΔE_L. In all cases, in the manner provided for in Equation 6-43, the transition between the intense, near core, In (4d), and the conduction band produces a perturbative "push" on the E_p for In systems.[30] However, this push should be relatively modest for In° and InSb, (due to Equation 6-36), and be countered by a push to higher binding energy by the valence band to conduction band transition.[30] In the case of InN, and particularly In_2O_3, the down field push of the $In(4d) \rightarrow$ conduction band transition is relatively large (due to the similarity of E_p and those particular E_j), and the resulting dispersion is ~3.5 (InN) and 4.0 (In_2O_3) eV, respectively.[30]

C. SiO_2-Si° and Al_2O_3-Al° Systems

In the case of oxide systems, it is tempting to suggest that any dispersions may be induced by the near core transition from O(2s) to the conduction band. This transition is

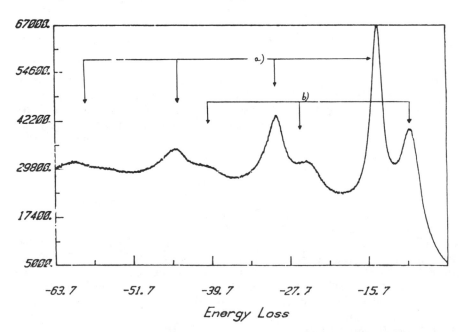

Figure 6-4 Loss pattern for Al° demonstrating Poisson distribution typical of free electron induced plasmons.

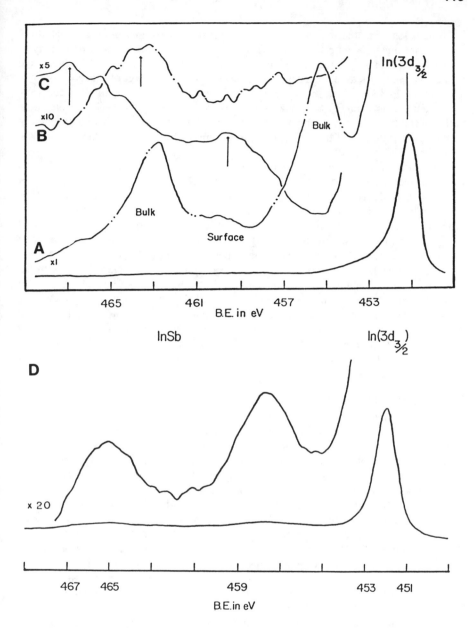

Figure 6-5 In(3d) loss lines for various indium compounds: A(In°), B(In₃O₃), C(InN), D(InSb).

approximately 27 to 30 eV for SiO_2 and Al_2O_3, in view of the respective positions of the two states involved. Based upon Equations 6-8 and 6-43, and the corresponding values for E_p, this should provide the ingredient for a significant perturbation. This, however, seems to be countered by the relatively weak photoelectron density provided by the O(2s), and one does not anticipate significant dispersion from this source.[28]

Thus, in the case of oxides with significant covalent bonding (e.g., SiO_2 and Al_2O_3), one is more inclined to attribute any dispersions (always reductions) of the plasmon-dominated losses to the reduction in "free" electron density in the valence band that results from this inherent covalency.[7,30] Thus, one should employ an electron density of N' in place of N in the plasmon calculation (Equation 6-4), where $m = N - N'$ is indicative of the covalently "tied-up" valency electrons. Evidence in support of this contention is given in the loss spectra of the oxides (in Table 6-3) that exhibit E_L that are generally significantly less than E_p. SiO_2 may be an interesting exception as it exhibits little variation (perhaps due to a countervariant, "push-up" dispersion provided by the valence band to conduction band transitions that, due to the extreme width of the SiO_2 bands, range from ~10 to ~22 eV).[28]

The character of the loss spectra for $Si°(2p)$ shows that it is dominated by plasmon influences, despite the band gap of ~1.1 eV. The ΔE_L for SiO_2-$Si°$ was utilized in a study of a $SiO_2/Si°$ layered system, with particular applicability at the interface (Figures 6-6 and 6-7).[28]

In a related study of the ESCA loss spectra provided by $Al(2s)$ and $Al(2p)$ during the oxygen-induced growth of the oxide on the surface of pure Al metal,[29,41] the substantial ΔE_L (= 8.2 eV) was used to follow the growth characteristics. However, as the oxide formed, the $Al°$ subsurface was still detectable because of the (~40-Å deep) depth of field of the ESCA. Thus, a variety of mixed loss lines resulted[28,41] (see Figure 6-8). Examination of these lines indicated a predominance of 15.8 eV E_L for $Al°$ and 24.0 eV E_L for Al_2O_3. Further loss peaks shifted up-field from the $Al(2p)$ of the metal by ~24 eV were detected suggesting the possible detection of extrinsic-only losses produced by $Al°(2p)$ photoelectrons during their passage through the outer layer of Al_2O_3.[29] (The extrinsic-only character results from the localization of the hole of the $Al°$ photoelectron in the metal matrix.) Additional details were also observed, but such studies are still too poorly controlled for any definitive judgments.

D. GENERAL FEATURES

It is interesting to try to catalog the rationales for the different dispersions suggested for ESCA loss spectra of various compound systems. Some of these speculations are summarized in Table 6-5. Table 6-6 demonstrates another feature about the general nature of the loss spectra of these compounds. If these losses have primarily a plasmon origin (based on Equation 6-41), then it is apparent that their E_L values should be directly proportional to the electron density in the valence band. This suggests that systems that are isoelectronic, *and not affected by other factors*, should yield essentially the same E_L values. In Table 6-6 we see that this is the case for most of the $Ge°$ isoelectronic series, with the exception of KBr. The same argument, however, does not hold for the $Sn°$ series. It is suspected that the exceptions noted are due to significant, selective dispersive forces that increase with the increased band gaps as one radiates out and down in these Group IV isoelectronic series.[30] The fact that the I-VII compounds deviate substantially from the tetrahedral structure that predominates for the other compounds is another reason for this dispersion.

Perhaps one of the most interesting (and potentially the most useful) examples of the status of plasmon-only vs. dispersed plasmon losses occurs for carbon-based systems. A preliminary analysis of several carbon systems is presented below.

VI. AN ADDITIONAL, IMPORTANT ESCALOSS SPECTROSCOPY EXAMPLE — THE CARBON PROBLEM

A. INTRODUCTION

One of the most discouraging problems in traditional ESCA analysis is the lack of uniqueness found in the positions of the principal $C(1s)$ lines for many different carbon-based systems (i.e., those with $-C-C-$, $-C=C-$, $-C≡C-$, and $-C-H$ bonds).[32,42] Thus,

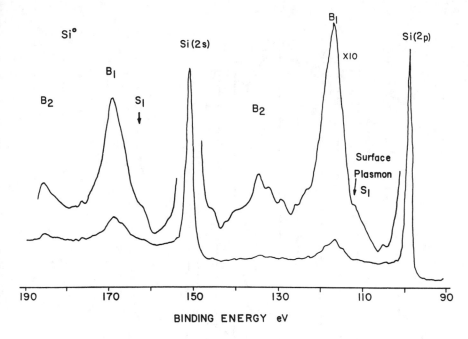

Figure 6-6 Loss spectra for Si(2s) and Si(2p) of Si° (note the E_L values and their Poisson repetition despite the Eg ~ 1.1 eV).

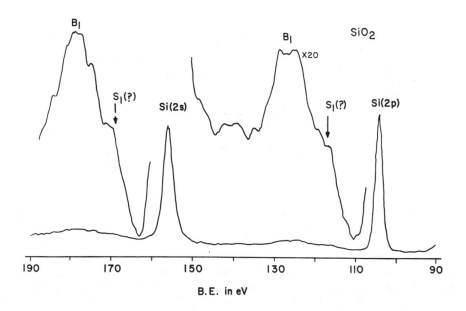

Figure 6-7 Loss spectra for Si(2s) and Si(2p) of SiO_2 (note the substantial increase in E_L over that for Si°, but significant reduction in loss peak intensity and lack of Poisson character).

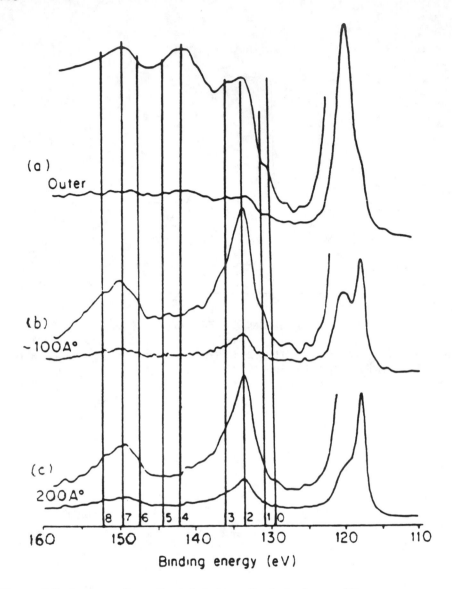

Figure 6-8 Loss results realized during growth of alumina on Al°.

whereas the (apparent) binding energy of the principal C(1s) peak may be employed to identify the presence of certain different organic functional groups (e.g., –C–O–, –C=O, – C–OH; as exemplified above in the AR-XPS study of polypropylene oxidation[43]), it cannot be readily employed to differentiate between, for example, polypropylene, polystyrene, and graphite. In fact, if asked to identify the bonding nature of the ubiquitous carbonaceous deposition that always occurs on air-exposed surfaces (referred to in the surface analysis literature as *adventitious carbon,*[32,44]) many researchers have classified it as a graphite-like substance, whereas others have suggested an adsorbed hydrocarbon polymer, with generally no evidence provided for either claim. There can be little doubt that this *constantly present* shortcoming is one of the more frustrating features of applied ESCA.

The principal reasons for this chemical shift problem would appear to be twofold and related. First, most of the systems in question are, at minimum, modest insulators and,

Table 6-5 **Loss splittings assuming plasmon character with dispersion**

System	ΔE_G	f_i^a	Cal. ε_P	Exp. E_L	Comment
Group IV systems					
$C_{(C_xH_y)}$	5.2[b]	—	31.1	22.0[c]	LoCo ↓
Si	1.2	—	16.6	17.8	VB→
Ge	0.7	—	15.6	16.5	VB ↑
Sn	0.1	—	14.1	14.0	M⟨–⟩
Pb	0.0	—	13.5	13.5	M⟨–⟩
Select III-V Systems					
GaAs	1.4	0.32	16.3	16.2	Ga(3d)→C.B. ↓ V.B. ↑
GaSb	0.8	0.27	15.3	14.8	Ga(3d)→C.B. ↓
					V.B. ↑
InN	1.7	0.5	19.0	15.5	In(4d)→C.B. ↓ Ion ↓
InSb	0.2	0.33	13.9	13.4	In(4d)→C.B. ↓ V.B. ↑
Select II-VI Systems					
ZnSe	2.8	0.62	15.6	16.0	Zn(4d)→C.B. ↑ V.B. ↑
ZnTe	1.2	0.60	14.5	15.5	Zn(4d)→C.B. ↑ V.B. ↑
CdS	2.5	0.68	NA	18.5	Cd(4d)→C.B. ↑ ↑
CdSe	1.8	0.68	14.7	17.0	Cd(4d)→C.B. ↑ ↑
CdTe	1.4	0.68	13.6	16.5	Cd(4d)→C.B. ↑ ↑
Selected Oxides					
BeO		0.62	29	29	None
ZnO	3.3	0.65	NA	—	—
Al_2O_3	7.5	0.80	27.0	24.0<	O(2s)→C.B. ↓ Ion ↓
Ga_2O_3	—	—	—	21.2	O(2s)→C.B. ↑ Ion ↓
In_2O_3	3.6	—	22.5	18.5	In(4d)→C.B. ↑
					O(2s)→C.B. ↓ Ion ↓
SiO_2	9.0	0.57	23.0	23.0	O(2s)→C.B.
GeO_2	6.0	0.73	NA	—	Ion ↓
SnO_2	3.8	0.78	26.0	20.0	O(2s)→C.B. Ion ↓

Note: LoCo = localized covalent bonding; Ion = localized ionic bonding; M = metal; V.B. = valence band; C.B. = conduction band; f_i = ionicity; ↑ = effect increases E_L.

[a] From Levine, B. F., *J. Chem. Phys.*, 59, 1463, 1973.
[b] Diamond.
[c] Approximate value for polymeric hydrocarbon.

therefore, due to charging produce somewhat uncertain C(1s) binding energy positions. Even if charging is curbed, however, most researchers have suggested that the chemical shift (if any) for C(1s) between, for example, graphite and an adsorbed hydrocarbon, is very small, perhaps too small to register. Hence, while some have suggested a C(1s) binding energy for graphite systems of 284.2 eV and that for a hydrocarbon polymer at ≈284.8 eV, and still others have recently described an apparent progressive shift of C(1s) from ~284.6 to ~285.5 eV for hydrocarbon polymers with increasing complexity, it is still common to shift all carbon-carbon and carbon-hydrogen C(1s) binding energies to a single value, recently adjusted to 284.7 ± 0.5 eV.[32,43,44] This lack of distinction, of course,

Table 6-6 **Loss splitting (E_L) in eV ± 1.0 for
isoelectronic series based on Ge_2 and Sn_2**

	E_p	E_L
Ge	15.6	16.5
GaAs	16.3	16.2
ZnSe	15.6	16.0
KBr	11.1	13.7[a]
Sn	14.1	14.0
InSb	13.9	13.4
CdTe	13.6	16.5[a]
RbI	10.9	NA[a]

[a] Plasmons, for these cases, may be too weak to be
primary producers of these peaks.

tends to make ESCA a less than adequate tool for investigations in the rapidly growing fields of organic composites and films. It is the purpose of the present discussion to show how ESCALOSS may be employed to provide at least a partial answer to some of these questions.

Based upon the methodology outlined above, a series of ESCA spectra were recorded for a variety of hydrocarbon and carbon-only systems, with particular emphasis on their core loss peaks. These results were produced using several different types of ESCA spectrometers. It is significant to note that reasonably reproducible loss results were achieved with all of these systems. Hence, the results described below are not unique to particular ESCA instruments. We have personally observed, however, that the ease with which these results are generated and the ability to differentiate certain subtleties in intricate use situations seem to be aided by use of a monochromator-based ESCA unit. Total analysis also usually requires the generation of a good survey scan (to examine for ubiquitous impurities, such as O, N, Na, Cl, etc.) and a good high-resolution scan of the C(1s) peak. The latter should be run with the use of an electron flood gun to remove charging shifts, an approach that is not necessary in the case of the loss spectral analysis, due to the independence of the loss splitting, E_L from charging.[22,32,43]

In addition, we found that valence band spectra scanned from about 0 to 30 eV may provide some delineation of the chemical character of these carbon systems.[6,32,44] A recent use of this type of study has been provided by Galuska et al.[45] Often, however, this valence band method may be even more difficult to register and control than the loss peaks. Fortunately, only one of these types of spectra is generally necessary to achieve analytical results. (One version of appropriate valence band data for certain carbon-based systems is described later in this text.)

A variety of different C(1s) loss lines also have been examined, and each has their own particular merit. Since the analysis to be described pivots on the determination of the loss splitting, E_L defined as the difference between the principal no-loss C(1s) line and the peak of the (rather broad) first, major, (perhaps) plasmon-dominated loss line, it is useful to include a spectral run that includes both of these features in the same spectrum. In addition, we have often recorded spectra that run up field to 350 eV or more in order to determine the presence of any secondary loss lines. Such lines were included because plasmon-dominated losses are found to arise in evenly spaced, repeated (harmonic oscillator-imposed) intervals, which fall off exponentially in intensity (i.e., with what is referred to above as a Poisson character;[6,7] see Figure 6-4). Hence, spectra of this type were recorded for a variety of carbon only and/or C–H- only compounds to examine for plasmon character in their loss spectra.

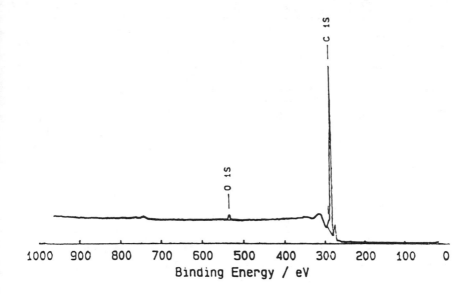

Figure 6-9 Survey scan for wafer of clean (but air-exposed) graphite.

Initial detection of some of the characteristic XPS loss peaks described herein (for graphite and related compounds) were reported earlier by several of the pioneering ESCA groups (e.g., that of Siegbahn[46] and Shirley[47]). In the latter case, they have suggested two separate prominances in the broad, primary loss peak of graphite that surrounds the 30-eV F_L point. We have not been able, however, to totally distinguish these weak peaks; therefore, we have suggested a singular transition of $E_L \sim 30$ eV.

Non-XPS-generated losses for graphite and related material have also been considered by others. These have included studies in X-ray generated appearance potential spectroscopy by Park et al.[48] and high-energy EELS investigations by Weissmantel et al.[49] and Liang and Cundy.[50] The carbon loss realized in Auger analyses of select systems were also considered by Luries and Wilson.[51] All of these features have been recently surveyed by Tsai and Bogy.[52]

B. GRAPHITE

As part of this analysis, several pressed wafers of chemically pure graphite were examined in various ESCA systems. Unpressed, pure, graphite powder also was examined. The state of graphite purity was gauged both by bulk specification and surface analyses (in the latter case, by employing the chemical detection of ESCA). In all instances, if air exposed, these materials were found to suffer very moderate (<2%) surface oxidation (see the survey scan, Figure 6-9). It should be emphasized that to varying degrees this is the case for *all* air exposed solids![43,53] Examination of systems cleaned of this slight oxide (and also of graphite more extensively oxidized, see Chapter 8 in this book) suggests that this level of oxidation does not play a significant role in the subsequent general analysis.

The characteristics of the C(1s) ESCA loss spectra for graphite are illustrated in Figure 6-10 and Table 6-7. The experimental patterns produced closely replicate that suggested for a carbon-based material that is a near total producer of plasmon-type loss excitations. Further, based upon the loss results obtained on various other polycrystalline and/or powdered materials, we do not expect the previously mentioned lack of total, macroscopic structural integrity to influence the position of the loss peaks. Thus, we find an $E_L \simeq E_p$ of $\sim 30.0 \pm 1.0$ eV (the large uncertainty results from the broad nature of the loss

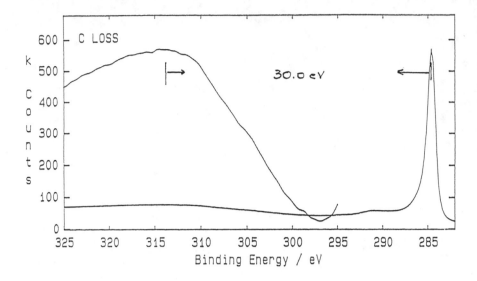

Figure 6-10 C(1s) loss splitting, E_L, for relatively clean graphite.

peak, see Figure 6-10). It is suspected that a very pure, structurally sound, single crystal of graphite, oriented to permit maximum ESCA analysis perpendicular to its basal plane, should produce a substantial enhancement in resolution, but will not alter these general conclusions. This feature is under investigation.

The present samples of moderately clean graphite also exhibit relatively intense, reasonably well-resolved second loss peaks at ~345 eV, Figure 6-11 (i.e., split from the first loss peak by ~30 eV). All of these results suggest that the aromatic π structure of graphite, with its density located primarily at or near the Fermi edge, is experiencing a near total collective (plasmon-like) loss behavior. The general lack of intensity in the resulting loss lines would further suggest that the sp^2 covalent part of the graphite system only provides weak "support" to this collective loss behavior.[32]

C. CARBON-BASED SYSTEMS DEVOID OF AROMATIC STRUCTURE — POLYPROPYLENE

The loss spectra of a variety of carbon-based materials that do not feature any aromatic π character were also examined with several ESCA systems (Section 6.VI.A). One of the most detailed of these studies was that for very pure, catalytically produced films of polypropylene[32,43,44,53] (see the section on angular resolution). This

Table 6-7 **Characteristic loss splitting for carbon containing systems (note correlation between graphite and E_P for carbon)**

Carbon species	Loss splitting E_L	$\Delta E_L i \rightarrow^{j\ a}$
Graphite	30.0 ± 1.0 eV	0
Polypropylene	21.0 ± 1.0 eV	9.0 ± 1.0 eV
Adventitious carbon	22.0 ± 1.0 eV	8.0 ± 1.0 eV
Diamond[b]	32.8 (23.0) ± 1.0 eV	−2.8 ± 1.0 eV

[a] Take graphite as i, other species as j.

[b] Based upons results from Reference 47. All other results from our experiments.

has since been expanded to include a similar analysis of high-density polyethylene.[32] Both of these systems produce ESCA loss results that seem to be representative of those for many carbon-based materials without the interjection of interlaced π networks. In these cases, the valence band region displays the expected "pull back" of density from the Fermi edge, as the π electron region of the graphite hybridizes into the balance of the covalent system. This results when the π bond in the propylene monomer is converted to form the sp^3 bonds of the hydrocarbon polymer (see Figure 6-12). As the carbon $3p_\pi$ density pulls away from the Fermi edge and "rolls back" into the valence electron density region occupied by the covalent (σ) balance of the carbon valence band density, several additional features evolve. First, the charging shift (which was muted or entirely absent for graphite) begins to emerge for the nonconductive, hydrocarbon polymer. Second (and more to the point of our present discussion), there is a substantial change in character and shift in the C(1s) loss spectra.

Hence, the C(1s) loss peaks for polypropylene are significantly attenuated in size compared to graphite and shifted in energy to an average E_L of 21.2 ± 1.0 eV. These features are summarized in Table 6-7 and Figure 6-12, including the $\Delta E_L{}^A(1 \rightarrow 2)$ case, where A indicates carbon, and 1 and 2 stand for graphite and either polypropylene or polyethylene, respectively. In this case, the ΔE_L is 8.8 ± 1.0 eV. The C(1s) loss spectra of several other hydrocarbon-dominated polymers, films, coatings, and powders have been examined by ESCA; in all cases these materials were found to produce loss spectra with ΔE_L of ~9.0 eV with respect to graphite.[32]

A variety of other carbon-based materials with varying degrees of conjugated, π, character between polypropylene and graphite have also been examined with ESCA, and their C(1s) loss spectra reveal E_L values between 31 and 22 eV.[32,44] Often several peaks have been produced, and there is a suggestion that the ΔE_L may increase in a progressive fashion as the degree of conjugation is enhanced. Proof of the latter, however, awaits future evidence.

We also suggest here that the ~9.0-eV reduction in E_L is not the only shift that can be experienced by carbon as it loses the (essentially) infinite π character of graphite. Diamond is the most covalent of all carbon systems, but its electronic system is quite different from that of the aforementioned hydrocarbons. Thus, whereas polymeric hydrocarbons represent a series of sp^3 bonds that repeatedly terminate at the hydrogens, diamond is an "interlaced", carbon-only sp^3 system with some delocalization of electrons throughout the lattice. This delocalization is not as "total" as that in the two-dimensional, π lattice of graphite, but the diamond interlace extends over a complete three-dimensional network. As a result, diamond seems to produce two distinct loss peaks when measured by XPS (see, for example, Reference 47). One peak (1) produces an E_L of ~23 eV, while the other (2) is at 32.8 eV. An interpretation of these results suggests that (1) may be due to localized electron effects, whereas (2) results from the aforementioned delocalized character. The shift in the latter from E_p may be due to a band-to-band perturbation of the free electron plasmon.

D. ADVENTITIOUS CARBON

One of the least desired, but most persistent, parts of nearly every surface study (no matter what the problem) is the ubiquitous presence of carbon deposits which are difficult to remove and even more difficult to keep off. These deposits form to varying degrees on surfaces when they are not properly maintained in extreme UHV. These carbonaceous species were originally attributed to the vacuum pump oil, but it is now known that they arise from numerous sources, and are significantly deposited on all air-exposed surfaces. Because of the persistent nature of these species, they have been generally labeled as *adventitious carbon* (AC).[54]

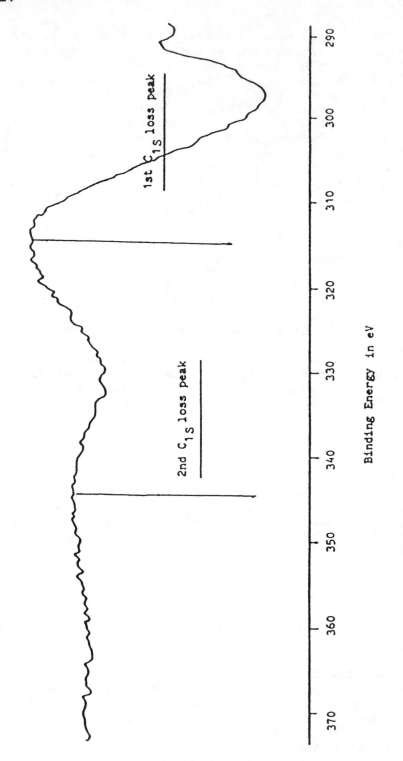

Figure 6-11 Poisson character of ESCA loss splitting for graphite.

Binding Energy in eV

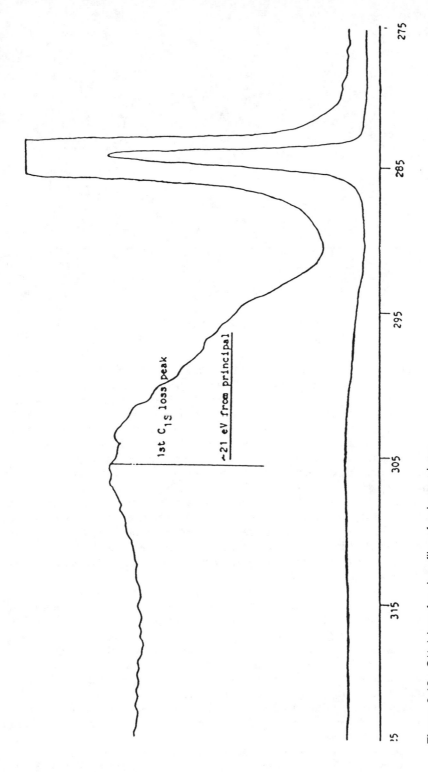

Figure 6-12 C(1s) loss for clear film of polypropylene.

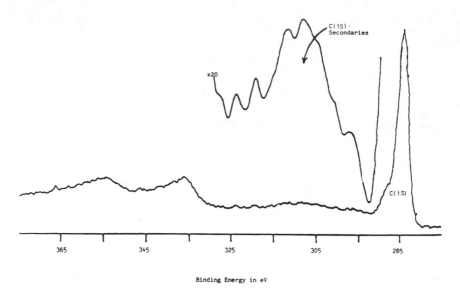

Figure 6-13 C(1s) loss pattern for typical systems coated with adventitious carbon (a pressed wafer of a catalyst).

Generally, the ESCA spectra produced by these deposits form a rather consistent pattern that demonstrates that 75 to 90% of the deposited species are either C–H– or C–C-containing compounds; the balance are various oxidized systems, usually carbonyls. The designation AC is often reserved for the former, and its binding energy is generally fixed by some means at 284.6 ± 0.4 eV. This value is frequently employed to establish a general (if somewhat precarious) binding energy scale (see Chapter 2).[22] Because of the uncertain origins of these materials, and the lack of differentiation of the principal C(1s) peak, designation of the exact type of species involved in AC, as mentioned above, is uncertain.

We have employed ESCALOSS to examine a large number of adventitious carbon systems deposited on a variety of materials with diverse histories. In *all* cases, the loss spectrum produced was found to closely replicate that of polypropylene (see Figure 6-13). Thus, the characteristic E_L for most adventitious carbon seems to be 22.0 ± 1.0 eV. *This strongly suggests that the C–C and C–H parts of AC are hydrocarbon-based species, with little or no graphitic materials present.*[32]

The insidious nature of many of these AC deposits has been examined and documented by Barr in a companion study of various surfaces.[22,43,55] It should be noted that these carbon systems seem to form continuously, but are generally lacking in sufficient chemical bonding strength to prevent displacement by other, more voracious bonding species. Thus, assuming that over relatively long periods of time most of the growth of adsorbants on a surface is vertical, then the AC products are continuously being pushed toward the outer, evolving surface by any chemically attached film of "growing" species, such as an oxide.[43,53] *It is also interesting to note that AC species are even found on the outer surface of carbon films.* To illustrate, we have noted that air-exposed graphite, if examined by angular resolution, exhibits, in addition to a small amount of carbon oxides, an outermost surface that has a very thin deposition of AC. The latter is detected, at grazing incidence, as an appearance of a small ~305.5 loss peak, E_L ~ 22 eV, and a shift of the main loss peak down field from 314.5 eV. This also may explain why our E_L for graphite at standard incidence is somewhat smaller than E_P.

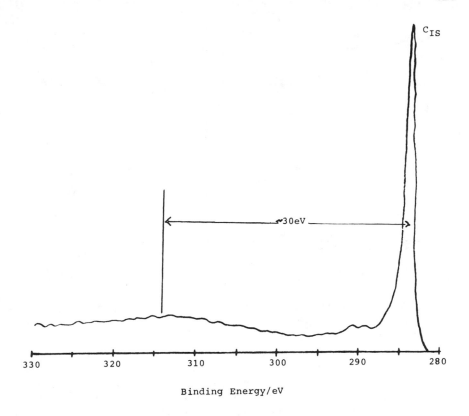

C_IS

$\leftarrow$ ~30eV $\rightarrow$

330   320   310   300   290   280

Binding Energy/eV

Figure 6-14 C(1s) loss spectrum for pretreated carbon paper before surface alteration to form fuel cell (note 30 eV graphite-like E_L).

In addition, one should realize that it is a common experience to find a broad loss peak associated with all photoelectron lines of most air-exposed solids with loss splittings in the region of ~22 eV. It is our contention that these noncarbon containing materials with "dirty" surfaces are simply reflecting (extrinsic) losses in the AC overlayer, rather than loss transitions of the materials themselves.[1]

E. UTILITY

The ultimate proof of the analytical potential of the C(1s) loss spectra method rests, naturally, on its utility in research and applied situations. Several research groups in an association with that of the author (TLB) are presently exploiting this method in a number of areas, two of which are outlined here:

1. Certain potential forms of fuel cells may be prepared by using carbon-based supports onto which metal catalysts, such as platinum, are implanted. Particular examples of this type of system have been prepared and tested in the laboratories of Professor Giordano (U. Messina) and surface analyzed with ESCA by the research group of Barr.[33] The base for these systems was a pure, graphitic, carbon paper. The evolving materials were then examined with ESCA at select stages during their involved production. The ESCA, C(1s) loss analysis for the carbon paper base (Figure 6-14) revealed a pattern strongly suggesting graphite, with only slight alterations. During an early stage in the production of the fuel cell, this graphite material was coated with a fluorinate hydrocarbon polymer and subjected to lengthy heat treatment. The resulting, relatively thick,

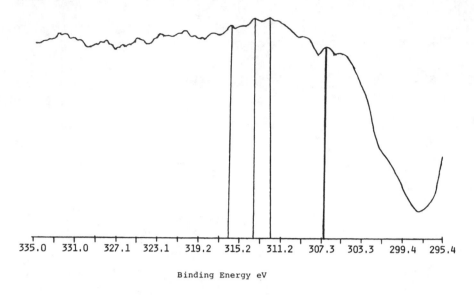

335.0 331.0 327.1 323.1 319.2 315.2 311.2 307.3 303.3 299.4 295.4

Binding Energy eV

Figure 6-15   C(1s) loss spectrum realized following FEP polymer coating and physical alteration (note introduction of 21.8 eV E_L chracteristic of hydrocarbon polymer plus retention of graphite loss due to spotty coverage).

polymeric film was apparently rather spotty, with substantial uncovered regions. Thus, whereas the C–C–, C–H part of the principal C(1s) line suggested no change, the C(1s) loss spectra for this system (Figure 6-15) supported these suggestions of splotchy coverage by revealing peaks that suggest the simultaneous presence of both graphitic and hydrocarbon polymer materials. It should be noted (Figure 6-15) that, as expected, these peaks are shifted somewhat from exact correspondence with the pure precursors, and there is also the apparent presence of several additional, "intermediate" peaks in the general manifold. The latter effects are perhaps indicative of the alterations in chemistry and interfacial properties resulting from the coating and subsequent treatment. Further treatments of these materials produce equally revealing scenarios, with ESCA analysis assisting in detecting changes in the fluorine and platinum chemistry, as well as the coverage.[33] The utility of the C(1s) loss spectra proved to be a key point in the total analysis.

2. A second, practical use of the C(1s) loss analysis has occurred during studies of aluminum alloy composites, doped with various amounts of graphitic or non graphitic carbon, in order to provide instructive variations in the tribological properties of these systems. These materials are being prepared in the laboratories of Rohatgi et al.[56] of UWM with ESCA analysis conducted in the laboratory of Barr.[57,58] Once again, ESCA C(1s) loss spectra are playing a crucial role. Hence, the loss spectra were employed to examine the character of the carbonaceous species that were found deposited on, or segregating to, the surface of these composite systems. For example, C(1s) loss spectra were employed to differentiate between segregated graphitic deposits and the presence of non graphitic carbons, including the ubiquitous, AC that also is generally present.[33,57,58] Most important, however, the spectra were used to determine when wear treatment begins to degrade graphite and destroy its lubricating properties.[57,58]

F. RATIONALE

In all of these analyses, use is made of the nature of the loss spectra for carbon-containing systems. In this regard, we have noted that if the potential π density in the carbon system is maximized (graphite), it becomes an interlaced collective network of electrons that form a near perfect, free-electron plasmon with $E_L \simeq E_P \simeq 31$ eV. As the interlaced π network is reduced and pulled back from the Fermi edge, the "collective" electron density N in Equation 6-4 is reduced by this covalent formation. The degree of π network reduction in polypropylene is apparently ~30%, and that for diamond is suggested to be even larger. Possible perturbative involvement of the largest part of the valence band (located at ~17.5 eV) in band-to-band transitions of the type described in Equation 6-43 must also be included in the creation of E_L values in the 22- to 19-eV range.[6,33]

REFERENCES

1. **Bloch, F.,** Double electron transitions in X-ray spectra, *Phys. Rev.,* 48, 187, 1935; **Feinberg, E. L.,** Ionization of the atom due to decay, *J. Phys.,* 4, 423, 1941; **Levinger, J. S.,** Effects of radioactive disintegrations on inner electrons of the atom, *Phys. Rev.,* 90, 11, 1953; **Sachenko, V. P. and Demekhin, V. F.,** Satellites of X-ray spectra, *Soviet Physics JETP,* 22, 532, 1966; **Carlson, T. A. and Krause, M. O.,** Electron shake-off resulting from K-shell ionization in neon measured as a function of photoelectron velocity, *Phys. Rev. A,* 140, 1057, 1965; **Krause, M. O., Carlson, T. A., and Dismukes, R. D.,** *Phys. Rev.,* 170, 37, 1968; **Carlson, T. A. and Nestor, C. W., Jr.,** Calculation of electron shake-off probabilities as a result of X-ray photoionization of the rare gases, *Phys. Rev. A,* 8, 2887, 1973. See, for example, **Aberg, T.,** *Phys. Rev.,* 156, 35, 1967; **Carlson, T. A. and Krause, M. O.,** Atomic readjustment to vacancies in the K and L shell of Ar, *Phys. Rev. A,* 137, 1655, 1965; **Carlson, T. A., Krause, M. O., and Moddeman, W. E.,** *J. Phys. C,* 20, 102, 1971; **Carrol, T. X. and Thomas, T. D.,** Shake-up spectra and core ionization potentials for formaldehyde. The role of electron spin in shake-up, *J. Electron Spectrosc. Rel. Phenomena,* 10, 215, 1977.
2. **Chang, J. J. and Langreth, D. C.,** Deep-hole excitations in solids. I. Fast electron-plasmon effects, *Phys. Rev. B,* 5, 3512, 1972; **Chang, J. J. and Langreth, D. C.,** Deep-hole excitations in solids. II. Plasmons and surface effects in X-ray photoemission, *Phys. Rev. B,* 8, 4638, 1973. **Langreth, D. C.,** Boin-Oppenheimer principle in reverse: electrons, photons and plasmons in solids-singularities in their spectra, *Phys. Rev. Lett.,* 26, 1229, 1971. **Langreth, D. C.,** Singularities in the X-ray spectra of metals, *Phys. Rev. B,* 1, 431, 1970.
3. **Gadzuk, J. W. and Sunjic, M.,** Excitation energy dependence of core-level X-ray photoemission-spectra line shapes in metals, *Phys. Rev. B,* 12, 524, 1975. **Gadzuk, J. W.,** Plasmon satellites in X-ray photoemission spectra, *J. Electron Spectrosc. Rel. Phenomena,* 11, 355, 1977.
4. **Manne, R. and Aberg, T.,** Koopmans theorem for inner shell ionization, *Chem. Phys. Lett.,* 7, 282, 1970.
5. **Fadley, C. S.,** Peak intensities and photoelectric cross sections in the sudden approximation, *J. Electron Spectrosc. Rel. Pheonmena,* 5, 895, 1974; **Fadley, C. S.,** *Electron Spectroscopy: Theory, Techniques and Applications,* Vol. 2, Brundle, C. R. and Baker, A. D., Eds., Academic Press, New York, 1978, 1.
6. **Kowalczyk, S. P., Ley, L., McFeeley, F. R., Pollack, R. A., and Shirley, D. A.,** *Phys. Rev. B,* 8, 3583, 1973; **Pollack, R. A., Ley, L., F. R., McFeeley, F. R., Kowalczyk, S. P., and Shirley, D. A.,** Characteristic energy loss structure of solids from X-ray photoemission spectra, *J. Electron Spectrosc. Rel. Phenomena,* 3, 381, 1974; **Ley, L., McFeeley, F. R., Kowalczyk, S. P., Jenkin, J. G., and Shirley, D. A.,** Many-body effects in X-ray photoemission from magnesium, *Phys. Rev. B,* 11, 600, 1975; **Kowalzyk, S. P., Ley, L., Martin, R. L., McFeeley, F. R., and Shirley, D. A.,** Relaxation and final-state structure in XPS of atoms, molecules and metals, *Discuss. Faraday Soc.,* 7, 1975. See also **Pollack, R. A.,** Ph.D. thesis, University of California, Berkley, 1972.

130

7. **Bohm, D. and Pines, D.,** *Phys. Rev.,* 82, 625, 1951; **Pines, D. and Bohm, D.,** A collective description of electron interaction. II. Collective vs. individual practicle aspects of the interaction, *Phys. Rev.,* 85, 338, 1952; **Bohm, D. and Pines, D.,** *Phys. Rev.,* 92, 609, 1953; **Pines, D.,** *Phys. Rev.,* 92, 626, 1953. **Pines, D.,** Collective energy losses in solids, *Rev. Modern Phys.,* 28, 184, 1956. **Pines, D.,** *Elementary Excitations in Solids,* Bejamin Publishers, New York, 1964.

8. **Nozières, P. and Pines, D.,** Electron interaction in solids. General formulation, *Phys. Rev.,* 109, 741, 1958. **Nozières, P. and Pines, D.,** Electron interaction in solids. Collective approach to the dielectric constant, *Phys. Rev.,* 109, 762, 1958. **Nozières, P. and Pines, D.,** Electron interaction in solids. The nature of the elementary excitation, *Phys. Rev.,* 109, 1062, 1958. **Nozières, P. and Pines, D.,** Correlation energy of a free electron gas, *Phys. Rev.,* 111, 442, 1958.

9. **Mahan, G. D.,** *Phys. Status Solidi,* B55, 703, 1973.

10. **Gadzuk, J. W.,** in *Photoemission and the Electronic Properties of Surfaces,* Feuerbacher, B., Filton, B., and Willis, R., Eds., John Wiley & Sons, New York, 1978, chap. 5.

11. **Sunjic, M. and Sokcevic, D.,** Inelastic effects in X-ray photoelectron spectroscopy, *J. Electron Spectrosc. Rel. Phenomena,* 5, 963, 1974; **Sunjic, M. and Sokcevic, D.,** On the problem of 'intrinsic' vs. 'extrinsic' scattering in X-ray photoemission from core levels of solids, *Solid State Comm.,* 18, 373, 1976.

12. **Doniach, S. and Sunjic, M.,** *J. Phys. C,* 3, 285, 1970. **Gadzuk, J. W. and Doniach, S.,** A soluble relaxation model for core level spectroscopy on adsorbed atoms, *Surface Sci.,* 77, 427, 1978.

13. **Schaich, W. L. and Aschcroft, N. W.,** *Phys. Rev. B,* 3, 2452, 1970; **Lundqvist, B. I.,** *Phys. Kondens. Material,* 9, 236, 1969; **Sunjic, M. and Lucas, A. A.,** *Phys. Rev. B,* 3, 719, 1971; **Hedin, L. and Lundqvist, S.,** *Solid State Phys.,* 23, 1, 1969; **Lundqvist, S. and Wendin, G.,** *J. Electron Spectrosc. Rel. Phenomena,* 5, 513, 1974; **Cederbaum, L. C. and Domcke, W.,** *J. Chem. Phys.,* 64, 603, 1976.

14. Statements to this effect have appeared quite often in the literature and text books.

15. **Powell, C. J. and Swan, J. B.,** Effect of oxidation on the characteristic loss spectra of aluminum and magnesium, *Phys. Rev.,* 118, 640, 1960; **Powell, C. J. and Swan, J. B.,** *Phys. Rev.,* 115, 869, 1959; 116, 81, 1959. **Callcott, T. A. and McRae, A. U.,** *Phys. Rev.,* 178, 769, 1969; **Smith, N. V. and Spicer, W. E.,** *Phys. Rev. Lett.,* 22, 769, 1969. **Pireaux, J. J., Ghijsen, J., McGowan, J. W., Verbist, J., and Caudano, R.,** Surface plasmon spectroscopy: photoelectron spectra from adsorbates on aluminum, *Surface Sci.,* 80, 488, 1979. **Sokcevic, D., Sunjic, M., and Fadley, C. S.,** Strength of intrinsic plasmon satellites in XPS from adsorbates: dispersion and lifetime effects, *Surface Sci.,* 82, 383, 1979.

16. **Barr, T. L.,** XPS analysis in Pt-metal catalysis, *ACS Div. Petrol. Chem.,* 33, 699, 1988.

17. **Parmigiani, F., Kay, E., Bagus, P. S., and Nelin, C. J.,** *J. Electron Spectrosc. Rel. Phenomena,* 36, 257, 1985.

18. **Platzman, P. M. and Wolff, P. A.,** Waves and interactions in solid state plasmas, in *Solid State Physics,* Vol. 13, Ehrenreich, H., Seitz, F., and Turnbull, D., Eds., Academic Press, New York, 1973.

19. **Ludeke, R.,** Electronic Properties of the (100) surfaces of GaSb and InAs and their alloys with GaAs, *IBM J. Res. Dev.,* 22, 304, 1978.

20. **Froitzheim, H., Ibach, H., and Mills, D. L.,** *Phys. Rev. B,* 11, 4980, 1975. **Ibach, H. and Mills, D.,** *Electron Energy Loss Spectroscopy and Surface Vibrations,* Academic Press, New York, 1982.

21. **Ibach, H.,** *Phys. Rev. Lett.,* 24, 263, 1970; **Ibach, H.,** *J. Vac. Sci. Tech.,* 9, 713, 1972.

22. **Barr, T. L.,** *J. Vac. Sci Tech. A,* 7, 1677, 1989.

23. **Hedin, L. and Johansson, A.,** *J. Phys. B,* 2, 1336, 1969.
24. **Bohm, D. and Gross, E. P.,** *Phys. Rev.,* 75, 1851, 1864, 1949.
25. **Anderson, P. W.,** Infrared catastrophe in Fermi gases with local scattering potentials, *Phys. Rev. Lett.,* 18, 1049, 1967.
26. **Nozières, P. and De Dominicis, C. T.,** Singularities in the X-ray absorption and emission of metals. III. One-body theory exact solution, *Phys. Rev.,* 178, 1097, 1969.
27. **Barr, T. L., Greene, J. E., and Eltoukhy, A. H.,** *J. Vac. Sci. Tech.,* 16, 517, 1979; **Natarajan, B. R., Eltoukhy, A. H., Greene, J. E., and Barr, T. L.,** Mechanism of reactive sputtering of indium. II. Growth of indium oxynitride in mixed N_2-O_2 discharges, *Thin Solid Films,* 69, 217, 1980.
28. **Barr, T. L.,** An XPS study of Si as it occurs in adsorbents, catalysts and thin films, *Appl. Surface Sci.,* 15, 1, 1983.
29. **Barr, T. L.,** Applications of electron spectroscopy to heterogeneous catalysis, in *Practical Surface Analysis,* Briggs, D. and Seah, M. P., Eds., John Wiley & Sons, Chichester, England, 1983, chap. 8.
30. **Barr, T. L., Kramer, B., Shah, S. I., Ray, M., and Greene, J. E.,** ESCA studies of the valence band and loss spectra of semiconductor films: ionicity and chemical bonding, *Mat. Res. Soc. Proc.,* 47, 205, 1985.
31. **Kubiak, C. J. G., Aita, C. R., Tran, N. C., and Barr, T. L.,** Characterization of sputter deposited Al-nitride and Al-oxide by X-ray photoelectron loss spectroscopy, *Mat. Res. Soc.,* 60, 379, 1986.
32. **Barr, T. L. and Yin, M. P.,** ESCA induced loss spectroscopy as a means to determine the surface chemistry of C-C and C-H containing systems, *J. Vac. Sci. Tech. A,* 10, 2788, 1992.
33. **Barr, T. L., Yin, M. P., and Giordano, N.,** ESCA studies of the processing of C/Pt fuel cell electrodes: with emphasis on XPS induced loss results, *J. Chem. Soc. Faraday,* to be submitted.
34. **Meldner, H. W. and Perez, J. D.,** Observability of rearrangement energies and relaxation times, *Phys. Rev. A,* 4, 1388, 1971.
35. **Müller-Hartmann, E., Ramakrishnan, T. V., and Toulouse, G.,** Localized dynamic perturbations in metals, *Phys. Rev. B,* 3, 1102, 1971.
36. **Jones, W. and March, N. H.,** *Theoretical Solid State Physics,* John Wiley & Sons, New York, 1973, 76.
37. **Kittel, C.,** *Introduction to Solid State Physics,* 4th ed., John Wiley & Sons, New York, 1971, chap. 9.
38. **Fuggle, J. C., Fabian, D. J., and Watson, L. M.,** Electron energy loss processes in X-ray photoelectron spectroscopy, *J. Electron Spectrosc. Rel. Phenomena,* 9, 99, 1976.
39. **Barr, T. L.,** ESCA studies of metals and alloys: oxidation, migration and dealloying of Cu-based systems, *Surface Interface Anal.,* 4, 185, 1982.
40. **Natarajan, B. R., Eltoukhy, A. H., Greene, J. E., and Barr, T. L.,** Mechanism of reactive sputtering of indium. I. Growth of InN in mixed Ar-N2 discharges, *Thin Solid Films,* 69, 201, 1980.
41. **Johnson, E. D.,** Preparation and Surface Characterization of Planar Transition Aluminas, Ph.D. dissertation, Cornell University, Ithaca, NY, 1985.
42. **Briggs, D.,** Applications of XPS in polymer technology, in *Practical Surface Analysis,* Briggs, D. and Seah, M. P., Eds., John Wiley & Sons, Chichester, England, 1983, chap. 9.
43. **Barr, T. L.,** *Corrosion 90,* Paper 296, NACE, Houston, 1990.
44. **Yin, M. P., Koutsky, J. A., Barr, T. L., Rodriguez, N. M., Baker, R. T. K., and Klebanov, L.,** *Chem. Mater.,* 5, 1024, 1993; Yin, M. P., Ph.D. thesis, University of Wisconsin, Milwaukee, 1991.

45. **Galuska, A. A., Maden, H. H., and Allred, R. E.,** Electron spectroscopy of graphite, graphite oxide and amorphous carbon, *Appl. Surface Anal.,* 32, 253, 1988.
46. **Siegbahn, K.,** private communication.
47. **McFeeley, F. R., Kowalczyk, S. P., Ley, L., Cavell, R. C., Pollack, R. A., and Shirley, D. A.,** *Phys. Rev. B,* 9, 5268, 1974.
48. **Park, R. L., Houston, J. E., and Laramore, G. E.,** Proc. 2nd Int. Conf. on Solid Surfaces, 1974, Japan; *J. Appl. Phys., Suppl. 2,* 757, 1974.
49. **Weissmantel, C., Bewilogua, K., Schurer, C., Breuer, K., and Zscheile, H.,** *Thin Solid Films,* 61, L1, 1979.
50. **Liang, W. Y. and Cundy, S. L.,** *Philos. Mag.,* 19, 1031, 1969.
51. **Luries, I. G. and Wilson, J. M.,** *Surface Sci.,* 65, 476, 1977.
52. **Tsai, H.-C. and Bogy, D. B.,** *J. Vac. Sci. Technol. A,* 5(6), 3287, 1987.
53. **Barr, T. L.,** *J. Phys. Chem.,* 82, 1801, 1978.
54. **Briggs, D. and Seah, M. P., Eds.,** *Practical Surface Analysis,* 2nd ed., John Wiley & Sons, Chichester, England, 1990.
55. **Barr, T. L.,** Electron spectroscopic analysis of multicomponent thin films with particular emphasis on oxides, in *Multicomponent and Multilayered Thin Films for Advanced Microtechnologies: Techniques, Fundamentals and Devices,* Anciello, O. and Engemann, J., Eds., NATO/ASI Ser. E, Vol. 234, Kluwer Academic, Amsterdam, 1993, 283.
56. **Rohatgi, P. K., Dahotre, N. B., Liu, Y., Yin, M. P., and Barr, T. L.,** Tribological behavior of Al alloy-graphite and Al alloy-microcrystalline carbon particle composites, in *Cast Reinforced Metal Composites,* Fishman, S. G. and Dhinga, A. K., Eds., American Society of Metals, Chicago, 1988, 367.
57. **Rohatgi, P. K., Liu, Y., Yin, M., and Barr, T. L.,** A surface analytical study of tribodeformed aluminum alloy 319-10 vol.% graphite partical composite, *Mat. Sci. Eng. A,* 123, 213, 1990.
58. **Rohtgi, P. K., Liu, Y., and Barr, T. L.,** *Metallurgical Trans. A,* 22, 1435, 1991.

Chapter 7

Old Problem Areas — New ESCA Analysis Methods

I. THE CHARGING SHIFT AND FERMI EDGE REFERENCING AS TOOLS

A. FUNDAMENTALS OF THE CHARGING SHIFT

Charging occurs in ESCA because a nonconductive sample does not have sufficient delocalized, conduction band electrons available to neutralize the charged centers that build from clustering of the positive holes created with the photoelectron and/or Auger electron ejection.[1] As a result, a positive outer (or Volta) potential builds near the materials surface, producing a retardation or "drag" on the outgoing electrons.[2] This retardation appears in the electron spectrum as an additional positive shift, either subtracting from the uncharged kinetic energy or, correspondingly, adding onto the "normal" binding energies of the outgoing electrons. Compared to most of the features in these experiments, charging shifts are often relatively slow in their establishment; thus, in the time frame of a normal experiment, charging has both dynamic and static components.[3] Charging also depends upon the macro-surface and bulk morphology of the measured system, as well as its chemistry and microstructure.[4,5] As indicated, charging does not arise for "good" conductors (or relatively narrow band-gapped semiconductors, see Chapter 2) because these systems possess, at room temperature, sufficient (internally and externally provided) conduction band electrons to neutralize the aforementioned "excess" charge.[1,6,7] In the ESCA process itself, when direct X-ray impingement onto the sample surface is utilized, there is often sufficient stray, low-energy electrons from Bremsstrahlung and other processes to partially neutralize these charge centers. On the other hand, when a monochromator is employed, indirect, crystal scattered X-rays are focused onto the sample, and these, along with other features, provide a relatively electron-clean environment around the sample. Thus, most of the aforementioned stray, neutralizing electrons are eliminated, and as a result a substantial, nearly complete, charging effect emerges. The latter includes the various differential charging features that may result from different components and unique morphological variations.[4,6,8] Hidden in this dichotomy are a variety of additional major problems that may arise, particularly pertaining to the lack of establishment of a valid *Fermi edge and differential charging*. These difficulties tend to make energy referencing for insulators a persistent and quite formidable problem,[4,5,9–11] particularly when the system being examined is a doped catalyst with poor dispersion.[11] For this reason, it is common for some researchers to avoid all use of absolute binding energies for insulators and wide band-gapped semiconductors. This we find to be an extreme measure that may be avoided in many (but not all) circumstances. In fact, at times this problem may be favorably overshadowed, since in some cases the presence of Fermi edge decoupling and differential charging may be employed as a useful auxiliary tool.[6] Although this "tool" has obvious limitations, it has exhibited a surprising versatility in practical, as well as basic, problems.

B. DIFFERENTIAL CHARGING

As suggested above, the collective property generally labeled as *differential charging*[4,6,12] arises because of a number of factors, of which four of the most common, interrelated forms, are considered herein: (1) the photoelectron spectroscopy sampling depth; (2) the depth of field of any neutralization device (such as an electron flood gun) that may be utilized to adjust (and perhaps remove) any charging; and, most importantly, (3) the sample morphology, including layered systems; and (4) clusters. Previously, we briefly outlined the causes and nature of the charging shift itself (see Chapter 2). It was also pointed out that part of the general problem area often considered under the collective heading of "charging" is actually due to either the lack of the existence of a true Fermi edge for an insulator (or wide band-gapped semiconductor) or to the often-realized inability to conductively couple the Fermi edge of a part of a sample under investigation to the Fermi edge of the other components in the system. In particular, the prevention of coupling between a conductive component and the spectrometer was shown to interfere with the determination of conventional, valid binding energies. Both charging and the inhibition of Fermi edge coupling result from the nonconductive nature of all or part of the samples under study. However, charging only occurs following emission (or introduction) of a charged particle (e.g., photoelectron), whereas Fermi edge decoupling occurs both before and after photoelectron emission, and may actually be partially rectified by the latter process.[5,13] In any case, although they are two separate features, charging and Fermi edge decoupling are often collectively detected as part of the same measurement. Hence, since they are both subject to uncertain, difficult-to-control dependencies on the morphological as well as the chemical parameters of the system under study, one must employ great care in trying to reach anything other than very general suppositions about such features.

As we described in more detail elsewhere,[6] it is most useful to construct general models as representative cases of how these properties should interrelate, and then use these models to try to establish which ones are applicable to the several specific examples considered. In Figure 7-1 we have, therefore, constructed one of these models.

An important point of warning must be heeded at this point. As many experienced ESCA users are now aware, the realization of charging shifts and their removal is very much system dependent. As mentioned above, those practitioners employing systems with conventional, direct focus X-ray units generally experience far less charging than do those users with monochromators. In addition, many charge shift removing devices are less responsive than others. In fact, some just do not work at all! It is generally true that flood guns employed with conventional, direct focus X-ray XPS systems are only partially successful. This apparently results from some aspect of the competition between the flood gun electrons and those inherent to the Bremsstrahlung processes. All of these less than adequate means for manipulating the effects of Fermi edge detachment and charging will leave the user in a position where she/he will be unable to realize the full thrust of the features to be described in this section. The author cautions everyone of this point and suggests that those interested should join in lobbying instrument manufacturers to "get it right". It should be remembered that in science the only things worse than unrealized results are "instrumentally hidden or adversely altered results".

C. MORPHOLOGICAL CONSIDERATIONS
1. Layered Structures
The first problem of potential interest concerns the results when layers of insulators are deposited in a rather uniform fashion on top of relatively flat surfaces of conductors or narrow band-gapped semiconductors. The chemistry at the resulting interface is perhaps the most crucial problem of interest to electron spectroscopists for these types of systems. Typical ESCA examples were provided in the SiO_2/Si^o interface studies of Vasquez and Grunthaner[14] and Hollinger et al.[15] The implementation of these studies required the careful registration of the binding energies of all the species, including those that are insulators, (see, for example, Figure 2-1b). As described from results of a similar ESCA study of related SiO^2/Si^o layered systems,[4] we found that charging and Fermi edge

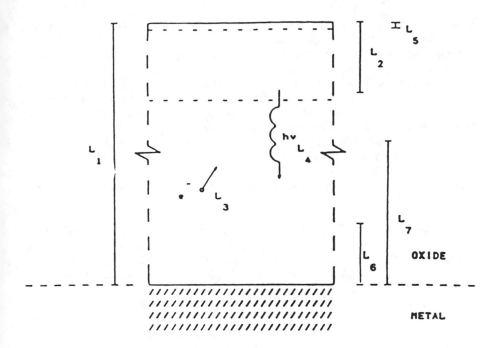

Figure 7-1 Representative renditions of surface situations expected during ESCA analysis: atomically flat (oxide) surface (unrealistic) on metal [also presented are representative depths and fields; $L_1 \equiv$ oxide thickness, $L_2 \equiv$ photoelectron analysis depth (low energy), $L_3 \equiv$ path of undetected (deep) photoelectrons, $L_4 =$ depth of X-ray photons, $L_5 \equiv$ flood gun electrons, $L_6 \equiv$ Schotky barrier penetration of electrons from metal substrate].

decoupling play major roles in the analyses of these systems, and these features must be properly registered, not only to develop accurate binding energies, but also because they can be useful in the total analysis.

In this regard, it is of interest to examine some results of a study in which we conducted a stagewise ESCA analysis of various levels in a silica film (film thickness of approximately 500 Å, thermally grown by Lagally [University of Wisconsin-Madison] on top of a film of Si°), the details of which have been presented elsewhere.[4] For our present purposes we will concentrate on several of the results exhibiting progressive charging.

Initially, when we examined the outer surface and the evolving ESCA spectra from within the silica layer, we observed a charging behavior that resembled, in many ways, that obtained for a relatively thick silica wafer (as shown above in Figure 2-3b). Perhaps the major difference between the two is that the total charging experience by the silica layer at a thickness of approximately 200 Å is always less than that realized by the thick wafer (we will return to this later). It may be assumed that both silica systems are relatively uniform in structure and chemistry. The general features of these results suggest, therefore, that even relatively uniform samples do not necessarily charge uniformly; rather, they often seem to experience a differential field in which the preponderance of the detected Si(2p) photoelectrons charge more substantially than the balance. Based upon the arguments mentioned previously,[4,6,16] we suggest that the large protrusion on the high binding energy side of the Si(2p) (the place where the charging field is largest) is produced by that part of the silica at or near the surface of the wafer. In this vein, the plateau that spreads out toward lower binding energies (less charging) is produced by the detected subsurface parts of the wafer. Subsequent application of an electron flood gun* (Hewlett-Packard 18623A) rapidly removes this charging shift (Figure 2-3b), but does so in stages that strongly suggest that the (grazing incidence) neutralization electrons are also responding in a differential manner, removing first the surface charging sites closest to the flood gun source, and only reaching less accessible charge centers after a substantial increase in electron density (current). Application of (electron driving) energy to the flood gun electrons also tends to produce results which indicate that the phenomenon of charging and its removal produce interesting, potentially useful, depth-of-field effects.[6] Obviously, if this statement is true, the most useful cases will result when the system is not a single species or uniform composite, but rather when significant discontinuities, such as surfaces, interfaces, or cluster growth are considered. An interesting example of the interfacial situation occurs when the aforementioned silica layer has been reduced to <30 Å, so that both it and the relatively flat Si^o substrate are simultaneously detected by the ESCA instrument.

Representative analyses of these types of systems are displayed in Figure 2-3a. A feature of major concern is, of course, the possible detection of suboxide (SiO_x [x < 2]) species, whose ESCA spectrum should be located in the $Si(2p)$ binding energy region between that for the SiO_2 and the Si^o. For the "flood gun off" case, the shoulders in this $Si(2p)$ region bare striking resemblances to those detected by Vasquez and Grunthaner[14] and Hollinger et al.[15] (see Figure 2-3a). It is tempting to assume, therefore, that all of these intermediate peaks result from SiO_x (x < 2). Subsequent flood gun application, however, indicates that at least some of the oxidized species are still experiencing charging. Thus, in this case, one must be careful not to ascribe suboxide status to ESCA-detected species that may, in fact, be silica components suffering only modest charge shifting. A representative example of the removal of some of the charging for this system is presented in Figure 2-3a.[4,6] In this case, we should note that the uncertain status of the chemical nature of the oxides is not the only interesting feature. We also find here a dramatic example of differential charging where the relatively wide band-gapped silica species are found to respond to the flood gun current, whereas the contiguous sublayer of narrow band-gapped Si^o is essentially unaffected.[4,6] This is thus not only a dramatic example of differential charging, that reflects differences in chemistry, but also one that suggests useful morphological information. In fact, as we shall describe below in more detail, this differential charging behavior alone strongly suggests that the system under observation is a relatively thin layer of silica on top of a contiguous thick film of Si^o with (perhaps) some suboxides in between. Few other techniques exist that can provide this morphological information for the undisturbed ~20 Å interface in conjunction with the detailed chemistry provided for this system by the balance of the ESCA analysis.

Although obviously limited, related differential charging analyses have been utilized in a number of basic and applied interfacial situations. For example, we have detected similar examples of differential charging during the formation of passivating oxides on metals and alloys, e.g., during the oxidation of α-brass[17,18] and also during corrosion of a Cu_{98} Be_2 alloy.[18,19] All of these cases are typified by the covering of either a contiguous conductor or semiconductor by an insulator or relatively wide band-gapped semiconductor.

2. The Si^o Implanted in SiO_2 System[4,6]

This particular system is configured as shown in Figure 7-2. Note that both Si^o and SiO_2 are simultaneously detected in the ESCA. The results are characterized by several interesting features. First, significant amounts of the Si^o are visible without sputtering and therefore the $Si(2p_{3/2}-2p_{1/2})$ splitting is easily detected. Second, the charge-shifted silicon

---

* Oriented almost parallel to the sample surface, in order to exploit the attractive potential of the positive charge for the relatively soft flood gun electrons.

oxide peak manifold exhibits mainly SiO_2, with only a small amount of SiO_x (x < 2). This is true because of the physical nature of this system, where most of the SiO_2 and Si° observed are *not* at an interface. The orientation of the principal constituents is also well characterized by the energy splitting revealed during flood gun analysis (Figure 7-2). The "flood gun off" value is only slightly charge shifted. Application of the flood gun first induces shifts in both the silica and Si° regime, with a more extensive shift in the former. Further increase in the flood gun current provides a noticeable additional shift in the silica peak with only a slight effect of the Si° region (Figure 7-2). In addition, there is a noticeable change in the relative intensities of the peaks involved with a corresponding flood gun induced "flow" of finite peak density into the region between the two principal peaks. The latter may result from charge-shifted SiO_x (x < 2) but, due to its flood gun behavior, we are inclined to attribute much of the detected density shifts to different physical orientations of silica. This will be expanded upon below. It should be noted that these suppositions were supported by a variety of other spectral features, including a detailed ESCALOSS spectral analysis in a manner related to that described in Chapter 6 of this book. In any case, application of the flood gun to remove most of the charge shift finally produced values for the binding energies of the $Si(2p)$ of the principal manifold of SiO_2 that are typical for that system (see, e.g., Figure 7-2).[4,6]

The most interesting aspect of this system is the behavior of the Si° peak during flood gun analysis. Unlike the previously described case of SiO_2 on top of a Si° crystal (Figure 2-3a), the present case of a Si° sliver exhibits a noticeable charge-shift compensation. It is suggested that this occurs because the Si° (now lodged in the SiO_2 insulator) has lost good Fermi level contact to the spectrometer, i.e., the Fermi edge of the Si° is floating (see Figure 2-2). The Si° is a contiguous unit, however (i.e., not mixed into the silica matrix), so that its flood gun shift behavior does not follow that of the much poorer surface conductor, SiO_2. The latter, while not the major species in the detection range of the ESCA, is also contiguous, in this case, all the way across the spectrometer range. Thus, in this case, most of the results obtained for SiO_2 during flood gun analysis generally

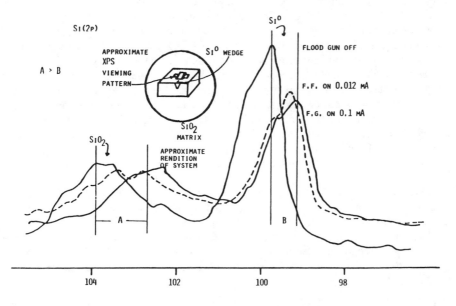

Figure 7-2 $Si(2p)$ binding energy study for Si° wedge **in** SiO_2 matrix. Note effect on all silicon peaks of different flood gun settings. These results should be compared to those for a thin film of SiO_2 **on** Si° (see also Figure 2-3).

mimic that produced by the outer surface of the previously described SiO_2 film, and also by the SiO_2-pressed wafer system model (see Figures 2-3a and b, and 7-2). It is apparent that the present wedged system is an excellent example of a moderate (surface) conductor (Si^o) lodged in a surface insulator (SiO_2).[4,6,16] These results demonstrate once again that the charging behavior of a particular system examined under ESCA can provide unique morphological information as well as detailed chemical analysis about such materials.

D. ADDITIONAL MORPHOLOGICAL CONSIDERATIONS
1. Depth of Fields

One of the most interesting and potentially useful features of charging arises when the fields described in our first model become partially overlapped; as a result, they may expose the selective morphological features or chemical behaviors of several constituents in a mixed system (Figure 7-1). Since this process is predicated upon the variations of charging or Fermi edge decoupling with location of the constituents in the matrix under investigation, it is considered herein to be a form of differential charging.[6,16]

To illustrate this behavior, a series of experiments were performed by Kao in conjunction with Merrill and the author[6,20] in which a thin film of alumina was grown on top of a clean surface of Al^o.* The alumina overlayer which grew naturally (at STP) in air [producing, no doubt, a mixture of Al_2O_3, AlOOH, and $Al(OH)_3$ products] or in oxygen (to yield only Al_2O_3), eventually produced a near-uniform film that generally terminated at a thickness of about 15 to 25 Å. The latter was determined by noting the relative ESCA, [Al(2p)] visibility of the alumina, and subsurface Al^o in a conventionally aligned ESCA system. ESCA spectra were then recorded for the various principal lines of this system using a flood gun, first off and then on, with a progressive increase in both beam current and voltage. A representative selection of the results for the Al(2p) line is displayed in Figure 7-3.[6,20]

A detailed analysis of these results is not obvious, but the following comments may offer a reasonable scenario. Comparing the charging experienced by the present thin film of alumina to that for a representative pressed wafer of an insulator (see, e.g., the SiO_2 in Figure 2-3b), we see, as indicated above, that the ready presence of the Al^o substrate in the present case must markedly reduce the general charge shift.[21] Part of the alumina layer is apparently still charging, as is suggested in the modest peak shifts produced with the application of (just) flood gun current (with no driving voltage). Some of this charging alumina may, however, be overcompensated by the application of 0.4-mA current, as it seems to shift the alumina peaks to points that remove the differentiation between the modest Al^o detected and this (perhaps) overshifted Al_2O_3. In any case, the broad peak centered at about 74.6 eV suggests that a charging shift was present in Figure 7-3a, and that the shift has been essentially removed through application of only flood gun current; however, the broad nature of the peak also suggests the detection of different species (and perhaps differential charging). Application of various relatively large flood gun voltages definitely establishes the presence of (flood gun-induced) differential charging in the alumina layer, with very interesting results. Thus, the results in Figure 7-3 suggest that some of the charging species (approximated by segment 3 in this figure) are shifting with the applied voltage (almost volt for volt), whereas other parts of the system (segment 2) while influenced by the flood gun are apparently only partially free of the influence of other "conflicting" fields. Still other parts of the alumina layer do not appear to experience any response to the flood gun at

---

* To be sure, the microstructural characteristics of the Al^o and resulting Al_2O_3 almost certainly played significant roles in the *details* of the results to be described, but it is suggested that the *general* features are independent of these parameters. Thus, except to note that these microfeatures were monitored and will be considered in other studies, we ignore them in the present case.

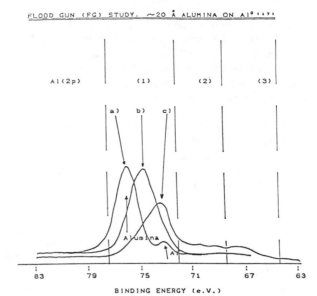

Figure 7-3 Flood gun (FG) study — ~20 Å alumina on Al°: (a) flood gun off; (b) flood gun, 0.4 Ma; (c) flood gun, 0.4 Ma, 8.0 eV.[20]

all (segment 1), *as if they were composed of good conductors*. The nature of this conflict is no doubt very complex, but if we assume that the simplified model shown in Figure 7-1 has some validity one may speculate that the principal conflict is between the neutralization current from the flood gun and the electrons provided by the space-charge effect to the adjacent alumina by the conductive Al°. Thus, in this simplified model we may assume that the alumina immediately adjacent to the electron-providing Al° (segment 1) is not charging and is therefore unresponsive to the flood gun. The degree of response of the balance of the alumina to the flood gun thus grows (as does the negatively induced shift) as one goes up in the alumina layer away from the conductive Al° until a point is reached at about 20 Å, at which the response of the alumina is dominated almost entirely by the flood gun. As demonstrated elsewhere,[6,20] a thick wafer of alumina produces an Al(2p) that would shift a relatively singular peak down field in the flood gun, approximately 1 eV for each volt added.

All of this suggests that the apparent destruction of the normal structure of the photoelectron peak for a thin insulating overlayer on top of a conductor [in this case the Al(2p) of alumina over Al°] has an *interesting potential utility that results from the establishment of an apparent differential charging in the insulator that varies continuously with the separation of the insulator from the conductive substrate.* Hence, it may be possible in select cases to use ESCA not only to detect the chemistry of constituents, but also to exploit the effect of differential charging to locate (and perhaps map) the evolving (z-dimensional) positions of these constituents. A crude preliminary example of this possibility is shown in Figure 7-4, where the differential charging shifts of a carbon deposition laced throughout the alumina layer is seen to follow almost the identical differential charging behavior of that displayed by the matrix itself. In this vein, it is not inconceivable that this procedure may be employed to follow the evolution of select segregations (and diffusions), perhaps locating such key features as points of meeting and the nature of any reaction between two or more species migrating in an insulating matrix.

A number of other geometrical features of insulators have produced ESCA results which suggest the possibility of similar controlled differential charging (or Fermi edge decoupling) effects[6,16,22] (as, for example, seems to be the case for the previously described situation of a Si^o sliver lodged in a SiO_2 matrix). *All of these features suggest a potential utility that requires a full, initial realization of charging, followed by its controlled manipulation.*

2. Dispersions and Clusters

In view of the previous discussion that suggests that some forms of differential charging may arise due to the geometrical characteristics of an insulator, it should not be surprising that the state of dispersion of one component (dopant) in another (matrix) may also often yield unique situations in differential charging. In fact, now that we know that the binding energies produced by a species will differ slightly depending upon whether that entity is at its surface or in a bulk matrix[23,24] (see Section VII in Chapter 9), we should expect that the characteristic charging response exhibited by any component in a mixed environment will depend primarily upon whether it is in electrochemical equilibrium with the Fermi level of some referencing conductor. This feature will, in turn, be dependent upon whether the species in question (A) is (1) either a continuous film or (2) mixed into another matrix, (B). Further, if (2) applies, then the critical question concerns the state of dispersion of A in B. Dispersion is a very useful and important concept in catalysis or colloids, but one that is very difficult to define in scientific terms. In general, we may assume that A is well dispersed in B if the concentration [A] << [B] and the particle size of A is very small (e.g., generally less than 20 to 30 Å). It is further necessary that the A units be uniformly distributed throughout the B matrix. Under these circumstances one may also assume that *the A units have no direct electrochemical equilibrium with one another,* i.e., *the Fermi edge (if it exists for such small units) of one unit, A1, can only couple to another, A2, through the band structure of B.* Obviously, this type of arrangement of a dopant, A, is special, and requires special physical conditions to maintain. Further, if one "abuses" these conditions, some form of migration and possibly clustering (perhaps with surface segregation) of dispersed A units may take place. As a result, clusters (perhaps relatively finite sized >50 Å crystallites) of A may develop. These two diverse situations should be visualized strewn throughout a matrix such as that depicted in Figure 7-1. Only if A and B are both reasonably good conductors will charging not play an important, often unique, role in these types of situations. One of the key situations in which charging becomes most interesting occurs when B is an insulator or wide band-gapped semiconductor, and A is a conductor or narrow band-gapped semiconductor. In

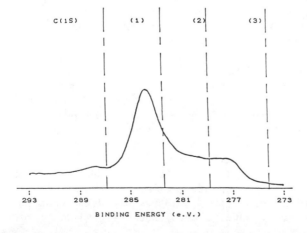

Figure 7-4 Flood gun result — carbon on Al^o and in alumina (flood gun, 0.4 Ma, 8.0 eV).[20]

such cases, one can also expect the possibility of a variety of interesting differential charging situations.

Consider, for example, the case where A is a conductor and B an insulator. This is the classic situation in metals catalysis (e.g., Pt in Al_2O_3). We recently presented a detailed description of ESCA results for some typical mixtures in which 0.2 to 0.75 wt % platinum are mixed into an alumina matrix by standard catalytic means.[22] The latter usually implies ion exchange of certain Pt salts, followed by an oxidative and then reductive step. Although examined in the previous study at all stages, we are herein primarily concerned only with the results following the reduction of the catalyst and whether the material is either well handled or physically abused.

3. The Optimal Reduced Pt-γ-Al$_2$O$_3$ System

All of the catalytic studies related to this system strongly suggest that optimal behavior requires that the Pt° particles produced in the reductive step are well dispersed in the Al_2O_3 matrix. When this occurs, the catalyst performance, microscopic examination, and X-ray diffraction all suggest that the platinum is uniformly placed in the alumina in "key" sites, in the form of small platinum particles, generally no more than several atoms on a side, with essentially no crystallites having diameters of more than 20 Å. When ESCA is employed to examine this system, one observes substantial charge shifting that is readily removed with application of a flood gun in a manner that appears to be identical to that for a wafer containing only γ-Al_2O_3. As suggested above, *the various platinum particles seem to charge shift with the alumina as if they were simply a continuation of the macroscopic feature of that matrix.* Unfortunately, this tends to hide the Pt from ESCA view. Thus, after the purported total removal of charging, the relatively intense Pt(4f) lines are entirely covered by the far more intense Al(2p) (see Figure 7-5a). Until recently, this feature precluded acceptable surface chemical detection; however, we recently published the first example of a detailed study of the behavior of the much weaker Pt(4d) lines under these conditions.[22] In these cases, the Pt(4d$_{5/2}$) for the reduced Pt/Al_2O_3 system was found repeatedly at 314.0 ± 0.5 eV, suggesting *the formation of uniformly dispersed Pt°*. This point is emphasized because of its relationship to surface-to-bulk shift studies (see Section VII in Chapter 9).

Thus, it appears that uniformly mixed, tiny Pt particles charge (and discharge) with the alumina (matrix) support. Therefore, one might suspect that some other type of behavior might occur if the metals are, instead, forced to cluster into relatively macroscopic-sized crystallites. Electron microscopists[25] and catalyst scientists[26] have demonstrated that such clustering is often induced in these systems when they are thermally abused (these abuse processes may often be simulated by such treatments as steam heating). We have discovered that, if the aforementioned Pt catalyst systems are abused, both the Pt and alumina spectra still exhibit extensive charging; however, during the process of charge removal (using devices like a flood gun) the degree of (discharge) shift induced in the Pt(4f) was significantly different from that for the Al(2p). As a result, when the Al(2p) line reached the point previously established as that of charge neutrality (with approximately the same setting as that for the pre-abused catalyst), the photoelectron lines for the platinum were in quite different positions. In fact, the new positions were such that the Pt metal crystallites produced Pt(4f) lines "pushed" from 1.5 to 3.0 eV below their "normal" Pt° positions. They are pushed so low, in fact, that the Pt(4f$_{7/2}$) lines now are free from the previously described Al(2p) blockage. A figurative example of this process is shown in Figure 7-5b, and a practical example for an extensively abused Pt/Al_2O_3 system is presented in Figure 7-6. The newly detected platinum clusters are still reduced to the metallic state, but are symbolized as Pt°' to indicate their difference from the previously revealed Pt°.

The newly produced cluster system would thus suggest the presence of differential charging.[6] In fact, the real cause may be a combination of differential charging and the

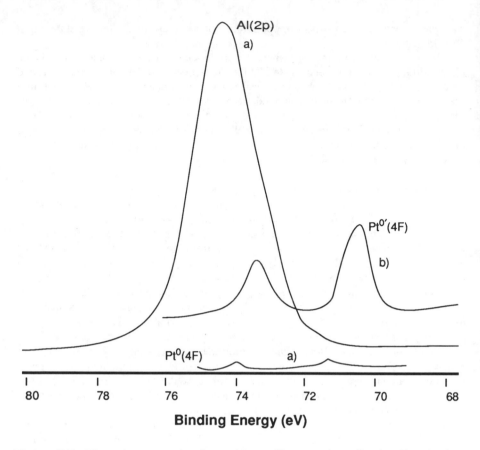

Figure 7-5 Figurative example of metal crystallite growth — floating Fermi edge case: (a) dispersed metals; (b) following metals agglomeration.

previously described Fermi edge decoupling, and the soon to be described "cluster shifts"[4,6,22] (see Section VII.F in Chapter 9). Thus, the finite-sized Pt crystallites, generally now 50 to several hundred angstroms in diameter, would appear to be capable of some of the solid-state properties of a conductive metal. If this is true, they have a Fermi edge, but are decoupled from the corresponding edge of the spectrometer due to the surrounding alumina (insulating) sea. The latter, which is still continuous, thus behaves as it did before crystallite growth; however, the Pt is now electronically "floating", and following charge shift removal achieves its own Fermi edge, zero point, totally independent of that for the spectrometer.[4,6,22] Since binding energies are still produced based on the spectrometer Fermi edge, the binding energies for these Pt crystallites are "unique". For reasons that are not entirely understood, the resulting (spectrometer-zeroed) values for most situations like the ones described, (i.e., conductive crystallites immersed in an insulating matrix) seem to yield a "negative" binding energy shift, see also Chapter 9. The relative degree of this negative shift also seems to depend upon the size and shape of the crystallites formed. In addition, shifts of this type may be realized through additions of another dopant, as seems to be (selectively) the case when various alkali cations are added to a Pd° doped-alumina system (Figure 7-7). The latter situation may reflect the forcing by the alkali atoms of electrons into the palladium valence band to produce Pd^{-x} ions.

Based upon these results, a rather unique tool may be suggested. Obviously, the value of this charge shift approach will depend upon its controllability and versatility. In order

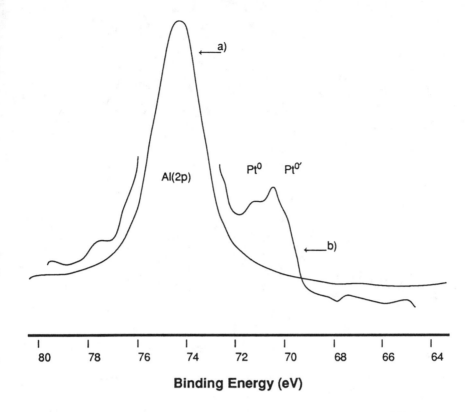

Binding Energy (eV)

Figure 7-6 Example of Pt crystallite growth: (a) reduced pre-abuse, Al(2p) covers Pt(4f) Pt° dispersed in Al_2O_3; (b) extensively abused Pt/Al_2O_3 catalyst [$Pt^{0'}$(4f), with $4f_{7/12}$ ~70 eV now detected].

to thoroughly establish the validity of these methods, much more carefully devised model studies are necessary.

II. ESCA-INDUCED VALENCE BAND SPECTRA

A. INTRODUCTION

Despite its prominence as a chemical tool, X-ray photoelectron spectroscopy (XPS) has seen only marginal use in valence band spectroscopy. There are several reasons for this; perhaps the most important of these are (1) the general, overlapping complexity of many valence bands; (2) the complicated nature of the photoelectron cross sections of these valence regions; (3) the general observance that these valence band cross sections are much weaker than the most favorable core levels; and (4), perhaps most important, the fact that these valence band photoemissions may be achieved with generally greater resolution using ultraviolet (UPS) and/or synchrotron radiation sources. Despite these drawbacks, we feel that there are a number of features that make XPS-generated valence band spectral analysis useful addendums to the more conventional core level spectra and that, in a number of cases, these valence band results are the critical features.

B. TECHNICALITIES OF THE METHOD

It should be readily apparent that most of the arguments presented earlier in Chapters 2, 4, and 5 apply to valence band analysis in exactly the same manner as they do to core

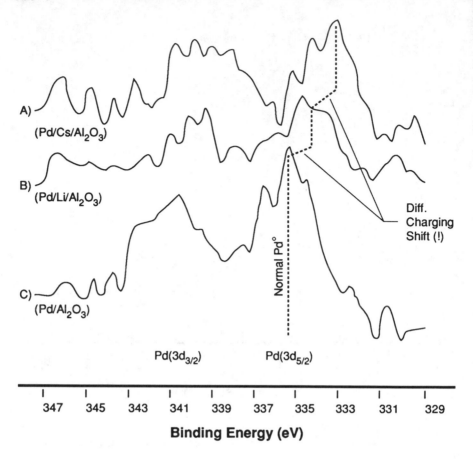

Figure 7-7   Examples of Pd$^{o'}$ created by increased alkalinity.

level studies. Most of the spectroscopic considerations are the same. However, due to the relatively weak cross sections of valence bands, significantly more patience is needed in these types of studies, and one must ensure that the XPS spectrometer being used is operating with maximum stability. In addition, there are several select but important features that may make valence band analysis unique compared to the generation of core level results. To note:

1. It was suggested above that one may often produce useful core level results for nonconductive materials without physically removing any charging. In the case of valence band studies, however, due to the poor cross sections and complex nature of the peaks involved, it is often vital that optimal techniques be exercised to remove (if possible) all of the charging shifts before these valence bands are generated.

2. The relaxation and loss spectral features that are described in Section IV of Chapter 2 and of Chapter 6 may be significantly different in the valence band region compared to core level studies. Several reasons contribute to these differences, particularly the deep vs. shallow hole concept described earlier in Section IV of Chapter 6. It is important that analysts be cognizant of these facts, particularly noting that valence band loss spectra often differ from their core level equivalent, but at the same time keeping in mind that loss features generally occur in both cases.[27-29]

3. Caution should be exercised in conducting certain XPS valence band studies using (conventional) spectrometers with non monochromatic sources. In particular (as will

be described below), one of the more interesting new forms of analysis is to try to monitor the production or destruction of states in the band gap for nonconductive materials, even those with rather finite (3- to 10-eV) band gaps.[30-32] It should be noted that most of the materials involved produce relatively broad bands that often stretch up field in a range from 5 to 15 eV. One should also recall that photoelectron states detected with conventional X-ray sources will produce X-ray satellite peaks at various points down field (at lower binding energy). (For example, in the case of an $Al_{K\alpha}$ source, the largest of these satellites occurs about 9 eV lower in binding energy.) In many valence band situations, therefore, these satellites would be placed directly in the area where the generally weak defects and surface states occur. Corrections employing satellite subtraction methods for these delicate analyses may be very difficult, and may also conflict with the defect analyses.[30-32]

C. XPS-VALENCE BAND ANALYSES OF CONDUCTIVE SYSTEMS

Examples of the determination and use of valence band spectra produced in X-ray photoelectron spectroscopy (XPS) date from the origins of the spectroscopy,[7] with particular emphasis on gas phase analysis.[33] It was, however, primarily in the work of the groups of Shirley[27,34] and Wertheim[28,35] in which XPS valence bands of solids were first seriously examined as potential substitutes for (or at least important supplements of) core level results, and also as alternatives to results obtained in UPS and the X-ray spectroscopies.

The Shirley group, for example, systematically succeeded in demonstrating the utility of XPS valence bands in studies of the differences between various forms of solid carbon.[36] They also showed how a combination of valence band spectra and near core lines could be employed to make critical observations about the sequential bonding variations often found in compound semiconductors.[37]

At almost the same time, the Wertheim group made a very careful study of the nature of spectroscopically generated valence bands for certain transition metals, comparing the XPS results realized to similar data obtained using UPS, and particularly the X-ray spectroscopies.[35] In these studies, it also was demonstrated that there seemed to be a rather unique, direct correlation between the XPS-generated densities of state and those obtained in the best one-electron band structure calculations. Some shifts and additional densities were found in the XPS data due to various many-electron and final state effects, but these were shown to be generally less of a problem for the XPS results than they were for the X-ray spectroscopies and UPS. With proper handling and interpretation of the cross-sectional features and "lifetime" effects, the Wertheim group concluded that the XPS-valence band approach can be very useful. It is also informative to note the systematic changes in band structure that this group were able to realize for binary (x/y) alloys with progressive changes in the (x to y) ratio.[38] *It was made apparent in these studies that the simultaneous capability to realize key core level lines with the valence band data could be very useful in confirming an analytical interpretation.* In order to improve the approach, Hüfner et al.[38] suggested a need for better resolution (while still maintaining the advantages of a high energy source!).

D. VALENCE BANDS OF NONCONDUCTIVE COMPOUNDS
1. Transition Metal Oxides

Wertheim et al. rapidly extended the previously mentioned studies to include investigations of both simple[35] and mixed oxides[28,35] of these same transition metals. They were particularly concerned with the structural alterations induced by the changes in chemistry and the realization of new many body effects. Simultaneously, it was realized that the substantial reduction in both intensity and resolution of the resulting XPS-valence band densities of state would present a formidable problem in later use of the technique.[28,33,34] The enhancement in the degree of overlap of the valence band leading edge with that of the Fermi level as the oxide became more metallic and the striking similarity between these results and the valence band structures predicted by Goodenough[39] were both very encouraging. In particular, it was noted (as suggested by Goodenough) that the part of transition metal oxide band structures dominated by $O(2p)$ remains relatively fixed (at about 5 eV below E_F, the pseudo-Fermi energy) for a series of third period (3d) oxides (see Figure 7-8). On the other hand, the corresponding (predominantly) metallic, somewhat localized, 3d bands were found at different positions in the resulting $[O(2p)]$ gap, first moving toward E_F (as the number of 3d electrons are increased), then drifting away from E_F at Cu, and finally "jumping" all the way out of that gap to about 11 eV below E_F for Zn where the d band is completely filled. (We will return to the potential utility of these shifts later in this chapter.)

Surprisingly, these potentially useful results have not seemed to elicit a major effort by others to utilize these approaches. As a result, except for several key cases, there have been few XPS-valence band studies of oxide systems that followed these initial efforts. The charging problem obviously played a role in this neglect. In addition, it would appear that the lack of peak intensity and generally modest XPS instrumental resolution forced many investigators to turn to synchrotron radiation sources, particularly for studies of initial oxidation.[40] Unfortunately, as indicated, the latter studies often sacrifice corroborating core level results.

One of the key XPS investigations that did evolve out of the early valence band work of the Shirley and Wertheim groups was that of transition metal oxides by Fiermans et al.[41] This investigation was concerned primarily with core level results, but also included an extensive, informative analysis of the valence bands for ZnO, ReO_3, Cr_2O_3, V_2O_5, NiO, and related compounds. These studies seemed to substantiate many of the conclusions of the earlier investigations, and also included an interesting comparative discussion of the ionicity arguments as developed for III-V and II-VI materials by Kowalczyk et al.,[37] (see below).

2. Valence Band Studies of Tetrahedral Compound Semiconductors

If one examines the valence band spectra realized by almost all tetrahedrally oriented, s-p electron dominated systems, one will find a recurring theme based around a three-peaked manifold.[42] A representative example is the case of InSb (as exhibited in Figure 7-9).[43] Detailed XPS studies by the Shirley group[37] and later by Ley[42] support these conclusions. Although some differences have been found with the theory[41] and UPS results,[44] the generality of these ESCA results has been verified. The reason for this generality can be seen in basic band structure calculations,[45] or even through simple molecular orbital (MO) arguments,[46] where four valence orbitals were found generally to yield three bands, particularly in the common cleavage direction (III) (see Figure 7-9). Shoulders are readily detected in many specific cases, suggesting a variety of "complications", including band shifts with changing orientations. In fact, specific examples have been detected in which two of the particular peaks, P_{III} and P_{II}, shift from the values reported by Ley[37,42] for single crystals of GaSb cleaved along the (III) axis ($\Delta E_{III-II} = 3.6$ eV), to values near the reported X1 and X3 points ($\Delta E_{III-II} = 2.4$ eV), respectively, when GaSb films are forced into a (100) orientation as a result of their sputter deposition onto (100) GaAs.[43] Similarly produced (100) InSb, on the other hand, yields ESCA splittings similar to those of (111) InSb. The supporting evidence suggests that in the latter case the structural integrity of the InSb may be altered by the sputtering, probably producing a polycrystalline mix, with preferred (111) orientation.[43]

Examination of the molecular orbitals that can be constructed as a cursory basis for these tetrahedral systems[46] suggests that extensive s-p hybridization occurs for the lighter diamond-structured (group IV) systems (C and Si). As one proceeds down the periodic table (to Ge and Sn), the hybridization dissipates, and the orbital that represents the lowest

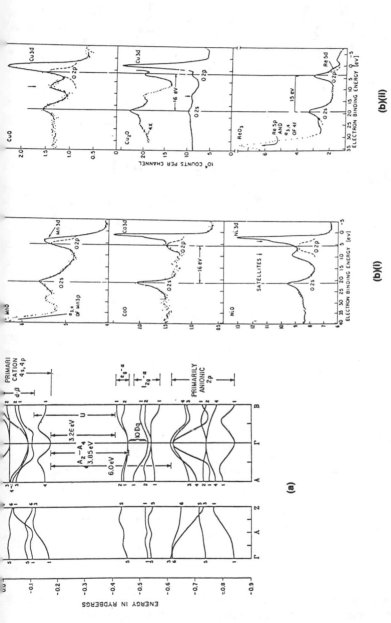

Figure 7-8 (a) Representative valence-band development for transition metal oxide (from Reference 39) (calculated energy bands of ordered MnO₂ along the three directions from the origin to the symmetry points on the boundary of the Brillouin zone [various bands are identified in the right-hand column]); (b) XPS-generated valence bands for select transition metal oxides. Note positions of O(2p) and metal 3d bands (from Reference 35) (i), XPS valence-band spectra of MnO, CoO, and NiO ([position of satellites in NiO deduced from the Ni(3p) spectra are indicated by arrows; the position of O(2p) is indicated where not directly visible in the spectra]; (ii), XPS valence-band spectra of CuO, Cu₂O, and ReO₃ [satellite position in CuO and Cu₂O obtained from Cu(2p) obtained from Cu(2p) are indicated by arrows].

148

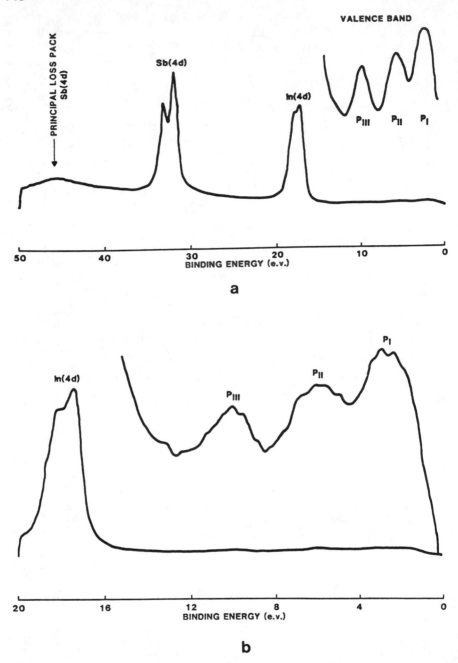

Figure 7-9 ESCA valence band results for (III) InSb: (a) 0 to 50-eV scan demonstrating integrity and quantification; (b) high-resolution 0 to 20-eV scan, demonstrating valence band triple peaks and their splitting.

valence band, Γ ($k = 0$), splits out from the balance, becoming more and more s-like. Eventually one refers to the electrons in this band as an inert (Sidgwick) pair.[46] Note that the second band [symbolized by X_3 in the (100) direction] also becomes predominantly s-like. However, its behavior is somewhat more complicated than the inert band that yields the higher binding energy peak. In addition, Kowalczyk et al.[37] have discovered that this same "detachment" of the s-like bands (yielding a splitting symbolized by ΔE_s) increases noticeably when one considers the band structure of the zinc blend systems, A_nB_{8-n} (A ≠ B) that are isoelectronic with a particular diamond-structured, group IV system, e.g., ΔE_sGe < ΔE_sGaAs < ΔE_sZnBe < ΔE_s KBr. This pattern of increasing ΔE_s is suggested by Kowalczyk et al.[37] to denote increased ionicity.

A variety of related valence band spectra have been generated, by research groups connected to the author, of mixtures of $(III-V)_x \cdot (IV_2)_{1-x}$ in metastable equilibrium[47] (see, for example, Figure 7-10). Borrowing on the arguments of Onodera and Toyozawa[48] and Kowalczyk et al.,[37] these results have been used by our groups to describe the nature of the bonding situations involved, suggesting in some cases the presence of amalgamated[43,49] and, in other cases, persistent[50] mixtures.

3. Valence Band Studies of Silica and Alumina

XPS-generated valence bands were employed to investigate some of the properties of the surface and near-surface regions of select silicas and aluminas.[31] Results that are similar to those obtained in previous band structure calculations[51,52] and measurements with X-ray spectroscopy and XPS[53] were achieved for several well-formed systems (Figures 7-11 and 7-12). These results included the detection of sub-band features that permit delineation of the bonding (B) and nonbonding (N) electron density regions of the bands. The determination of the relative widths and positions of the total band, and certain sub-band features, have permitted the suggestion of the relative covalency/ionicity of these systems (see Table 7-1). In this regard, alumina is shown to be significantly more ionic than silica.[31] Several of these systems also were subjected to ion alteration until a steady state was reached. The silicas exhibited little or no change in chemistry during this process, although their structures were obviously modified (Figure 7-13). In addition, the ion beam was shown to create localized, defect states on the leading edge of the silica bands, and perhaps small discrete defect state of the E' or F type (Figure 7-13). Ion alteration of alumina did not produce detectable quantities of these defect states, but did alter the chemistry, as well as the structure of the system (Figure 7-14). A reduction produced in [O/Al] was accompanied with a transfer of electron density from the nonbonding (N) region to the bonding part of the valence band. This, and other changes, suggest the formation of suboxides, particularly Al(II)O. There seems to be the possibility of π-bond creation in the latter. It also should be noted that similar suboxide production may be induced through extreme thermal abuse of certain aluminas.[31,54]

4. Valence Band Studies of Germanium Oxides
a. Sputter-Deposited GeO₂

Typical valence band spectra for a freshly grown thin film of oxygen-maximized, germanium oxide (produced by the research group of Professor C. R. Aita) are displayed in Figure 7-15.[32] Note that, while the resolution of this valence band is more accurately rendered in these 0- to 20-eV scans (Figure 7-15), 0- to 50-eV scans (not shown) were also generated. The latter have the virtue of (1) verifying the key features in the previous, more-detailed valence band and (2) simultaneously displaying the key shallow core peaks for Ge(3d) and O(2s). The quantitative ratios of the latter yield a [O/Ge] surface of about 2.0 ± 0.1 for the present material.[32]

Several specific features of the resulting valence band spectrum (Figure 7-15a) should be carefully considered. First, it is our contention (based upon analogy and subsequent results) that the XPS-generated spectrum for true, total GeO_2 (at the present resolution) would approximately follow the pattern contained in the dotted lines of Figure 7-15a. One should note, in particular (1) the total width of this band (about 9.0 eV at half maximum);

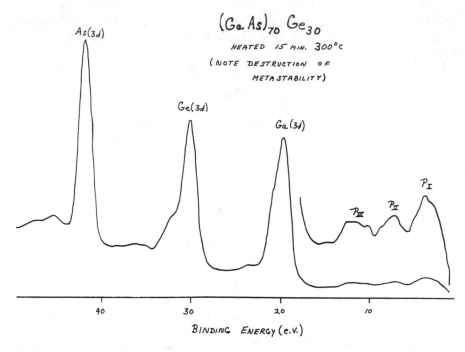

Figure 7-10 Typical XPS valence band spectrum for select $(III\text{-}V)_{1-x}$ $(IV_2)_x$: $(GaAs)_{70}$ $(Ge)_{30}$. Note that the three peaked valence band suggests presence of only GaAs.

(2) the presence of three readily discernible sub-bands; and (3) the positions and separations of those sub-band peaks from one another.[32]

Analogies have also been drawn between these features and previously presented related results for the aforementioned, more completely described bands for SiO_2 and Al_2O_3 (and other group IIIA and IVA oxides).[30,31] Several key numerical values for these systems are tabulated in Table 7-1. Attempts have been made to reference these values to a common Fermi edge — a procedure that is not easily achieved for these surface insulators. Suffice it to say that it is suggested that greater heed be paid to the trends (e.g., total width, subpeak separations, etc.) than to the absolute numbers realized in this uncertain procedure.[4,6,55]

It should be noted that our rendition of the GeO_2 valence band contains most (but not all) of the total spectrum found in Figure 7-15a.[32] The resulting GeO_2 features include, in particular, a set of three characteristic subpeaks (B_I, B_{II}, and N) of slightly different widths and heights. (The latter features are realized, of course, because of the unique combination of valence band peak density and photoelectron cross-section for these systems.) Three features of the latter are omitted in the former. These may be loosely depicted as

1. The small satellite peak (or shoulder) that protrudes from the experimental band on the leading edge side (i.e., into the band gap!). This peak (or peaks) seems to exhibit binding energies from somewhere between 3.75 to 2.50 eV. In this case, we are assuming that our "truncated" valence band spectrum for GeO_2 refers to a totally intrinsic system with a Fermi edge equal to approximately 1/2 the band gap, E_g; then E_g is approximately 5.8 to 6.4 eV for GeO_2, and the aforementioned peaks produce a side band neck in this band gap shifted in excess of 1 eV.

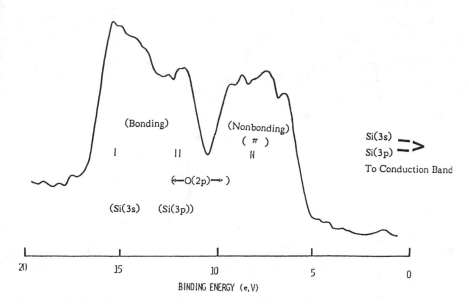

Figure 7-11   Representative XPS valence band spectrum for α-SiO₂. Note band width, positions of edges, and character of three sub-band peaks.

Figure 7-12   Representative XPS valence band spectrum for α-Al₂O₃. Note key band features.

2.   The somewhat obscured shoulder on the high binding energy side of the peak labeled as N. The latter [labeled herein as B(π)] seems to be a broad protrusion, peaked at ~7.0 eV. The rationale for not including it as part of the "normal" XPS band structure for GeO₂ is described later.

3.   There is also the suggestion that the peak structure for the other two sub-bands, B_I and B_II, may be somewhat sharper for GeO₂ than as experimentally produced in Figure 7-15a. We realize that enhanced spectrometer resolution should produce some of these results, but we are actually implying that the result shown in Figure 7-15a is also broad because these sub-band peaks (particularly B_I) are somewhat modified from that for *true* GeO₂.[32]

b.   Defects and Suboxides, Production and Removal

It should be apparent that the explanation offered above for the results provided in Figure 7-15a and Table 7-1 with respect to GeO₂ imply that we are ascribing some origin (other than GeO₂) to those parts of the band structure experimentally detected, but omitted

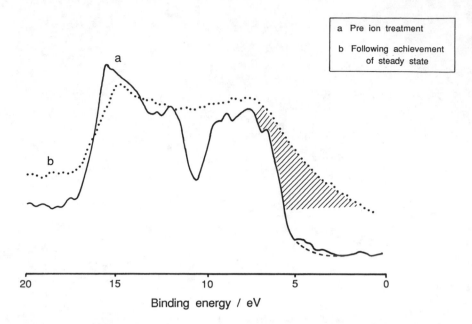

a Pre ion treatment

b Following achievement
   of steady state

Binding energy / eV

Figure 7-13 XPS valence band demonstration of steady-state ion alteration of α-SiO₂: (a) pre-ion treatment; (b) following achievement of steady-state.

from, inside the dashed curves. In view of the informative work of others investigating the *bulk* properties of germanium oxides,[56,57] it seems natural to associate part of our non-GeO₂ band features to the presence of extensive (*surface oriented*) defect structures and/or suboxide species. Indeed, Cohen and Smith[58] have demonstrated the ready production of oxide defects for GeO₂. In addition, Weeks et al.[57,59] have attributed the results of their fused germania/kinetics/ESR studies to the presence of atomic order and suboxide defect structures. On the other hand, Takano et al.[60] (using XPS) and Kawazoe et al.[61] (using ESR and other techniques) have suggested the presence of suboxide-like systems that seem to result from the production of selective mixtures of GeO₂ and small, isolated groups of Ge°. In addition, several authors have investigated the possibility for production of E' (an oxygen vacancy occupied by one electron)[62] and peroxide-type[63] centers in GeO₂ glasses. Utilizing ESR, Watanabe et al.[64] feel that they also have detected the former in GeO₂/SiO₂ optical glasses. Most recently, Aita et al.[65,66] have employed UV-visible and infrared spectroscopies to examine a variety of sputter-deposited germanium oxide systems. In these studies, they discovered that the band gap shrinks from that of GeO₂ as oxygen is removed,[65,66] producing apparent defect centers; however, subsequent ESR studies did not detect any paramagnetism — suggesting the possible creation of F-like centers,[65] or mixed oxide-metal cases[66] rather than those of the E' type. One should also consider the prospect of the formation of surface-oriented microcrystalline or paracrystalline structures as suggested by Phillips.[67]

As outlined above, we detected the creation of what appears to be a defect state (or states) in the band gap of that part of the valence band attributed to GeO₂ during a sputter deposition of the latter.[32] If these states are indeed due to the presence of some type of defect, then they may also be enhanced during some aspect of the ion sputtering process. It seemed reasonable to suspect that ion bombardment (in an atmosphere devoid of oxidation state maximizing O₂) may be the culprit that induces the deformation. In order to examine this point, surfaces of sputter-deposited germania were interacted with moderate-energy (pure) Ar⁺. This process was continued sequentially, with interspersed

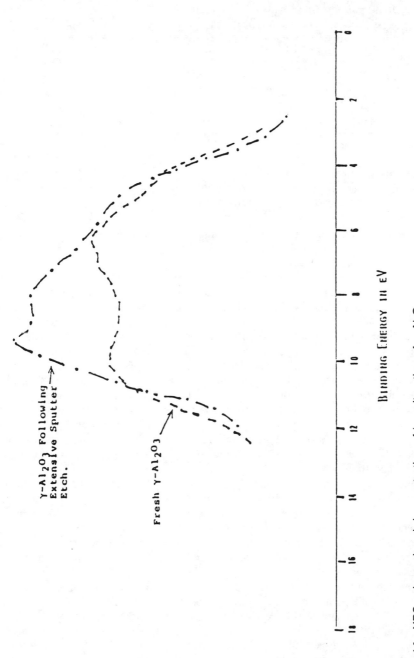

Figure 7-14 XPS valence band demonstration of ion alteration of α-Al_2O_3.

Table 7-1 **XPS valence band results related to covalency/ionicity for select group IVA and IIIA oxides**

Oxide	Band gap $(2 \times (E_{LE} - E_F))$	Band width (1/2 height)	E_c covalency factor (In_2O_3 as O)	Ionicity factors		
				E_I (SiO_2 as O)	f_i Our results	f_i Levine[c]
SiO_2	10.0	10.4	4.4	0.0	0.50[a]	0.57
GeO_2	6.0	9.6	3.6	1.9	0.65	0.62[b]
SnO_2	3.5	7.7	1.7	2.9	0.74	0.78
PbO_2	3.0	N.A.	N.A.	3.6	0.78	N.A.
Al_2O_3	7.5	8.9	2.9	1.8	0.64	0.79
In_2O_3	4.0	6.0	0.0	3.9	0.80	N.A.

Note: N.A. = not available.

[a] Approximate SiO_2 ionicity set to 0.50.
[b] Average of GeO_2 Quartz (0.51) and GeO_2 Rutile (0.73).
[c] From Levine, B. F., *J. Chem. Phys.*, 59, 1463, 1973. With permission.

valence band monitoring until no further change in the valence band was detected. The end result of this process is illustrated in Figure 7-15b.[32]

One should note that several dramatic changes have occurred between the results of Figures 7-15a and b. First, there is a very substantial growth in the size of the density of states in the GeO_2 band gap. This growth occurs in the general range of the shoulder detected for the pre-altered system (Figure 7-15a), but one must not rush to the conclusion that the origin of this enhanced density and the latter are necessarily the same.[32]

In addition to the substantial enhancement of XPS detection of the electron-occupied density of states in the band gap, several other changes were found for the valence band of the ion-altered germanium oxides. The sub-band states, B_I and B_{II}, seem to change somewhat compared to those for the preceding [GeO_2 dominated (?)] system. Consideration of the significant uncertainties that exist in these types of spectra, however, would suggest fairly reasonable reproduction of the B_I region in Figure 7-15b compared to the dotted region in Figure 7-15a. The same is generally true for sub-band B_{II}, except that there is a suggestion that the post-ion altered system (Figure 7-15b) exhibits some enhanced filling of the density region at higher binding energy (i.e., between approximately 9 and 10.5 eV).

Region N in the valence band, on the other hand, exhibits an obvious pronounced, and (we believe) quite important, shift in occupied density. This shift suggests a substantial decrease in the occupied density on the leading edge side of the N sub-band, with the apparent filling [transfer (?)] of density into that region up-field, between B_{II} and N, designated here as $B(\pi)$. As with the detected variable density of states in the GeO_2 band gap, the density between B_{II} and N (outside the dotted lines) is assumed due to some species other than GeO_2![32]

In attempting to identify the causes of these shifts, it is most important to note that, following ion alteration, there may be a slight (but hardly detectable) shift in the binding energies of the key core level peaks, *but no evidence at all of the creation of Ge°*. On the other hand, the key [O(2s)/Ge(3d)] ratio drops precipitously from values of ~2.0 to ~1.4. This confirms (as suspected) that the ion alteration is significantly and selectively bleeding the altered region of oxygen [but apparently without a (noticeable) production of Ge°].[32]

A new set of questions needs to be considered. For example: (1) what will happen if oxygen is returned to this germanium oxide system and (2) are the two effects (a) creation

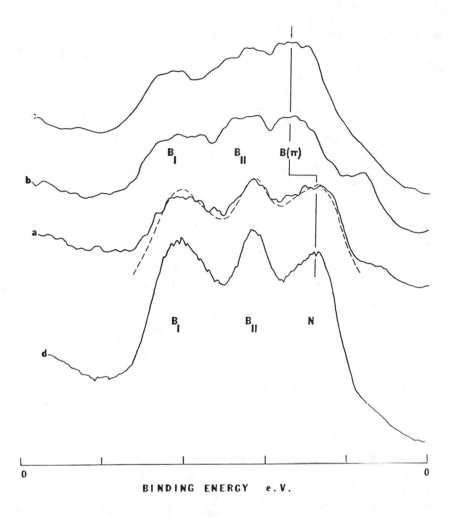

Figure 7-15 XPS valence band study of various modified germanium oxide systems: (a) fresh sputter deposited; (b) steady-state ion altered; (c) moderate oxide growth; (d) GeO$_2$ produced in O$_2$ at 400°C. Dashed line depicts "idealized" GeO$_2$.

of states in the band gap and (b) production of new states in the occupied density region, due to the production of the same or different oxygen-deficient species?

In order to test the latter points, the ion-altered system was gradually (but consistently) fed oxygen in a slow room temperature bleeding process in a preevacuated container. Following extensive exposure under those conditions, the system was reexamined using ESCA, and the spectrum in Figure 7-15c was produced. In this case, the aforementioned, near-discrete peak(s) in the band gap were almost entirely destroyed. The states inside the GeO$_2$ band system [particularly the B(π) detected between B$_{II}$ and N] were, on the other hand, if anything, even more intense than following ion alteration. Close scrutiny of the leading edge of the valence band density suggested the presence of something outside the limits of the predetermined GeO$_2$ region — but in this case, a (continuous) exponentially decreasing set of *tail states*. Once again, the core level spectra exhibited little or no alteration, except in the [O/Ge] that grew to ~1.65. The conclusion reached from this result is that *the causes of the observed discrete (?) band gap states and the effects causing the novel states inside the GeO$_2$ band are not the same.*[32]

In order to further verify these results, we subjected the same altered germanium oxide film to *extensive* reinvestment of O$_2$ by heating it to 400°C in flowing O$_2$. Following this treatment, the core level results once again exhibited little change (except in the case of [O(2s)/Ge(3d)], which increased to ~2.0. The valence band results presented in Figure 7-15d, however, did change. In this case, the principal sub-bands reverted to three peaks (Figure 7-15d) that are quite distinct from one another and are split by almost exactly the same values as those predicted in the dotted rendition of the original system (Figure 7-15a). In the present case, however, the (forced) oxygen "rich" system exhibits almost no evidence of any states in the gaps between the three principal sub-band peaks, B$_I$, B$_{II}$, and N. Thus, we are now essentially completely free of the B(π) states. In the band gap, we also find a substantial (*but not complete*) reduction in occupied density. The slight but obvious shoulder state found in the gap is assumed to be directly related to the same type of band gap states found in the original material (Figure 7-15a).

In conclusion, we argue (as suggested above) that the triple-peaked band structure, B$_I$, B$_{II}$, and N, is that due to GeO$_2$. [Note that we have not discerned whether the resulting germania is tetrahedral or rutile in structure; certain other ESCA-based considerations do suggest a mixture in the present case (see below)]. The discrete defect states detected are difficult to explicitly identify; however, their connection to oxygen depletion suggests state(s) of the E′ or F type.[32,57,65] The continuous tail states seen in Figure 7-15c may be a corresponding Cohen-Fritzsche-Ovshinsky state.[68] We suggest that the system resulting from selective oxygen removal [with band structure, as typified by a combination of B$_I'$ (near B$_I$), B$_{II}$, and B$_\pi$] is indicative of a collection of suboxides — perhaps dominated by Ge(II)O.[32] As in the case of the previously described Al(II)O,[31] we further suggest that the creation of these suboxides may result in the presence of total covalent-enhancing, π-bonding (thus, the shift in valence band — see below).[32] This should not be surprising, since π-bonding is definitely exhibited at the top of group IVA (carbon), and the formation of stable +2 states is still indicated near the bottom, Sn(II)O and Pb(II)O.

5. Valence Band Studies for Indium Oxides

In Figures 7-16 and 7-17, we display the valence band results for a typical pre- and post-ion altered indium oxide containing material.[30] Certain key numerical results also are presented in Table 7-2. Based upon these results, coupled with the core level values and the aforementioned quantification,[69] one should be able to draw some reasonable suppositions as to the composition of the materials in question.

a. Pre-Ion Altered Indium Oxide

The valence band displayed in Figure 7-16 and the corresponding results in Table 7-2 are rather nondescript in themselves; however, they may be informative if compared to other related valence band data.[30] For example, the total band width of 6.0 eV is the narrowest of all the group IIIA and IVA oxides yet measured (i.e., see Table 7-1). In addition, the protrusion at ~3.9 eV that forms the leading edge peak of the indium oxide valence band is the lowest binding energy of its type for all these oxides. The relatively square structure of the total band, without central dip, may be construed to be a continuation of the pattern initiated by alumina[31] and, to some extent, continued by gallia (Ga$_2$O$_3$).[70,71] One should also note, in particular, that the band is relatively flat across its entire density top with only a small (but noticeable) upward slope as one proceeds from high to low binding energy.[30]

b. Ion-Altered Indium Oxide

Following ion bombardment to the point of an apparent steady state, the indium oxide valence band resembles Figure 7-17.[30] As seen in this result (and Table 7-2), there are only modest alterations *of the band structure* from that of the pre-ion bombarded indium oxide. Thus, the total band width is still ~6.0 eV and based on the total extent of the band there still may be a peak at ~3.9 eV. Despite the general similarities of Figures 7-16 and 7-17, however, close scrutiny reveals certain factors that are different. For example,

1. There seems to be, in the ion-treated band, a moderate peak forming at ~4.4 eV that may not be present in the pre-ion altered material.
2. The total band structure is (following ion alteration) more definitely sloping upward from trailing to leading edge.
3. *There are (at least) two rather discrete peaks present in the band gap near the leading edge of the band. These were definitely not present before ion alteration.*

c. Comparisons of Indium Oxides to Related Oxides

In view of these considerations, it would seem that one of the most interesting aspects demonstrated in Figure 7-18 is the potential interrelationships of the valence band of In_2O_3 to those for other previously described group IIIA and IVA systems.[30-32,71] Some of the more pertinent comparative results are summarized in Table 7-1. In this regard, we note that the valence band for In_2O_3 differs from those for Al_2O_3 and Ga_2O_3 in a significant, but regular, progressive fashion.[30,70] Thus, whereas both Al_2O_3 and Ga_2O_3 exhibit two distinct sub-band regions (designated for Al_2O_3 as B and N), the valence band for In_2O_3 seems to be constricted into one, essentially continuous, nearly flat band [at least, when detected with a spectrometer having the resolution of that used (Hewlett-Packard 5950A)]. The total band width of In_2O_3 is also constricted to a value substantially less than that for Al_2O_3 (Table 7-1). In addition, the leading edge of the former is noticeably shifted to a much lower binding energy that is relatively close to its pseudo-Fermi edge (i.e., the midgap, zero reference).

These constricting and shifting patterns of the principal oxide valence bands (as one goes from the top to the bottom of a periodic group) are also generally followed by the equivalent oxides of group IVA, with the added proviso that the corresponding IIIA oxides are noticeably more constricted and shifted. Thus, for example (Table 7-1), india has a narrower valence band than SnO_2, and the former also is shifted to lower binding energy.[30-32,70,71]

All of this, we suggest, indicates that the bonding chemistry for india is more ionic than that for the other oxides shown in Table 7-1. The ionic effects referred to are suggested by us in other, more-detailed sources[70,71] to be due to a rather complex combination that presupposes substantial ionicity for a system in which the oxide valence band is entirely composed of $O(2p)$ orbital density [i.e., the metal, e.g., $In(5s)$ and $In(5p)$, orbitals contribute to *shift* the density, but do not contribute substantially to the density itself]. The reason for the supposition of only $O(2p)$ density is to insure that the ionic effects on the valence band are uniform throughout the band.

We have demonstrated that these ionic effects may be reasonably partitioned into an ionic field (E_i) that presupposes 100% ionic behavior and the ionicity (fi) that recognizes that no system (including In_2O_3) is entirely ionic.[30] The ionic field has been shown in an apparently reasonable model[72] to include several separable terms, including the field created by the O^{-2} ion itself, plus the contributions of the ions that surround the "representative" O^{-2}. The latter are most easily expressed as a Madelung-type potential development (V^M). Broughton and Bagus[73] have calculated the V^M values for many oxides using a spherical, point charge model. (Unfortunately In_2O_3 was not among the oxides calculated.) One can see from Table 7-3 that the calculated Madelung terms are large, and generally counter, in size and charge, the aforementioned O^{-2} point charge. It should be noted (Table 7-3) that the V^M values that exist for both group IIIA and IVA oxides exhibit a progressive decrease as one goes down the periodic table. When properly coupled with the previously suggested uniform influence of these terms on all parts of the valence

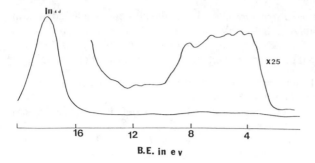

Figure 7-16 XPS valence band spectrum of thin film-deposited indium oxide.

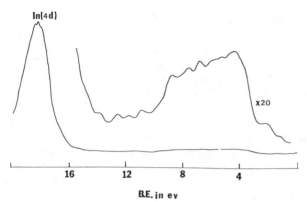

Figure 7-17 XPS valence band of indium oxide following steady-state ion alteration. Note (1) slight shift in band structure and (2) presence of moderate discrete peaks in band gap.

Table 7-2 **Key valence band data realized from XPS study of In_2O_3**

	(eV)
Band gap	~4.0
Lead "N" peak	3.9
Band width	6.0
Peak for In(4d)	18.0
	(2.0)

bands, one finds that these Madelung potentials suggest a progressive shift of the leading edge of the valence band toward the pseudo-Fermi edge as one goes down and to the left in this part of the periodic table.[30,70,71]

The terms considered thus far in the predefined ionic field, E_i, are, however, incomplete. In order to develop a more complete description, factors that treat both initial state repulsions (e.g., Pauling type double repulsions,[74] etc.) and final state relaxation effects[72] should also be included. [Mahan has demonstrated[29] that materials with finite band gaps (such as the oxides discussed here) will exhibit sufficient hole localization during the ejection of photoelectrons from their valence bands to produce relaxation shifts.] Unfortunately, calculations of these repulsions and relaxations only exist for small oxides such as BeO and MgO.[72] These results, and the progressions that are predicted for larger oxides, suggest that these effects are small and tend to cancel each other. Thus, despite the incomplete nature of our analysis, we do not anticipate that the inclusion of the missing features will alter the progressions predicted by the Madelung terms for the ionic field.[30,32,70,71,73]

The ionicity, on the other hand, is generally defined by assuming a Coulson-like[75] covalent-ionic partitioning of the wave function that represents the total valence band density, as

$$\Psi_{total} = a\Psi_{covalent} + b\Psi_{ion} \tag{7-1}$$

where Ψ is assumed normalized, i.e.,

$$a^2 + b^2 = 1 \tag{7-2}$$

and the ionicity, f_i, may be defined by the proportionality:

$$f_i \propto b^2 \tag{7-3}$$

and the ionicity (f_i) and covalency (f_c) are related through

$$f_i + f_c = 1 \tag{7-4}$$

The ionicity (f_i) should act to mitigate the ionic nature predicted by the ionic field, which presupposes *total* ionic behavior. Thus, we define the actual ionic nature as the ionic character, E_I, where

$$E_I^A(O) = f_i^A(O) \cdot E_i^A(O) \tag{7-5}$$

In this relationship, O indicates that the only orbital considered to contribute to this (admittedly approximate) model is the oxygen 2p, while the symbol A is used to label the different oxides being considered.[30]

In the case of india and related oxides, we further assume that the sub-band density near the leading edge of the valence band (previously labeled as N) results exclusively from electrons that occupy non(covalent)bonding O(2p) orbitals. Since these orbitals are nonbonding, they do not experience covalent effects but should experience the total ionic character, E_I^A. Thus, in this approximate model we expect the leading edge of these oxide valence bands to shift with respect to the pseudo-Fermi edge, depending upon E_I^A. In this model, india's greater ionic character relative to that of alumina may be seen in the lower binding energy of the former's leading band edge (Figure 7-18a).[30]

If this model is reasonable, and these valence bands of different oxides are artificially shifted to align their (true) leading edges, then their differences in ionic character, E_I^A, will have, in effect, been removed. This means that any difference in band width that remains on the trailing edge side following this shift must be due to their differences in covalency (see Figure 7-18b for the case of india and alumina, Figure 7-19 for the case of alumina and silica, and Figure 7-20b for the corresponding treatment of silica, germania, and SnO$_2$).[70,71]

A summary of these factors is given in Table 7-1. These results indicate that india follows a progression in which the ionicity of the maximum valent oxides of groups IVA and IIIA (and perhaps groups IIA and IA) appears, in general, to increase from top to bottom and from right to left in the periodic table.[71,72]

We submit that, while In$_2$O$_3$ is considered to be largely ionic (see Table 7-1), it retains some degree of finite covalency (≈ 0.2) that must influence its reactivity, adsorptivity, etc.[30] This covalency appears larger than that calculated by Levine[76] for MgO (0.16). In

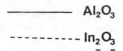

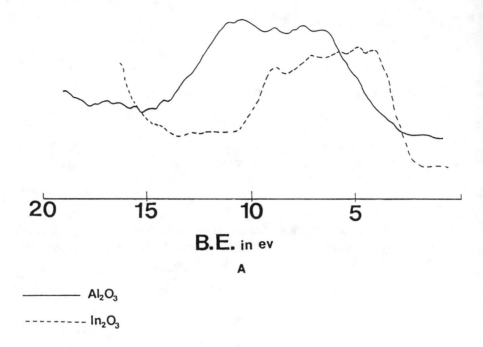

A

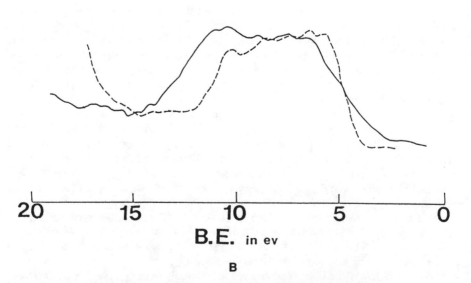

B

Figure 7-18 XPS valence band overlay of india and alumina: (A) band positions for common (quasi) Fermi edge; (B) same bands with leading edges aligned. Note ionicity and covalency shifts.

Table 7-3 **Representative structural factors for select oxides**

	Structural type	Lattice constant a	Madelung constant A	Point charge Madelung potential V^M
SiO	Quartz	4.91	4.44	30.8
GeO_2	Quartz	4.98	4.44	
	Rutile	4.40	4.82	25.1
SnO_2	Rutile	4.74	4.82	24.6
Al_2O_3	Rhombohedral (α-Al_2O_3)	4.76	25.03	26.4
Ga_2O_3	Rhombohedral (α-Al_2O_3)	4.98	25.03	25.3
In_2O_3	Cubic/CsCl	10.12	1.76	(?)
BeO	Wurtzite	2.70	1.64	28.7
MgO	Cubic/NaCl	4.81	1.75	23.9
CaO	Cubic	4.81	1.75	—

addition, it has been determined that an alkaline earth oxide such as BaO is significantly more ionic than In_2O_3;[70] a feature verified by a variety of XPS result.[70,71] Further we have discovered that this feature is of substantial importance in the chemistry of the new high T_c superconducting oxides, see Section VIII in Chapter 8.[77,78]

d. Suppositions Regarding Ion-Altered Indium Oxide

As described above, following ion alteration, indium oxide seems to exhibit only marginal changes. Thus, core level binding energies and line widths primarily suggest the "cleaning" of In_2O_3, rather than its modification. The [O/In] drops slightly, but this is felt to indicate primarily the removal of surface-oriented (air-induced) hydroxides and adsorbed oxygen, rather than the production of extensive suboxide species.[30,69]

The XPS-induced valence band results are even more revealing than the core level features. The former also suggest that the principal attribute realized following ion alteration of the indium oxide to an apparent steady state is the realization of relatively clean In_2O_3. It should be apparent that, without annealing in O_2, the structure of this system is undoubtably extensively compromised by the ion beam; however, we believe that the bonding chemistry is largely retained following such treatment. This suggests that the true version of the total XPS-realized valence band for In_2O_3 may have a distinct upward slope as it proceeds from high to low binding energy. The total width of the band (~6.0 eV) and the binding energy positions of the leading and trailing edges (as given in Table 7-2 for the presputtered indium oxide) seem to be retained following ion alteration. Thus, it appears that these characteristics are also approximate renditions of features inherent to the In_2O_3 system.[30]

Certain features in the valence band region, however, are altered following ion treatment. For example, inside the band itself, ion treatment to a steady state appears to produce a slight (but perceptive) change, particularly in the region nearest the leading edge where a noticeable peak seems to occur at ~4.4 eV. This may suggest a "roll-back" type shift to higher binding energy,[30-32,79] particularly since there seems to be simultaneous reduction in the leading edge, non-bonding (N) type peak at ~3.9 eV (Figures 7-16 and 7-17). Such features may support the concept of a slight production of In(II)O and related suboxide species, as have been detected following the ion alteration of Al_2O_3[31] and GeO_2.[32] That this production of suboxides should not fall entirely to zero for In_2O_3 would seem consistent with our previous detection of other defect states.[30,79]

The most obvious alteration induced by ion treatment, however, appears to be the several discrete peaks induced inside the band gap itself. These peaks are relatively close to the leading edge of the valence band, suggesting the formation of particular oxygen-deficient defect sites similar to those of the E′ or F type. In appearance, they are quite consistent with similar peaks described above that are observed following the treatment of GeO_2.[32] There are strong indications that these discrete defect sites are more readily achieved during ion alteration of group IVA, rather than group IIIA, species (see below).[31,32]

6. Valence Band Studies of Ion-Altered Tin Oxides

We have also considered the behavior of Sn during similar ion bombardments.[80] Tin is readily susceptible to suboxide formation; following "energetic disturbance" of the maximum valent tin (IV) oxide, there appears to be a "roll-back" of the valence band density into the middle of the band and a reduction in [O/Sn] [perhaps resulting in the production of Sn(II)O, see Figure 7-21].[80] In addition, ion beam alteration of the tin system also may produce a substantial concomitant growth of some form of discrete point defects which, if formed, appears to create a relatively large shoulder peak on the leading edge of the Sn(II)O valence band (see Figure 7-21).[80] In the tin case, the presence of these low binding energy states may confuse the interpretations regarding the correct valence band spectrum for Sn(II)O.[81]

7. Valence Band Studies of Zeolites

The first detailed, reproducible, experimental results of ESCA valence band spectra of a variety of commercially important zeolites[4,13,82] have also been developed by our research group (Figure 7-22). The general chemical formulation of the zeolites studied was $(SiO_2)_x \cdot (M_{1/p}^{+p} AlO_2^-)_y \cdot ZH_2O$, where, $M^{+p} = Na^+$ for most of the cases studied. Comparisons were made with well-known experimental and theoretical (MO and band structure) data for various silicas and alumina precursor[51–53] and for various zeolites.[83] Numerous reproducible

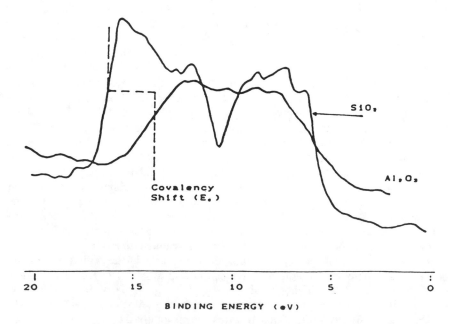

Figure 7-19 XPS valence bands for silica and alumina with leading edges aligned to half height of former. Note covalency shift.

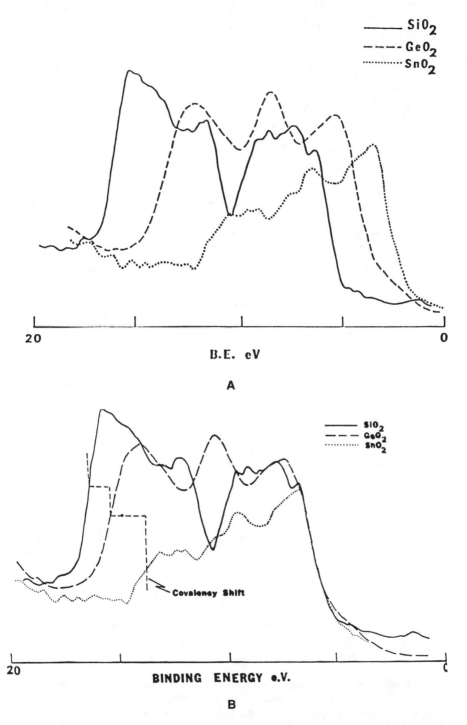

Figure 7-20 XPS valence band for select M(IVA)O$_2$ oxides: (A) referenced to common (quasi) Fermi edge; (B) bands aligned to leading edge of silica. Note covalency shift.

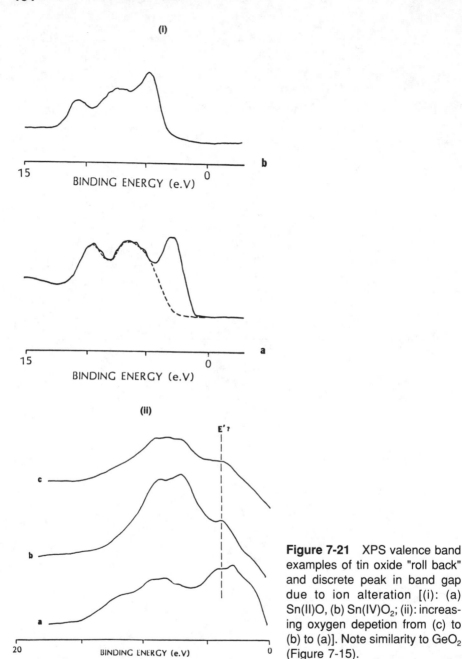

(i)

15　　　　　　　　　0
BINDING ENERGY (e.V)　b

15　　　　　　　　　0
BINDING ENERGY (e.V)　a

(ii)

E'?

c

b

a

20　　BINDING ENERGY (e.V)　　　0

Figure 7-21 XPS valence band examples of tin oxide "roll back" and discrete peak in band gap due to ion alteration [(i): (a) Sn(II)O, (b) Sn(IV)O_2; (ii): increasing oxygen depetion from (c) to (b) to (a)]. Note similarity to GeO_2 (Figure 7-15).

interrelationships were found involving most of the key sub-band features documented for the zeolites.[82] Distinct shifts and truncations were observed in the valence bands that appeared to be critical signposts of the bonding in these complicated mixed oxides (Figures 7-23 and 7-24).[82] Our findings suggested, for example, that zeolites more closely depict an amalgamation than a persistent-type mixture of the aforementioned precursors.[48,82] Analyses demonstrate that many of these features and changes seem to key in on the ratio of x to y in the general chemical formula given above, in a manner similar to that reported in

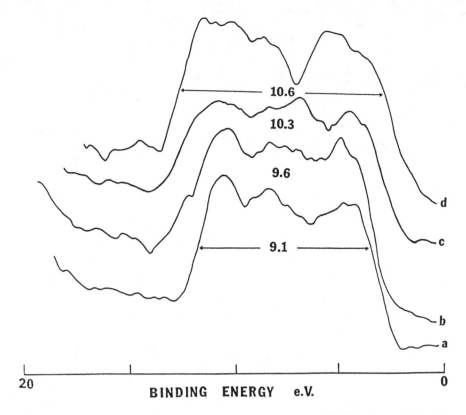

Figure 7-22 XPS valence band for zeolites with various [Si/Al] ratios with leading edges aligned: (a) NaA, [1/1]; (b) NaX, [1.2/1]; (c) NaY, [2.4/1]; and (d) mordenite, [5/1].

previous core level XPS studies.[82] Results also were obtained that suggested distinct, detectable, valence band differences that reflect major changes in the critical structures of these materials.[82] In particular, the data suggested that XPS valence bands may be employed as a means to determine novel groupings into (1) primarily cage-like zeolites (e.g., A, faujasites, and L) and (2) primarily chain-like (e.g., mordenite, ZSM-5, and silicalite) structures.[82] Our studies have shown these groupings and related features to be directly related to changes in the bonding of these zeolite systems and their particular relationships to ionicity/covalancy and acidity, see Chapters 2 and 8.[83]

These investigations were followed by ion alteration studies of some of these zeolites that were similar in type of those mentioned above for aluminas and silicas.[31] In the former case, a controlled, low energy, Ar+ ion beam was found to destroy the structural integrity of certain zeolites, apparently producing mixtures of the silica and alumina precursors which were then subsequently ion altered as described above for the pure, separate systems (see, e.g., Figures 7-25a and b).[84]

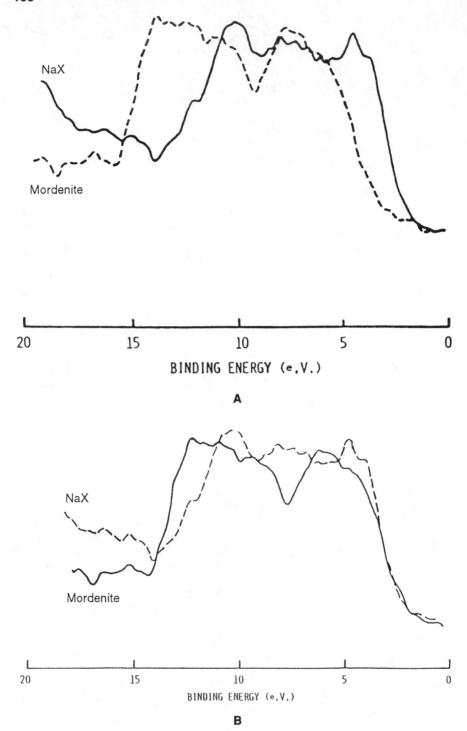

Figure 7-23 XPS valence band overlay for NaX and mordenite: (A) based on common (Fermi edge) zero; (B) aligned to leading edge of NaX. Note enhanced covalency of mordenite.

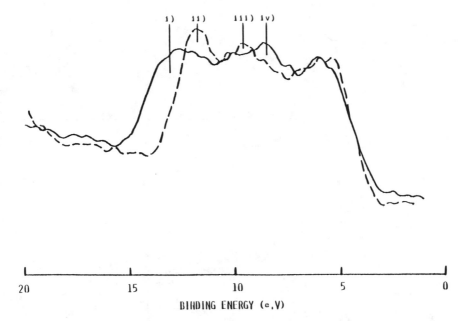

Figure 7-24 XPS valence band overlay of NaA and NaY aligned to leading edge of NaY. Note enhanced covalency in NaY. This should enhance acidity of O–H bonding in NaY.

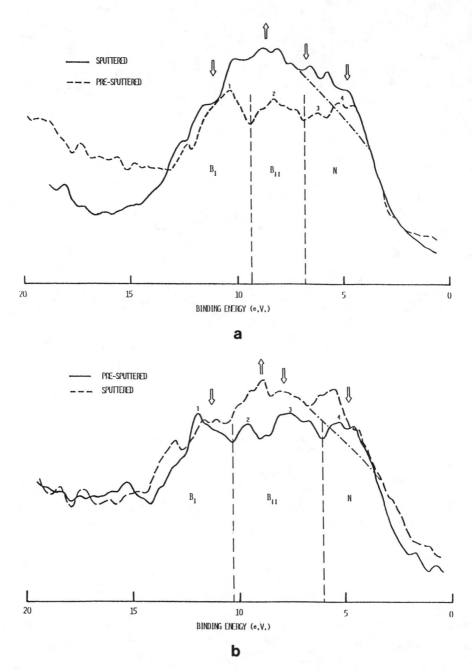

Figure 7-25 XPS valence band for select zeolites following ion alteration: (a) NaA; (b) NaY.

REFERENCES

1. **Lewis, R. T. and Kelly, M. A.,** *J. Electron Spectrosc. Rel. Phenomena,* 20, 105, 1980.
2. **Pollack, R.,** Ph.D. thesis, University of California, Berkeley, 1972.
3. **Swift, P., Shuttleworth, D., and Seah, M. P.,** *Practical Surface Analysis,* Briggs, D. and Seah, M. P., Eds., John Wiley & Sons, Chichester, England, 1983, app. 2.
4. **Barr, T. L.,** *Appl. Surface Sci.,* 15, 1, 1983.
5. **Barr, T. L.,** *Practical Surface Analysis,* Briggs, D. and Seah, M. P., Eds., John Wiley & Sons, Chichester, England, 1983, chap. 8.
6. **Barr, T. L.,** *J. Vac. Sci. Technol.,* 47, 1677, 1989.
7. **Siegbahn, K., Nordling, C., Fahlman, A., Nordberg, R., Hamrin, K., Hedman, J., Johansson, G., Bergmark, T., Karlsson, S. E., Lindgren, I., and Lindberg, B.,** ESCA — atomic, molecular, and solid state structure studied by means of spectroscopy, *Nova Acta Reginae Soc. Sci. Ups.,* Ser. 4, Vol. 20.
8. **Grunthaner, F. F.,** U.S. Patent 4,052,614, 1977.
9. **Barr, T. L.,** *Am. Lab.,* 10, 40, 65, 1978.
10. **Wagner, C. D.,** *Applied Surface Analysis,* Barr, T. L. and Davis, L. E., Eds., American Society for Testing and Materials, Philadelphia, 1980, 137; **Stephenson, D. A. and Binkowski, N. J.,** *J. Non-Crystalline Solids,* 23, 399, 1976.
11. **Fleisch, T. H., Hicks, R. F., and Bell, A. T.,** *J. Catal.,* 87, 398, 1984.
12. **Barth, G., Linder, R., and Bryson, C.,** *Surface Interfacial Anal.,* 11, 307, 1988.
13. **Barr, T. L. and Lishka, M. A.,** *J. Am. Chem. Soc.,* 108, 3178, 1986.
14. **Vasquez, R. P. and Grunthaner, F. J.,** *Surface Sci.,* 99, 681, 1980.
15. **Hollinger, G., Juguet, Y., and Duc, T. M.,** *Solid State Commun.,* 22, 277, 1977.
16. **Barr, T. L. and Svoboda, M. J.,** to be published.
17. **Barr, T. L. and Hackenberg, J. J.,** *Appl. Surface Sci.,* 10, 523, 1982.
18. **Barr, T. L.,** ESCA studies of metals and alloys: oxidation, migration and dealloying of Cu-based systems, *Surface Interface Anal.,* 4, 185, 1982.
19. **Barr, T. L.,** XPS as a method for monitoring segregation in alloys: Cu-Be, *Chem. Phys. Lett.,* 43, 89, 1976.
20. **Kao, C. C., Barr, T. L., and Merrill, R. P.,** to be published; **Kao, C. C.,** Ph.D. thesis, Cornell University, Ithaca, NY, 1988.
21. **Barr, T. L.,** to be published.
22. **Barr, T. L.,** *ACS Div. Petrol. Chem.,* 33(4), 649, 1988. **Barr, T. L., Yin, M. P., and Mohsenian, M.,** *The Role of Characterization in Catalyst Development,* Vol. 441, Bradley, S. A., Bertolacini, R. J., and Gattuso, M. J., Eds., ACS Books, Washington, DC, 203.
23. **Johansson, B. and Martensson, N.,** *Phys. Rev. B,* 21, 4427, 1980.
24. **Parmigiani, F., Kay, E., Bagus, P. S., and Nelin, C. J.,** *J. Electron Spectrosc. Rel. Phenomena,* 36, 257, 1986. **Wertheim, G. K., DiCenzo, S. B., and Youngquist, S. E.,** *Phys. Rev. Lett.,* 51, 2310, 1983.
25. **Dautzenberg, F. M. and Walters, H. B. M.,** *J. Catal.,* 51, 267, 1984.
26. **Sachtler, W. M. H. and Van Santen, R. A.,** *Adv. Catal.,* 26, 69, 1977.
27. **Kowalczyk, S. P., Ley, L., McFeeley, F. R., Pollack, R. A., and Shirley, D. A.,** *Phys. Rev. B,* 8, 3583, 1973; **Pollack, R. A., Ley, L., McFeeley, F. R., Kowalczyk, S. P., and Shirley, D. A.,** Characteristic energy loss structure of solids from X-ray photoemission spectra, *J. Electron Spectrosc. Rel. Phenomena,* 3, 381, 1974; **Ley, L., McFeeley, F. R., Kowalczyk, S. P., Jenkin, J. G., and Shirley, D. A.,** Many-body effects in X-ray photoemission from magnesium, *Phys. Rev. B,* 11, 600, 1975; **Kowalczyk, S. P., Ley, L., Martin, R. L., McFeeley, F. R., and Shirley, D. A.,** Relaxation and final-state structure in XPS of atoms, molecules and metals, *Discuss. Faraday Soc.,* 7, 1975. **Pollack, R. A.,** Ph.D. thesis, University of California, Berkeley, 1972.

28. **Hüfner, S., Wertheim, G. K., Buchanan, D. N. E., and West, K. W.,** *Phys. Lett. A,* 46, 420, 1974; **Hüfner, S. and Wertheim, G. K.,** *Phys. Rev. B,* 11, 678, 1975; **Campagne, M., Wertheim, G. K., Shanks, H. R., Zumsteg, F., and Banks, E.,** Local character of many-body effects in XPS from transition-metal compounds: Na$_x$WO$_3$, *Phys. Rev. Lett.,* 34, 738, 1975.

29. **Mahan, G. D.,** Photoemission from alkali halides: energies and line shapes, *Phys. Rev. B,* 21, 479, 1980; **Mahan, G. D.,** *Phys. Rev. B,* 22, 3102, 1980.

30. **Barr, T. L. and Liu, Y. L.,** An XPS study of the valence band structure of indium oxides, *J. Phys. Chem. Solids,* 50, 657, 1989.

31. **Barr, T. L., Chen, L. M., and Mohsenian, M.,** XPS valence band studies of zeolites and related systems. II. Silicas and aluminas, *J. Phys. Chem.,* to be published.

32. **Barr, T. L., Mohsenian, M., and Chen, L. M.,** XPS valence band studies of the bonding chemistry of germanium oxides and related systems, *Appl. Surface Sci.,* 51, 71, 1991.

33. **Gelius, U.,** *J. Electron Spectrosc. Rel. Phenomena,* 5, 985, 1974. **Siegbahn, K.,** Electron spectroscopy for solids, surfaces, liquids and free molecules, in *Molecular Spectroscopy,* Heyden & Sons, Ltd., London, 1977, chap. 15.

34. **Kowalczyk, S. P., McFeeley, F. R., Ley, L., Gritsyna, V. T., and Shirley, D. A.,** The electronic structure of SrTiO$_3$ and some simple related oxides (MgO, Al$_2$O$_3$, SrO and TiO$_2$), *Solid State Comm.,* 23, 161, 1977. **McFeeley, F. R., Kowalczyk, S. P., Ley, L., Cavell, R. G., Pollack, R. A., and Shirley, D. A.,** X-ray photoemission studies of diamond, graphite and glassy carbon valence bands, *Phys. Rev. B,* 9, 5268, 1974. **Kowalczyk, S. P., Ley, L., McFeeley, F. R., and Shirley, D. A.,** *J. Chem. Phys.,* 61, 2850, 1974. **Ley, L.,** *Photoemission in Solids,* Cardona, M. and Ley, L., Eds., Springer-Verlag, Berlin, 1978, chap. 1. **Kowalczyk, S. P., McFeeley, F. R., Ley, L., Pollack, R. A., and Shirley, D. A.,** X-ray photoemission studies of the alkali halides, *Phys. Rev. B,* 9, 3573, 1974.

35. **Wertheim, G. K. and Hüfner, S.,** XPS band structure of some transition-metal oxides, *Phys. Rev. Lett.,* 28, 1028, 1972; **Wertheim, G. K., Mattheiss, L. F., Campagna, M., and Pearsall, T. P.,** *Phys. Rev. Lett.,* 32, 997, 1974.

36. **McFeeley, F. R., Kowalczyk, S. P., Ley, L., Cavell, R. G., Pollack, R. A., and Shirley, D. A.,** X-ray photoemission studies of diamond, graphite and glassy carbon valence bands, *Phys. Rev. B,* 9, 5268, 1974.

37. **Kowalczyk, S. P., Ley, L., McFeeley, F. R., and Shirley, D. A.,** *J. Chem. Phys.,* 61, 2850, 1974.

38. **Hüfner, S., Wertheim, G. K., and Wernick, J. H.,** XPS of the valence bands of some transition metals and alloys, *Phys. Rev. B,* 8, 4511, 1973.

39. **Goodenough, J. B.,** Metallic oxides, in *Progress in Solid State Chemistry,* Vol. 5, Reiss, H., Ed., Pergamon Press, Elmsford, NY, 1971, chap. 4.

40. **Powell, R. A. and Spicer, W. E.,** Photoemission study of oxygen chemisorption on tin, *Surface Sci.,* 55, 681, 1976.

41. **Fiermans, L., Hoogeiwijs, R., and Vennik, J.,** *Surface Sci.,* 47, 1, 1975.

42. **Ley, L.,** *Photoemission in Solids,* Cardona, M. and Ley, L., Eds., Springer-Verlag, Berlin, 1978, chap. 1.

43. **Barr, T. L., Kramer, B., Shah, S. I., Ray, M., and Greene, J. E.,** ESCA studies of the valence band and loss spectra of semiconductor films: ionicity and chemical bonding, *Mat. Res. Soc. Proc.,* 47, 205, 1985.

44. **Eastman, D. E., Grobman, W. D., Freeouf, J. L., and Erbudak, M.,** *Phys. Rev. B,* 9, 3473, 1974.

45. **Stukel, D. J., Euwena, R. N., Collins, T. C., Herman, F., and Kortum, R. L.,** *Phys. Rev.,* 179, 740, 1969.

46. **Levin, A. A.,** *Solid State Quantum Chemistry*, McGraw-Hill, New York, 1977.

47. **Barnett, S. A., Ray, M. A., Lastras, A., Kramer, B., Greene, J. E., Raccah, P. M., and Abels, L. L.,** *Electronics Lett.,* 18, 891, 1982. **Newman, K. E., Lastras-Martinez, A., Kramer, B., Barnett, S. A., Ray, M. A., Dow, J. D., Greene, J. E., and Raccah, P. M.,** *Phys. Rev. Lett.,* 50, 1466, 1983.

48. **Onodera, Y. and Toyozawa, Y.,** *J. Phys. Soc. Jpn.,* 24, 341, 1968.

49. **Newman, K. E. and Dow, J. D.,** *Phys. Rev. B,* 27, 7495, 1983.

50. **Kramer, B., Tomasch, G., Ray, M., Greene, J. E., Salvati, L., and Barr, T. L.,** A high-resolution XPS study of the valence band structure of single-crystal metastable $(GaAs)_{(1-x)}(Ge_2)_x$, *J. Vac. Sci. Tech. A,* 6, 1572, 1988.

51. **Schneider, D. N. and Fowler, W. B.,** *Phys. Rev. Lett.,* 36, 425, 1976; **Chelikowsky, J. R. and Schlüter, M.,** *Phys. Rev. B,* 15, 4020, 1977; **Batra, I. P.,** in *Proc. Int. Conf. Phys. of SiO_2 and its Interactions,* Pantelides, S. T., Ed., Pergamon Press, Elmsford, NY, 1978, 65; **Lauglin, R. B., Joannopoulos, J. D., and Chadi, D. J.,** *Phys. Rev. B,* 20, 5228, 1979.

52. **Batra, I. P.,** Electronic structure of α-Al_2O_3, *J. Phys. C,* 15, 5395, 1982; **Ciraci, S. and Batra, I. P.,** Electronic structure of α-Alumina and its defect states, *Phys. Rev. B,* 28, 982, 1983.

53. **Thorpe, M. F. and Weaire, D.,** *Phys. Rev. B,* 4, 3518, 1971; **Balzavotti, A. and Bianconi, A.,** *Phys. Status Solidi B,* 76, 689, 1976; **Kowalczyk, S. P., McFeeley, F. R., Ley, L., Gritsyna, V. T., and Shirley, D. A.,** *Solid State Commun.,* 23, 161, 1977; **Flodström, S. A., Martinson, C. W. B., Bachrach, R. Z., Hagstrom, S. B. M., and Bauer, R. W.,** *Phys. Rev. Lett.,* 40, 907, 1978.

54. **Barr, T. L. and Yin, M. P.,** Studies of Pt metal catalysts by high-resolution ESCA, in *Characterization of Catalyst Development,* Bradley, S. A., Gattuso, M. J., and Bertolacini, R. J., (ACS Symp. Ser. 411), ACS Books, Washington, DC, 1989, chap. 19.

55. **Barr, T. L. and Lishka, M. S.,** ESCA studies of the surface chemistry of zeolites, *J. Amer. Chem. Soc.,* 108, 3178, 1986.

56. **Cohen, A. J. and Smith, H. L.,** *J. Phys. Chem. Solids,* 7, 301, 1958; **Purcell, T. and Weeks, R. A.,** *Phys. Chem. Glasses,* 10, 198, 1969; **Phillips, J. C.,** *Solid State Physics,* Ehrenrich, H., Seitz, F., and Turnbull, D., Eds., 37, 93, 1982.

57. **Weeks, R. A. and Purcell, T.,** *J. Chem. Phys.,* 43, 483, 1965; **Kordas, G., Weeks, R. A., and Kinser, D. L.,** *J. Appl.. Phys.,* 54, 5394, 1980; **Margruder, R. H., III, Kinser, D. L., Weeks, R. A., and Jackson, J. A.,** *J. Appl. Phys.,* 57, 345, 1985; **Jackson, J. M., Wells, M. E., Kordas, G., Kinser, D. L., and Weeks, R. A.,** *J. Appl. Phys.,* 58, 2308, 1985.

58. **Cohen, A. J. and Smith, H. L.,** *J. Phys. Chem. Solids,* 7, 301, 1958.

59. **Purcell, T. and Weeks, R. A.,** *Phys. Chem. Glasses,* 10, 198, 1969.

60. **Takano, Y., Tandoh, Y., Ozaki, H., and Mori, N.,** *Phys. Stat. Solid,* 130, 431, 1985.

61. **Kawazoe, H., Yamane, M., and Watanabe, Y.,** *J. Phys., Coll. C8,* 46, 651, 1985; **Kashiwazaki, A., Muta, K., Kohketsu, M., and Kawazoe, H.,** *Mat. Res. Soc. Proc.,* 88, 217, 1987.

62. **Feigl, F. J., Fowler, W. B., and Yip, K. L.,** *Sol. State Comm.,* 14, 225, 1974; **Yip, K. L. and Fowler, W. B.,** *Phys. Rev. B,* 11, 2327, 1975; **Griscom, D. L., Friebele, E. J., and Sigel, G. H.,** *Solid State Comm.,* 15, 479, 1974; **Griscom, D. L.,** *Phys. Rev. B,* 20, 1823, 1979; **Tasi, T. B., Griscom, D. L., Friebele, E. J., and Fleming, J. W.,** *J. Appl. Phys.,* 62, 2264, 1987.

63. **Stapelbroek, M., Griscom, D. L., Friebele, E. J., and Sigel, G. H.,** *J. Non-Crystalline Solids,* 32, 313, 1979; **Griscom, D. L. and Friebele, E. J.,** *Phys. Rev. B,* 24, 4896, 1981; **Edwards, A. H. and Fowler, W. B.,** *Phys. Rev. B,* 26, 6649, 1982.

64. **Watanabe, Y., Kawazoe, H., Shibuya, K., and Muta, K.,** *Jpn. J. Appl. Phys.,* 25, 425, 1986.

65. **Aita, C. R., Marhic, M. E. and Sayers, C. N.,** Proc. of 2nd Int. Conf. on the Effects of Modes of Formation on the Structure of Glasses, Differential and Defect Data, Vols. 53-54, Nashville, TN, 1987, 55.

66. **Abuhadba, N. and Aita, C. R.,** Growth and near-ultraviolet optical absorption characteristics of sputter deposited nominal germania, *J. Noncrystal. Solids,* 122, 305, 1990.

67. **Phillips, J. C.,** *Solid State Physics,* Ehrenrich, H., Seitz, F., and Turnbull, D., Eds., 37, 93, 1982.

68. **Cohen, M. H., Fritzsche, H., and Ovshinsky, S. R.,** *Phys. Rev. Lett.,* 22, 1065, 1969.

69. **Barr, T. L., Greene, J. E., and Eltoukhy, A. H.,** *J. Vac. Sci. Tech.,* 16, 517, 1979; **Natarajan, B. R., Eltoukhy, A. H., Greene, J. E., and Barr, T. L.,** Mechanism of reactive sputtering of indium II. Growth of indium oxynitride in mixed N_2-O_2 discharges, *Thin Solid Films,* 69, 217, 1980.

70. **Barr, T. L.,** *J. Vac. Sci. Tech.,* 49, 1793, 1991.

71. **Barr, T. L. and Bagus, P. S.,** Bonding patterns in oxides: XPS valence band chemical shifts, to be published.

72. **Broughton, J. Q. and Bagus, P. S.,** SCF studies of core-level shifts in ionic crystals. II. MgO and BeO, *Phys. Rev. B.,* 36, 2813, 1987.

73. **Broughton, J. Q. and Bagus, P. S.,** A study of Madelung potential effects in the ESCA spectra of the metal oxides, *J. Elec. Spec. Rel. Phenomena,* 20, 261, 1980.

74. **Pauling, L.,** *Nature of the Chemical Bond,* 3rd ed., Cornell University Press, Ithaca, NY, 1960.

75. **Coulson, C. A.,***Valence,* 2nd ed., Oxford University Press, London, 1961.

76. **Levine, B. F.,** Bond susceptibilities and ionicities in complex crystal structures, *J. Chem. Phys.,* 59, 1453, 1973.

77. **Barr, T. L., Brundle, C. R., Klumb, A., Liu, Y. L., Chen, L. M., and Yin, M. B.,** Novel bonding concepts for superconductive oxides: an XPS study, in *High Tc Superconducting Thin Films, Devices and Applications*, Margaritando, G., Joynt, R., and Onellion, M., AVS Series 6, AIP Conf. Proc. No. 182, New York, 1989, 216.

78. **Barr, T. L. and Brundle, C. R.,** On the bonding and electronic structure in high Tc superconducting oxides, *Phys. Rev. B.,* 46, 9199, 1992.

79. **Barr, T. L.,** Defects and suboxides in group A oxides: an XPS valence band study, to be published.

80. **Mohsenian, M.,** M.S. thesis, University of Wisconsin, Milwaukee, 1988; **Mohsenian, M. and Barr, T. L.,** to be published.

81. **Lau, C. L. and Wertheim, G. K.,** Oxidation of tin: an ESCA study, *J. Vac. Sci. Technol.,* 15, 622, 1976.

82. **Barr, T. L., Chen, L. M., Mohsenian, M., and Lishka, M. A.,** XPS valence band studies of zeolites and related systems. I. General chemistry and structure, *J. Am. Chem. Soc.,* 110, 7962, 1988.

83. **Sauer, J. and Zahradnik, R.,** *Int. J. Quant. Chem.,* 25, 793, 1984; **Derouane, E. G. and Fripiat, J. G.,** *J. Phys. Chem.,* 91, 145, 1987; **Sauer, J. and Engelhardt, G.,** *Z. Naturforsch.,* 37a, 277, 1982; **Sauer, J., Habza, P., and Zahradnik, R.,** Quantum chemical investigation of interaction sites in zeolites and silica, *J. Phys. Chem.,* 84, 3318, 1980.

84. **Barr, T. L.,** The nature of the relative bonding chemistry in zeolites: an XPS study, *Zeolites,* 10, 760, 1990.

Chapter 8

Areas of ESCA Application

I. INTRODUCTION

In this chapter we describe a representative array of the uses of ESCA to investigate problem areas in science and technology. Despite the obvious need for such considerations, the selection of the topics covered and the length of each section need to be clarified. It is here, perhaps moreso than anywhere else in this book that the criteria stated in the preface are employed. Thus, five basic arguments were used in arriving at the topics to be described: (1) We have taken the word "representative" to mean not only "popular", but also "diverse". Thus, the uses described cover a wide spectrum of interests and technologies. (2) At the same time, we have retained the original prejudices and "tilted the playing field" away from its more pristine surface physics side toward its more applied, surface chemistry end. This is, in part, a personal choice of the author, and although it, therefore, largely ignores an area of substantial contribution and growth; *it does actually concentrate on the area that has experienced the majority of the use of ESCA* (e.g., the IBM Corporation alone has more ESCAs than there are in all of the universities of some countries with major surface physics programs), and in addition it concentrates on the area that has been, in the past, the most neglected. This is, of course, not a total shift, nor is it the intention of the author to suggest to his readers that they should ignore the important contributions to ESCA science made in such areas as single crystal and chemisorption studies. The author accepts the contention that investigations in the latter areas *are the foundations* on which all surface science is built. However, it is also true that there are already a number of excellent texts that describe the techniques and impact of this fundamental surface physics (see, for example, the lists presented elsewhere), and it is not our intention to provide what would be, at best, a modest regeneration of these other sources. Instead, we suggest that any serious investigator will want to add some of these to her/his study. (3) Inside the many use areas that might still have been covered in the present text there is still a further prejudicial selection. In this case the author has exercised his own preferences and frankly concentrates on those areas of his personal interest. (4) In addition, we have tried to avoid certain repetitions within those interests. Thus, for example, there are discussions of the applications of ESCA in catalysis and corrosion science, but these presentations purposefully avoid most of the work described in the recent (1990) text *Practical Surface Analysis* edited by Briggs and Seah,[1a] i.e., in the chapters by Barr[1b] on Catalysis and McIntyre and Chan[1c] on Corrosion. We have also largely ignored the substantial area of application of ESCA to polymer science since this is also well treated in the aforementioned text by Briggs himself.[1d] The author does recommend that the serious reader examine Briggs and Seah, and the other useful sources listed in the appendix to complete his/her education in those areas. (5) Although we have presented sections on several fairly novel areas of ESCA application, a number of new areas have been omitted due to their extreme novelty and our lack of familiarity with their content. Two areas, in particular, need to be mentioned as they are growing rapidly and soon should assume quite prominent roles. They are ESCA studies of magnetic materials as exemplified by the studies of various IBM groups[2] and fullerenes as exemplified by the investigations of Ohno et al.[3]

It should also be mentioned that in many of the areas considered the author presents fairly detailed considerations of singular problems, rather than the more common, brief discussions of multiple problems. This is done to try to emphasize "how" rather than "what".

II. ESCA STUDIES OF ELEMENTAL OXIDATION

A. THE NATURAL PASSIVATION OF ELEMENTAL METALS: ORIGIN AND CONSEQUENCES

More than a decade ago the author introduced the concept of natural passivation (NP)[4-6] to describe the oxidative process by which (during exposure to air at STP) most elemental metals form relatively regular, thin films of oxides of progressive oxidation state, terminated by near monolayer films of hydroxides (and/or other related groups). The present arguments were generated through detailed applications of X-ray photoelectron spectroscopy (XPS or ESCA).[7] Certain aspects of these processes were, of course, borrowed from existing informative (non-ESCA) studies,[8,9] particularly the realization that most of these metals were being oxidized through some form of the classic Cabrera-Mott low temperature oxidation (LTO) mechanism.[10,11] Some of the generalities of our arguments are presented in Table 8-1. In particular, one should note the layering structure of the oxidation states formed by one class of oxides (A), whereas this feature does not occur for the other, (B). In addition, there appears to be a modest group of metals that do not readily form oxides under these conditions, whereas another group (or groups) seems to react with air so vociferously as to ignore LTO under these conditions, apparently suffering in air (at room temperature) diffusion-controlled (Wagner-type)[12] high-temperature oxidation (HTO).[8-11]

Several features that appear to be of significance were not described in these initial studies. These are presented in Table 8-2. One should note, in particular, that the

Table 8-1 **Elemental metals and their passivation oxides and hydroxides detected in this study**

Group	Metal	Bulk oxide (hydroxide)	Saturation (skin) oxide (hydroxide)	Zachariasen and Mott classification
A	Ce	Ce_2O_3, $Ce(OH)_3$	CeO_2, $CeO(OH)_2[Ce(OH)_4]$	
	Pd	PdO	(PdO_2) $Pd(OH)_4$	↑
	Fe	FeO	Fe_3O_4,Fe_2O_3[a] $[Fe(OH)_3$ or FeOOH]	Network
	Co	$CoO[Co(OH)_2]$	Co_3O_4, Co_2O_3,[a] CoOOH	modifiers
	Ni	NiO	Ni_2O_3, [NiOOH or $Ni(OH)_2$]	↓
	Cu	Cu_2O	$Cu(OH)_2$, (CuO)	
B	Si	SiO_2	SiO_2, $Si(OH)_4$	
	Al	Al_2O_3, AlOOH	Al_2O_3, $Al(OH)_3$	↑
	Zr	ZrO_2	ZrO_2, $Zr(OH)_4$	Network
	Sn	SnO_2	SnO_2, $Sn(OH)_4$	(glass)
	Mo	MoO_3 Mo_2O_5	MoO_3, mixed oxide-hydroxide	formers
	Y	Y_2O_3	Y_2O_3,[b] YOOH	↓
	La	La_2O_3 (LaO)	La_2O_3, LaOOH	
	Rh	Rh_2O_3	Rh_2O_3, RhOOH	

Note: Many of the hydroxides listed may in fact be hydrated oxides. Note the groupings based upon these results. Elements in parentheses are suspected constituents whose presence is largely based upon conjecture.

[a] The existence of the sesquioxides of Co, and perhaps Fe, are in doubt. However, the spectra resulting from the surface species believed to be Co_3O_4 may in fact be that due to islands of CoO and Co_2O_3.

[b] The Y_2O_3 detected in the saturation layer appears to be structurally different from that detected in the bulk of the passivation film.

Table 8-2 **Additional features of natural passivation**

1. In pure O_2 the same reaction sequences that take place in air occur at a somewhat different rate and without hydroxide termination.
2. During formation in air, adventitious carbon is also deposited.
3. The latter forms primarily a hydrocarbon layer (not a graphitic carbon layer as predicted by some researchers)!
4. The various carbon-containing species and the hydroxides (formed by deposited H_2O) form continuously during surface oxidation in air; generally, however, the deposited polymeric hydrocarbon, carbonate, and hydroxide species are constantly being forced up toward the evolving outer surface by the bulk oriented, 3d, lattice-forming oxides, this "push-up" process continues until termination of the oxide development, at which point the final, "stable" hydroxides and carbonates form the capping units, all of these species are further covered by a thin layer of adventitious hydrocarbon polymer.

formation of a hydroxide (and/or carbonate) layer is a continuous process, forming very thin films that are constantly being forced out to the outer surface by any remaining ability of oxygen to produce the generally more stable, oxide materials. The greater presence of oxides (particularly compared to hydroxides) on most typical air-exposed elemental metal systems is apparently as much a result of the enhanced lattice forming stability of the oxide, as it is the greater reactivity of O_2, compared to H_2O, for these M°.[13]

There are no absolute generalities for this process; however, if one allows for several notable exceptions, one may still provide a rather consistent picture in which we note that NP is a process by which many metals naturally form a series of oxides (and other oxidized products) that tend to protect these metals from further, generally deleterious (corrosive?) attacks.[4-6] Stated in terms of a specific example: if the surface of elemental copper were left totally exposed to the atmosphere, even one that is relatively free of pollution and contains only moderate humidity, e.g., ~30%, this metal would begin to corrode in a matter of minutes. It is the presence of the previously described NP layer that keeps metals, such as copper, in a state that for a substantial period of time suggests (to the often easily deceived eye) that little or nothing has happened.

It should be noted that the process that apparently produces LTO is, according to Cabrera and Mott (CM)[10,11] one that requires the introduction of a substantial electro-chemical field, F, provided by the extraction of electrons from M° by the (initially) chemisorbed O_2. This field often seems to produce a sufficient reduction in the normally insurmountable energetic barrier to thermal diffusion (V) that exists at this (relatively low) temperature, i.e.,

$$V - q\, a\, F \tag{8-1}$$

where, q is a standard charge, (a) is a unit dimension in the oxide lattice, and

$$F = \frac{u}{x} \tag{8-2}$$

In this case u is the potential difference established to lower the barrier to diffusion. One can see, however, that a field effect so introduced becomes smaller and smaller as the oxide film thickness, x, grows. Thus, as the process continues, the film thickness grows and the effective field reduces until a point is reached where the potential barrier to diffusion once again takes over, and the film growth ceases. At low temperature, therefore,

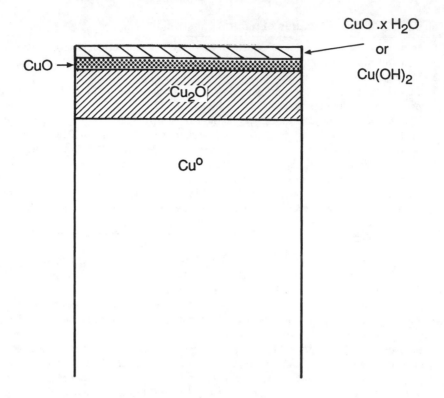

CuO .x H$_2$O

or

Cu(OH)$_2$

CuO →

Cu$_2$O

Cu°

Figure 8-1 Designation of the chemical species formed and approximate (~20 Å) layer structure following natural passivation of Cu° in STP air of moderate humidity.

this often results in only a very thin film. In the case of Cu°, we[4,14,15] (and Fehlner[11]) have found a room temperature oxidation film of ~20 Å thickness, Figure 8-1. Thus, as mentioned above, it is essentially undetected by the human eye.

It is interesting to construct a somewhat hypothetical rendition of these features, as in Figure 8-2. It is important to note that we have in this short description ignored the mechanism(s) for the establishment of these various types of fields, and/or oxide growths. Some consideration of these processes was provided by Cabrera and Mott[10] and amplified by Fehlner.[11] We have also speculated upon these mechanisms,[5,16] particularly as they relate to the induction of changes in oxidation state with oxide film growth. Several other features in Figure 8-2 need to be considered particularly the impact of places where the LTO curve exceeds room temperature.

B. VALENCE BAND STUDIES OF THE OXIDATION OF SELECT METALS
1. The Zn-O System
a. The Basic Problem

One of the easiest valence band systems to discern during studies of the process of the release (uptake) of oxygen is the zinc-oxygen system.

As indicated in Table 8-3, we have been able to produce (and reproduce) a series of core level spectra following ion etching of zinc systems that suggest the exposure of reasonably clean Zn°. The Zn°($2p_{3/2}$) binding energy of 1021.7 eV compares quite favorably with that of Lebugle et al.[18] (1021.82 eV). In our case sputter etching has successfully reduced the carbon contribution to the background and the oxide part of zinc to less than 1% (see above). Although the realization of this Zn(2p) binding energy may

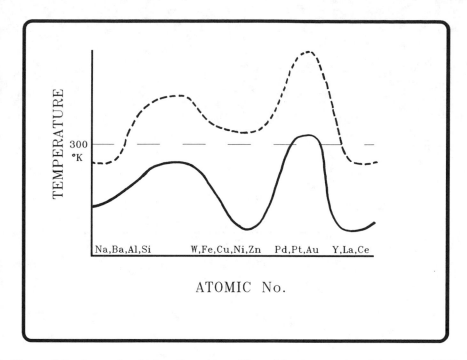

Figure 8-2 Approximate (nonlinear) rendition of the temperature barriers for HTO (– – – –) and LTO (——) in air.

Table 8-3 **Key core level binding energies in electronvolts (± 0.2) for this study**

	Zn	**ZnO**	**Cd**	**CdO**
Zn(2p$_{3/2}$)	1021.7[PS]	1022.0[PS]		
	1021.8[BS]	1022.1[BS]		
O(1s)	—	530.8[PS]	—	528.9[PS]
	—	N[BS]	—	528.6[GW]
Cd(3d$_{5/2}$)			404.5[PS]	404.2[PS]
			404.8[BS]	404.0[BS]

Note: PS = Present study; BS = Briggs, D. and Seah, M. P., Eds., *Practical Surface Analysis,* 2nd ed., John Wiley & Sons, Chichester, England 1990; GW = Gaarenstroom and Winograd;[27] N = not measured.

be viewed, in part, as indicative of the suppression of essentially all zinc oxide in favor of the Zn° metal, the best indication of this fact is carried in the narrow symmetric character of the 2p$_{3/2}$ line, and the well-formed structure of the Zn° KLL Auger lines.[15]

The general valence band produced by Zn° (Figure 3-7a) also makes a statement about the metallic (only) character of the material by exhibiting a relatively narrow 3d line pulled back from the Fermi edge to ~10.1 eV. (Note the results of the previous studies of Zn(3d) by Briggs[19a] and Brundle and Carley,[19b] and in our laboratory.[15]) In addition, the characteristic broad (free electron-like) 4s band may be readily discerned stretching from the apparent Fermi edge, E_F = 0.0 eV, to some point above 5 eV. Computer enhancement of the 4s part of this valence region indicates (Figure 3-7b) that the band stretches from a maximum point at ~1.8 eV to a point ~7 eV above the Fermi edge. As

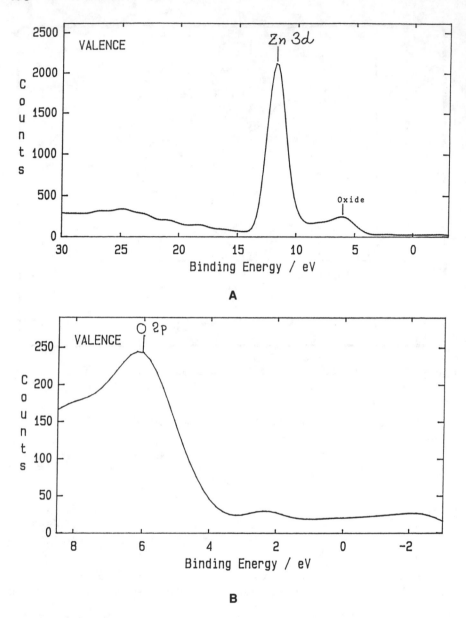

Figure 8-3 Valence band spectrum for representative ZnO system. (A) "General" spectrum, note O(2p) and shift of Zn(3d). (B) Computer-enhanced region showing split character of O(2p) and extensive removal of Zn(4s).

this spectrum is revealed, it indicates a gradual, but relatively consistent reduction in valence electron density with increasing binding energy. The exact width of the spherical, delocalized (metallic) free electron-type band cannot be determined from the XPS spectrum because the latter is subject to cross-sectional and other spectral features, and also its density at ~7 eV is crossed by the leading edge of the relatively localized (XPS revealed) 3d band.

The valence region of the zinc metal system is generally reputed to be $d^{10}s^2$. This is in contrast to the Ag° system that is supposed to be primarily $d^{10}s^1$.[20] It is interesting that the s band for silver metal also seems to dissipate in XPS, suggesting zero density at a point in front of its d band, i.e., at ~3.5 eV below its Fermi edge. This suggests that the simple free electron arguments are basically correct for these systems, i.e., the delocalized, s density region is "saturated" after two electrons are added. Thus, the addition of the second electron to a $d^{10}s^1$ system increases the occupied "depth" of the "well" in energy space, rather than doubling the density in the same energy region. Thus, Figure 3-7b applies to these results.

On the other extreme, ZnO produces a small (but apparently real) binding energy shift for the $Zn(2p_{3/2})$ to 1022.0 eV. All of this suggests the usual problems with charging and fixing binding energies.[21] We do suggest, however, that zinc probably does exhibit a small, but real, (positive) chemical shift from the metal to the oxide. It is perhaps much more revealing of oxide formation that the resulting Zn 2p lines broaden and the position and character of the Zn Auger change dramatically. These features have all been extensively exploited in a previous analysis to monitor the relative formation of various oxides during the oxidation of α-brass.[14,15]

Due to the effects of charging, the exact position of the O(1s) for ZnO is somewhat difficult to establish, but based upon our results it seems to be best positioned around 530.8 eV.[15] This suggests that (based upon our previous considerations[22]) ZnO is indeed a (modest) semi covalent oxide (e.g., O(1s) > 530.5 eV[22,23]). In fact, based on this result, ZnO should be roughly as ionic as MgO. In view of the filled shells and lack of overlap of Zn (4s) with Zn (4p) one might expect ZnO to be even more covalent than MgO (i.e., with its O(2p) pushed to relatively high binding energy[22–24]), but it is our suggestion that the presence of the near valent Zn(3d) actually acts as a "push state" moving the O(2p) to lower binding energy in ZnO, and keeping the O(2p) relatively close to the Fermi edge.[22–24]

The valence band region of the ZnO system exhibits the expected peak structure due to the interjection of O(2p) density, and also the expected removal of the aforementioned Zn(4s), Figure 8-3. The Zn(3d) for ZnO is somewhat broader than that for Zn° and slightly shifted upfield. The (XPS detected) O(2p) density (Figure 8-3b) seems to be divisible into two primary regions: one that we have labeled as B peaked at 7.5 to 8.0 eV and the other, labeled as N, peaked at 5.75–6.0 eV. The total band stretches up and down field from these peaks and is (relatively) much more intense than the Zn(4s) and, thus, even when

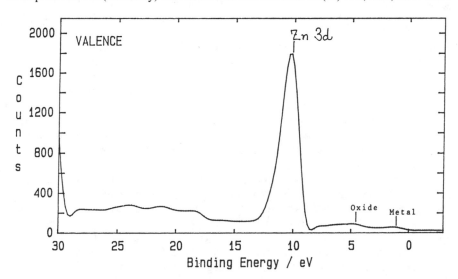

Figure 8-4   Zn-O valence band for partially oxidized case. Broad scan.

partially formed, O(2p) will easily cover the leading (higher binding) energy regions of the Zn(4s). In fact, there is ample evidence that the high binding energy side of the B part of the O(2p) is partially overlapped by the Zn(3d). On the low binding energy side the XPS-detected density of states for the O(2p)-dominated band for ZnO falls essentially to zero at ~3.25 eV, and remains flat through the (pseudo) Fermi edge.

b. Band Shifts During Oxide ↔ Metal Transformations

During the formation of an oxide, electrons are transferred almost entirely from select bands of the metals involved. Conversely, as the resulting oxide is transformed back to the zero valent metal, electrons must flow from the O(2p) band back into the appropriate metal valence bands. For the metals being considered herein the principal valence band involved is formed from an s-type band where density is adjacent to the metals' Fermi edge.

If we were concerned with the bulk of an elemental metals involved, then the band description would be relatively straightforward. Thus, for example, the 4s band of Zn is filled with two electrons and the band should be reasonably well depicted by the simple free electron model. This predicts a band that is spherical in k space and produces a density that rises from zero (at its point of highest binding energy) to its largest value adjacent to the Fermi edge. With some alterations due to the presence of certain modest second order effects and the influence of the appropriate photoelectron cross sections and some final state effects, Figure 3-7 resembles this prediction.

On the other extreme is the O(2p) band for ZnO. Once again, if one takes into consideration the qualifiers mentioned above, the band density described above should be a reasonable depiction of the band structure for bulk ZnO.

A major problem of interest occurs for the mixed case, that is, when the XPS system is able to detect simultaneously some ZnO and the near-surface Zn° being realized during a surface-induced reduction (or vice versa, see Figure 8-4). In these cases, the XPS system is detecting rather local (ZnO · Zn°) interfacial situations that are releasing or receiving electrons, and these localized environments may reflect the select regions of the s band that are receiving electron(s) (during reduction) and the select regions of the O(2p) band that are surrendering these electrons. Put in another perspective, one may ask "During reduction from the oxide to the metal, do the electrons first flow into the density region adjacent to the Fermi edge and then the s band builds with deeper and deeper density or does the s band evolve uniformly over its entire energy space, apparently stretching (for Zn°) from the Fermi edge to ~8 eV?"

In the case of the Zn-O system, the XPS determination of the region of origin for electron density transfer during oxidation/reduction is made difficult by the natural overlap of the key band regions, i.e., the Zn(3d) covers the trailing edge of the O(2p) and the latter covers the lower half of any Zn°(4s) between ~8 and 4 eV. This suggests some d band involvement in this process and indeed this is predicted. Most of the results, however, suggest that this is not a major factor. Two features, however, do appear to be occurring during the reduction cycle, see Figure 8-5. Thus, electrons are entering the Zn(4s) band; but initially do not seem to spread over the whole band, rather, the electrons appear to first fill states away from the Fermi edge, see Figure 8-5ii. The breadth of this density then grows as the system further evolves toward total Zn°. The electrons moving into the Zn(4s) (during reduction) all come from the O(2p), but qualitative shifts suggest that the density in the middle and front of the latter band are contributed first, followed by density from the trailing (high binding energy) side of O(2p). The reverse of this appears true during oxidation where the first electrons extracted from the Zn(4s) appear to come from density regions adjacent to the Fermi edge. In view of the relative ease with which electrons are removed from the Fermi edge of a metal, these results should not be entirely surprising.

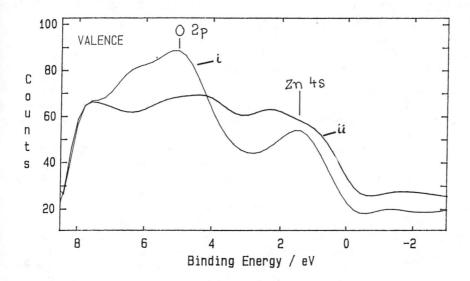

Figure 8-5 (i) Same case as presented in Figure 8-4 with computer-enhanced view of O(2p) and Zn(4s), (ii) typical result after further (but still incomplete) removal of oxygen.

One must keep in mind that we are herein referring to situations that carry a modest, but possibly significant, resemblance to the alloy cases described by Fuggle et al.[25] in which they demonstrated what they believe to be localized Martensson-type, interfacial, final state, relaxation shifts[26] in evolving d band densities as the relative concentration of these select binary alloys are changed. Our situation of a mixed, s band metal and oxide is hardly the same, but when the key features of the two species are created by relatively localized adjacencies the free electron character of the metal s band may be mitigated and localized relaxation induced shifts may become significant. We have not attempted to include any of the latter in the present analysis, but its possible contribution should not be arbitrarily excluded.

2. The Cd-O System
a. *General Results: Past and Present*
The results for cadmium are, on the one hand, merely a continuation of those established for zinc, while on the other hand the sparse nature of the previous XPS results reported for cadmium and its oxide[1a] are reason for a detailed investigation. The limited previous studies have suggested two anomalies. First, cadmium appears to be one of the metals that exhibits a negative (*measured*) chemical shift of its principal core lines in going from the metal (Cd°) to the oxide (CdO).[27] In addition, relatively early XPS studies also suggested an unusually low binding energy for the O(1s) of CdO.[27] The latter result has been disputed,[28] but our recent measurements of the oxide and metal support both of these contentions, see Table 8-3. A variety of explanations have been proposed for these atypical features, with most centering on the measurement induced relaxations experienced by all photoelectron peaks, particularly the core states. All systems have been shown to experience certain degrees of selective, extra-atomic relaxation shifts,[29] and Gaarenstroom and Winograd[27] have argued that the unusual binding energies of CdO results from its abnormally large extra-atomic relaxation. We have recently suggested that, whereas the latter is true, a significant part of the unusual binding energy character of CdO also results from the relatively large shifts induced in the O(2p) state by the

near-valent (~10 eV) 4d core state.[23] This perturbative "mixing", we suggest, "pushes" the O(2p) to lower binding energy than that commonly experienced by most metallic oxides (e.g., CaO and NiO). One of the goals of the present research is therefore to try to fix the binding energy range of the O(2p) band for CdO.

b. The Cadmium Oxide System

The core level binding energies produced during our study of CdO are displayed in Table 8-3, along with those of several other investigations. As indicated above, the relatively low O(1s) binding energy of 528.9 eV is of particular interest.

The valence band study of the oxidized system is intriguing, particularly because, if exposed to air instead of pure O_2, a shift in O(2p) is observed to accompany the corresponding shift in O(1s), but, whereas O(1s) increases to ~531.2 eV, suggesting formation of the hydroxide, the O(2p)-dominated valence band actually contracts to a relatively tight doublet with peaks stretching from approximately 5.7 to 4.7 eV. The oxide, on the other hand, produces a split peak valence band spectrum that suggests a peculiar behavior. In this case a somewhat singular peak at about 5.7 eV is joined by a broad (band) set with maximum plateau ranging from ~4.0 to 3.0 eV. This result is identified as Figure 8-6. It should be noted that, whereas the peak at 5.7 eV is near the O(2p) position exhibited by ZnO, the peaks from 4 to 3 eV are approximately 2 eV lower in binding energy than any O(2p) for ZnO. This shift is almost identical to that for O(1s) for CdO compared to ZnO.

c. The Cd° (Metal) System

Ion etching of a sheet of air-exposed cadmium metal eventually reveals a series of results that suggest pure Cd°. The latter conclusion is based upon the nearly complete disappearance of all oxygen (along with all other elements) and the generation of a singular $Cd(3d_{5/2})$ at 404.5 eV with line width of 0.9 eV. Further, and perhaps more expressive of the metal-only state, we find a sharp Cd°-only character for the MNN Auger lines[1,21b] and the various loss transitions.[21] Both of the latter features have distinct zero

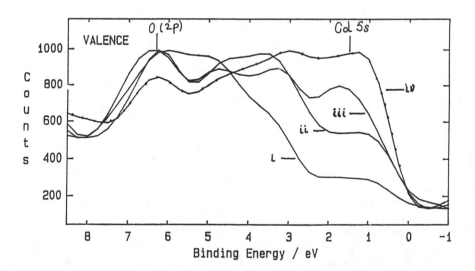

Figure 8-6 XPS valence band study of cadmium-oxygen system showing progressive involvement (or removal) of oxygen. (i) Well-formed oxide with slightly detected underlayer of Cd° metal, (ii) ↔ (iv) progressive removal of O(2p). Note rolling nature of growth of Cd°(5s). Note (by comparing with Figure 8-7b) that there is still some O(2p) retained in iv).

valent signatures. It is also true that the near valent Cd(4d) has a distinct Cd° signature that demonstrates a near perfect double not detected at this level of resolution for the oxide.

As expected, the 5s band for Cd° is almost identical in appearance to the 4s band for Zn°. The breadth of the Cd° band seems to approach 8 eV with the uncertainty caused in part by the overlap of the relatively weak higher binding energy part of the s band by the leading edge of the Cd(4d). As suggested in Figure 8-7, there may be some structure in

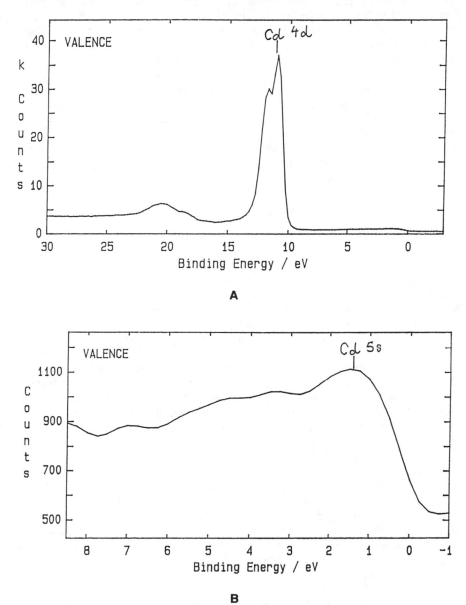

Figure 8-7 XPS valence band spectrum for Cd°. Note similarity to Figure 3-7 for Zn°. (A) Broad general scan. Note character of Cd°(4d). (B) Computer-enhanced version of Cd°(5s).

the Cd°(5s) band as it increases in size from ~8 to ~1.3 eV. The maximum at this point, however, is right where it is expected,[30] and suggests the validity of the free electron model. Basically, the results are similar to those of Pollack et al.[31] except the latter group found no internal band structure and a total width of ~7 eV.

d. The Mixed CdO-Cd° Systems

The reduction of oxide to cadmium metal or the converse oxidation produces the anticipated shift in density from the O(2p) to Cd(5s) band (and vice versa), see Figure 8-6. The general features of this process are similar to those detected during the related ZnO ↔ Zn° process. In the cadmium case, however, the major part of the O(2p) band is shifted down field (toward the Fermi edge) and therefore overlaps more with the metal s band, while the O(2p) band for ZnO overlapped more with the near valent d state. Once again in the case of the CdO ↔ Cd° it appears that the individual bands *do not* fill uniformly across the bulk-like band, but rather seem to build the band in a rolling fashion. Thus, the density peak near the Fermi edge for the metal is last to fill during the reduction process, and the first to deplete when the metal is oxidized. In view of the often-employed analogy of free electron bands and "bath tubs" (filling from the bottom and overflowing from the top) this should not be surprising. There also seems to be a rolling nature to the filling (removal) of electrons in the O(2p), but this is neither confirmed nor understood.

III. ESCA STUDIES OF INORGANIC SYSTEMS

The use of ESCA as a tool to try to describe the surface chemistry of inorganic compounds is another practice that dates from the origins of the technique. In addition to the many observations of the scientists in Uppsala,[7] one should consider a number of pioneering investigations particularly the voluminous tables of Jorgensson.[32] Such studies as these have provided many of the underpinnings of the science on which we build to this day. It is true, however, that many of these investigations were realized without full awareness of the difficulties resulting from the problems mentioned in our second chapter. Thus, the fact that most inorganic salts are insulators and their photoelectron core spectra are also endowed with complex substantial relaxation phenomena seems to have been minimized. However, rather than dwell on the problems, the reader should note how expressive these tabulations are and, in particular, how well their *comparative* numbers correlate with the better values of today. Once again, this is a testimonial for the simple utility of ESCA, and the apparent blind luck that surrounds the use of the argument presented in Chapter 2, Section IV. As a result of this success, the numerous "modern" inorganic results tabulated in Briggs and Seah[1a] may often be used with reasonable certainty of precision.

In this section, we will follow the practice of today and restrict our scope and describe the use of ESCA to explain just one type of material in inorganic identification. Following the practice often observed throughout the book we will restrict ourselves to inorganic oxides and, in particular, primarily those of the group A elements. In so doing, we can borrow heavily from results listed elsewhere in the book. As with other sections of this text, one should still exercise caution in that numerous problems and disputes still abound in this area, but the author feels that most of the numbers are reasonably precise and the all-important trends are correct.

A. GENERAL BONDING CONCEPTS OF SIMPLE OXIDE SYSTEMS

Recently our research group has conducted a variety of detailed studies of the surface chemistry of oxides. In general, these investigations have centered on the use of X-ray photoelectron spectroscopy (XPS or ESCA) as a tool to study primarily group A oxides, in both their simple (M_yO_z) and mixed ($R_wM_yO_z \cdot M'_y,O_z$) forms, where M_yO_z is typified by In_2O_3[1] and $R_wM'_yO_z \cdot M'_y,O_z$, by the synthetic zeolites, e.g., NaY = $(SiO_2)_{136} \cdot$ $(NaAlO_2)_{56} \cdot xH_2O$.[2] In the course of these investigations a number of patterns have become apparent in the revealed (surface) bonding chemistry that seem to have a collective generality. Historically, some aspects of these patterns have been suggested before[33] but most of the present observations and the total collective nature of our bonding arguments appear to be novel.

Most group A elements readily produce oxides with maximum, common valency [as examples consider: MgO for M(II)O and In_2O_3 for $M(III)_2O_3$]. It is often relatively easy to produce most of these oxides with near optimal stoichiometry, i.e., without significant (>0.1%) defects and suboxides, in their surface as well as in their bulk regions. If carefully prepared, some of these oxides also may be kept relatively (surface) clean, even if air exposed, e.g., SiO_2.[34] Other oxides, on the other hand, are subject to rapid and total outer surface conversion to carbonates and hydroxides, if not maintained under UHV conditions following production, e.g., BaO.[16,35] Unfortunately, sputter cleaning of the latter tends to produce a significant surface presence of the aforementioned defects and suboxides.[24,36,37] This difficulty has provided a substantial hurdle in attempts to study certain group IA and IIA oxides that is just now being scaled through the use of such techniques as ion beam deposition and UHV suitcase transport.[38]

In Table 8-4, therefore, we present a collection of some of the most pertinent (average) experimentally determined core level binding energies, $E^m(i)$, for many key group A oxides with maximum common valency. In addition, for comparative purposes we have included in Table 8-5 a similar compilation of the same type of measured binding energies for a number of maximum valent, group B (transition metal) oxides.

In all of these studies we have tried to be cognizant of the final state (relaxation) effects that also influence the results. In fact, we have utilized the arguments of Wagner et al.[39] and tried to compute the Auger parameters, α_i, for these materials, in order to employ the approximate relation:

$$\alpha_i \approx 2E_i^R \qquad (8\text{-}3)$$

however, this study has, for a variety of reasons, only been partially successful. It has, at least, confirmed our assumption that *the trends that arise from the measured numbers in Tables 8-4 and 8-5 are generally valid indicators of the initial state (chemical) differences for these oxides.*

In the present study, one must also be cautious not to summarily dismiss the effects of structural changes on these results. It should be noted, for example, that a progressive shift from tetrahedral to rutile to rock salt also follows the oxides in going from right to left (and down) in the periodic table. These structural changes also influence the experimental results and are, no doubt, intimately intertwined with any chemically oriented changes. In the following we will specify only the resulting chemical effects, which if totally argued would be amended to include the appropriate structural changes.

Before discussing our bonding arguments one should consider in Table 7-1 in Chapter 7 an additional, related set of data extracted from studies of the XPS-generated valence bands of selected members of the group A oxides. The detailed rational for the detection and use of these (relatively) difficult to obtain features (particular with XPS) are presented in Chapter 7.[23,24,36,37]

It is our supposition, therefore, that *a shift of the leading edge of these [O(2p) dominated] valence bands (N) toward the pseudo-Fermi edge, (reduction in band gap) is a direct measurement of increased ionicity. On the other hand, any growth in the total band width is primarily due to an increase in covalency of the oxide.* Since these two

Table 8-4 **"Key" XPS binding energies in eV for group A oxides**
1. $E(o_{1s}^=)$ 2. $E[M(oxide)]-E(M^°)$ 3. $E^I (M_xO_y)-E^I(M^°)$ 4. Near-valent "push" state[a]

	IA	IIA	IIIA	IVA	VA	VIA
	Na	Mg	Al	Si	P	S
1.	~529.7	~530.9	(531.2)	532.8	—	—
2.	0.8	0.9	(1.3–2.7)	3.9	5.4	6
		[NIO]				
3.	2.6	3.1	3.5	6.1	7.3	—
4.	—	—	—	—	—	—
					[SCO]	
	K	Ca	Ga	Ge	As	Se
1.	—	~529.8	530.6	531.3	531.4	—
2.	—	0.6	1.9	2.8	4.4	5
3.	—	2.75	4.05	4.8	6.0	6.7
4.	Δ	—	Δ(d)	—	—	—
	Rb	Sr	In	Sn	Sb	Te
1.	~529.0	~529.0	530.1	530.1	529.8	—
	[VIO]					
2.	—	—	1.1	1.4	2.4	4.0
3.	—	—	2.75	3.1	3.65	4.75
4.	ΔΔ	Δ	Δ(d)	—	—	—
				[NIO]		
	Cs	Ba	Tl	Pb	Bi	Po
1.	~529.4	~528.2	—	(531.4)	—	—
2.	—	0.4	—	0.8	—	—
3.	—	1.85	—	2.4	—	—
4.	ΔΔΔ	ΔΔΔ	ΔΔΔ(d)	Δ(d)	Δ(d)	—

Note: Highest (common) valent oxide; $\delta E = \pm 0.2$ eV. Binding energies of insulators fixed by several methods, including $E(C_{1s}) = 284.6$ eV for adventitious carbon. Numbers in parentheses suggest values in question. (d) Designates d state; otherwise push state is p type. ~ Indicates results achieved from impure sample. $E(O_{1s}^°)$ chosen as 534.0 eV.

[a] Deltas (Δ) are approximate separations of designated peaks from centroid of O(2p)-dominated valence band: 11 to 13 eV Δ; 8 to 10 eV ΔΔ; 5 to 8 eV ΔΔΔ.

(simplified) versions of chemical bonding vary inversely with one another, it is not surprising to expect the valence bands of certain oxides, e.g., In_2O_3, simultaneously to shift to lower binding energy and contract compared to certain other oxides, e.g., Al_2O_3.[36] This seems to be exactly what is observed as one considers oxides to the left and down in the periodic table from SiO_2, *suggesting the registration of a near regular, progressive increase in ionicity (decreasing covalency)*. As pointed out in Table 7-1 in Chapter 7, there seems to be a direct correspondence between the magnitude of the realized shifts and contractions, and the factors, $f_c(f_i)$, purported to register the degree of covalency (ionicity).

Representative examples of the kinds of shifts and contractions being described herein are also suggested in Chapter 2. Numerous other examples are presented elsewhere in this book.[33,34,36,37,40] A method has also been developed in which the leading edges of the valence bands are artificially shifted until they align. The rationale behind this forced alignment, Figure 7-23b, in Chapter 7, is based upon the premise that any enhancement

Table 8-5 "Key" XPS binding energies in eV for group B oxides

1. $E(O^{=}_{ls})$ 2. $E[M(oxide)] - E(M°)$ 3. $E'(M_xO_y) - E'(M°)$ 4. Near-valent "push" state[a]

	Sc	Ti	V	Cr	Mn	Fe	Co	Ni	Cu	Zn
1.	—	529.7	530.0	529.8	529.8	530.0	529.9	530.0	530.3	530.3
2.	—	4.9	4.7	4.6	~4	4.1	2.7	1.9	1.2	0.35
3.	—	5.35	—	—	(3.9)	6.1	3.95	2.45	2.25	2.15
4.	—	—	—	—	—	—	—	—	—	ΔΔ(d)

	Y	Zr	Nb	Mo		Ru	Rh	Pd	Ag	Cd
1.	529.3	529.9	—	530.4	[NIO]	—	530.0	530.1	528.3	528.6
2.	?	2.6	3.9	4.35	—	3.1	1.1	1.3	~0.3	~0.8
3.	—	—	—	5.35	—	—	—	1.45	0.75	0.4
4.	[VIO]Δ	—	—	—	—	—	—	—	ΔΔΔ(d)	ΔΔΔ(d)

	La	Ce		W	Re	Os	Ir	Pt	Au	Hg
1.	528.8	529.1	—	530.2	—	—	—	530.0	—	—
2.	?	?	—	4.5	—	—	3.0–3.5	—	—	—
3.	—	—	—	—	—	—	—	—	—	—
4.	ΔΔ	Δ	—	—	—	—	—	—	ΔΔΔΔ(d)	ΔΔΔΔ(d)

Note: Highest (common) valent oxide; δE = ± 0.2 eV. Binding energies of insulators fixed by several methods, including $E(C_{ls}) = 284.6$ eV for adventitious carbon. Numbers in parentheses suggest values in questions. (d) Designates d state; otherwise push state is p type. ~ Indicates result achieved from impure sample. $E(O°_{ls})$ chosen as 534.0 eV.

[a] Deltas (Δ) are approximate separations of designated peaks from centroid of O(2p)-dominated valence band: 11 to 13 eV Δ; 8 to 11 eV ΔΔ; 5 to 8 eV ΔΔΔ; 3 to 4 eV ΔΔΔΔ.

in ionic field for an oxide shifts the entire band down field uniformly, whereas, as indicated, a change in the degree of covalency primarily influences the breadth of the B side of an oxide band. Thus, an alignment of the "true" leading edge for two or more common oxides removes their differences in ionicity, while exposing differences in width that should be directly proportional to the size of their different covalencies. The word "true" is emphasized to note that this proposed alignment may often be encumbered by shifts in the mobility (leading edge) of the oxide valence band due to any varying presence of different types of defects.[34]

Consideration of the previously mentioned core level data suggests a number of additional features. In this regard, it should be noted that, if the O(2p)-dominated valence bands shift in a particular fashion, Figure 7-23a in Chapter 7, we should expect the oxygen core level peaks, e.g., O(1s), also to shift in *this same direction*, Table 8-4.[1b] (Note that obvious differences in relaxation effects between core level and valence band electrons generally negate direct comparisons of the *sizes* of these shifts.) The same covalency/ionicity arguments may thus be applied to explain the *patterns* exhibited by either type of spectra in Tables 8-4 and 8-5 and the valence bands in Chapter 7. In these examinations we see the reason for the common statement that "all ionic (metallic) oxides exhibit the same O(1s) binding energy of ~530 eV." In fact we find that *most* transition metal oxides, Table 8-5, and many group A oxides, Table 8-4, do exhibit measured O(1s) binding energies of 530.0 ± 0.4 eV, and this is accomplished while the corresponding binding energies of the interacting metals M (in a typical M_xO_y system) vary substantially. It is our determination, however, that these variations in the chemical shifts of the metals are primarily due to the drive of the different metal cations to create totally compensating positive fields about an oxygen species of almost uniform ionicity. These effects were argued in Chapter 2 where it was noted that these results are entirely consistent with the now-classic charge potential model for chemical shifts.[1] As pointed out by Goodenough,[41] in the case of the transition metal oxides, the oxygen binding energy positions are further locked into a narrow range by the presence and position of large d bands. Many of these effects are collectively seen in both Tables 8-4 and 8-5. In view of their common nature, those oxides that form the (530.0 ± 0.4 eV) group typified by most of the transition metals and many group A elements are classified herein as normal ionic oxides (NIO). Based upon our correlation with Levine's arguments,[42] we anticipate that all these oxides have ionicities roughly between 76 and 89%.

Examination of the listed O(1s) values and the resulting cation shifts suggests that there are also, as expected, several oxides in Table 8-4 that exhibit relatively significant, but varying, covalencies. These systems are labeled herein as semicovalent oxides (SCO) (in order to relate them to those even more covalent, omitted oxides formed by the elements on the right side of the periodic table). These SCOs are characterized by O(1s) binding energies >530.5 eV and, as expected, they exhibit corresponding increases in the size of their relative cationic shifts *and increases in the widths of their valence bands*.

The last designated group, labeled herein as very ionic oxides (VIO), is one that has not been properly recognized. Based upon the regions of Tables 8-4 and 8-5 that are suggested for members of this group, we see that a key characteristic of this type of oxide is that their O(1s) binding energies range from 529.5 to ~528.0 eV. In addition, it is our projection that the valence bands of these oxides (e.g., BaO and SrO) will be contracted and pushed several electronvolts lower in binding energy than that exhibited by a typical NIO (e.g., In_2O_3). For example, we suggest a valence band leading edge for BaO that initiates near 1.5 eV. If we follow the arguments presented above, and the results listed in Table 8-4, it is not surprising that these VIO materials are extremely (almost totally) ionic. However, if we view enhanced ionicity simply as a result of a continuation of the previously described movement of, in this case, all bonding electrons from metal to oxide, then several problems obviously arise. First, an increase of ionicity from, say, 86 to 95%

Table 8-6 **General oxides and their ESCA characteristics**

Type of oxide	O(1s) range	Band width (1/2 maximum)	~Ionicity
Semicovalent (SCO)	530.5–533.0	10–7.5	50–75
Normal ionic (NIO)	530.0 ± 0.4	7.5–6.0	76–89
Very ionic (VIO)	529.5–~528.0	6.0–5.0	>90–98

(typical of the transformation from a NIO to a VIO) seems to hardly explain the large binding energy shifts of core and valence states that may exceed 1.5 eV. Second, this does not explain why certain oxides of the heavier d^{10} transition metals seem also to fall into the VIO class, see Table 8-5. The answer to this dilemma, we feel, falls outside of the simple ionic shift argument presented so far, and may be of substantial importance in such fields as high T_c oxide superconductivity, see Section VIII of this chapter.[23,37,43]

Coupling these results together we suggest the general categories in Table 8-6 for all the oxides under consideration.

If we examine the electronic shell structure for the oxides from the top three rows of the periodic table, we find a ready distinction (energetic separation) between valence electrons and core level electrons. Proceeding down the table on its left side, however, a somewhat peculiar effect occurs when we reach the fourth row. A near-valent, metal, p, core state emerges that is close enough to the valence band for the latter oxides (e.g., K_2O) to exert a slight molecular orbital-like (mo) interaction with the O(2p)-dominated valence shell. As we explore further down the table, we find similar near-core states that are positioned even closer to the valence band, see Tables 8-4 and 8-5, and thus their interactive contributions in the previously mentioned molecular orbitals must grow and grow, i.e., these p core states have relatively large densities and their symmetries are such, that as the ΔE between them and the [O(2p)-dominated] valence band shrinks toward 10 eV (and less), the effect of the mo-type interaction should become more and more significant. Further, one can show that, as with most common mixings of two quantum states, the effect of this interaction will be to increase the separation of these two states by roughly the square of their interaction divided by the separation, ΔE. This means that the valence band densities *in these oxides* should be pushed even closer to their pseudo-Fermi edge by the presence of any large density near-core states that are of correct symmetry and close enough to experience significant overlap. Metallic p states should be the most effective at this. Thus the valence band "push" exhibited by BaO [as a result of the Ba(5p)] is relatively large and that for even Ce and Y may be of some significance. Equivalently placed d states may not be as effective at this as p states, but in some cases the d states may be even closer to the valence band, thus increasing their relative effectiveness. Thus, in the case of those transition metals with filled d shells, [($d^{10}s^1$) and ($d^{10}s^2$)], the d bands for their oxides are "popped out" of the band gap between the O(2p) and the Fermi edge, and moved to near-valent core level positions, e.g., ZnO.[30,41] However, these new, near-valent, d, core states are not far below the O(2p) and, once again, they move even closer as one proceeds down the periodic table. This suggests that some of the most shifted (very ionic) oxide systems should be AgO, CdO, and HgO, etc. and the evidence (Table 8-5) strongly supports this contention.

Thus, as described above, these progressive shifts in the XPS spectra of oxides should be experienced by both their valence band and core level peaks in a manner that may correlate directly with progressive changes in the covalency/ionicity of the metal-oxygen bond. The interrelationships of these patterns may provide information about more complex, applied oxide systems. For example, we have used these arguments to help describe such important features as the relative acidity in zeolites[34,40] and a possible mechanism for the superconductivity in the new high T_c oxides.[23,37,43]

B. BONDING IN MIXED OR COMPLEX OXIDES
1. Introduction to the Problem of Complex Oxides

There is a practice often utilized in the chemistry and materials science literature to employ a certain shorthand nomenclature to try to represent the relationship between many complex or mixed compounds, such as complex ($A_zM_sO_t$) oxides (e.g., Mg_2SiO_4), and the simple compounds often considered to be their progenitors (e.g., MgO and SiO_2). This practice includes (1) the use of generalized names to designate the family of origin, e.g., silica (SiO_2) is often described as the simplest member of the group of compounds called silicates; (2) in addition, it is common to use theoretical and/or empirical collective bonding arguments to describe the bonds exhibited by these systems, e.g., the bonding in SiO_2 and Mg_2SiO_4 are assumed to exhibit the *same* SiO_4^{-4} (silicate) form; and (3) it is also suggested that complex oxides are nothing more than different ratios or combinations of simple oxides, e.g., $Mg_2SiO_4 \equiv 2 \, MgO \cdot SiO_2$.

These shorthand forms for nomenclature and property description are often expressive of certain relationships, and perhaps origins, and they also seem to assist in simplifying compositional or structural considerations[44] and/or features related to solid phase diagrams.[45] (Even the author, for example, has employed this practice as a method to indicate the principal components in select zeolites as the Si/Al ratio is varied.[34,40,46]) The problem is that *chemically* the three collective arguments are often misleading, and may in many cases be simply incorrect.

The rationale behind the latter statement is based around the following observations and arguments largely realized through the use of ESCA. If, in fact, a complex oxide $A_zM_sO_t$ can be properly and completely represented as $pA_mO_n \cdot qM_xO_y$, where $p \cdot m = z$ and $q \cdot x + s$ and $p \cdot n + q \cdot y = t$ (e.g., $Mg_2SiO_4 \equiv 2 \, MgO \cdot SiO_2$), then it must be true that in forming this complex oxide the chemical bonding factors (bond lengths, bond angles, relative ionicities, etc.) between A and O and between M and O are essentially unchanged from those that exist in the separate "simple" oxides.* If the latter feature is indeed true, then it suggests that formation of $A_zM_sO_t$ is more closely associated with the creation of a "solid *solution*" of restricted composition, than a specific "mixed or complex *compound*."[47] It also suggests that following preparation the $A_zM_sO_t$ system theoretically may be separated out into its composite units, eventually yielding unaltered A_mO_n and M_xO_y. Obviously, the status of this feature may be critical to such fields as catalysis[48] and nanoclusters.[49]

Theoretically, an interesting adjunct to this problem has been analyzed by several research groups, one of which is that of Onodera and Toyozawa.[50] Arguing primarily from the basis of optical spectroscopy, these authors have noted that mixtures of two or more unique chemical units to form a more complex third species may be described quantum mechanically in terms of various gradations of two extreme cases, either *persistent* or *amalgamated*. Thus, the mixed system is deemed *persistent* if the quantum (spectroscopic) features of, for example, $A_zM_sO_t$, become simply a composite mixture of those for A_mO_n and M_xO_y with relative peak size dictated by the ratios of p/q in the previously mentioned formulation. If, on the other hand $A_zM_sO_t$ is an *amalgamation*, then its quantum (spectroscopic) features display sizable shifts from the peak positions of the simple oxides themselves to a new set of peaks that may be properly attributed to only a single chemical species. It should also be noted that a form of these arguments is consistent with the coherent potential model.[51]

It is an essential feature of the present argument that in most of the cases investigated *true* complex or mixed compound (e.g., oxides) formation is indicative of the amalgam-

---

* It should be kept in mind throughout the present discussion that either A_mO_n or M_xO_y (or both) may themselves be complex oxides, e.g., the case where $M_xO_y \equiv SiO_2$ and $A_mO_n \equiv Na_2Al_2O_4$ and $A_zM_xO_t \equiv$ various types of sodium zeolites.[34,40,46]

ation case.[46] The evidence for the latter statement has been extended by the author's research group beyond optical spectroscopy to an analysis of the X-ray photoelectron (ESCA) spectra of select complex oxides.

2. Further on the Mixed Oxide Bonding Concept

Numerous examples of mixed or complex oxide systems that verify the amalgamation argument have been described by us elsewhere, see also Chapter 2.[22] Perhaps the most expressive of these are in the case of the silica-aluminate systems that typify most zeolites[34,40,46] and clays presented in Chapter 2 and later in this chapter.[1b,52] In this regard the results shown in Table 2-5 and those exemplified in Figures 2-8 and 2-9 are typical examples.

What we describe herein are numerous situations in which one or more of the three methods of oxide designation described above have been employed to designate and/or describe the bonding characteristics of complex oxides, and in most of these cases these practices have been found by ESCA analysis to be incorrect (at least for the materials in question). In fact, the details of the resulting errors have been defined to the point where we have been able to describe the generalized nature of the (changing) bonding patterns involved.[22]

These generalized statements about oxide bonding were described in detail in Chapter 2 and are again summarized as follows:

1. If an oxide (M_xO_y) is "interacted" with another oxide (A_mO_n) such that a complex oxide (symbolized herein as $A_zM_sO_t$) results, then the bonding chemistries between M and O and between A and O are both altered from that exhibited by the respective simple oxides, and that alteration is *often* so substantial that it can be readily registered on any ESCA with reasonable high-resolution capabilities.

2. The resulting binding energy shifts generally form consistent patterns (e.g., see Table 8-4) that are such that if A is more electropositive than M then the shifts seem to suggest that the binding energies of A increase [*became more ionic (positive)*] following formation of the complex oxide, whereas those of the metal (or metalloid) M become less positive (*more covalent*) than is the case for the respective species in their simple oxides.[22,37,46]

3. Similarly, the binding energies exhibited by the oxygen [e.g., O(1s)] shift for the complex ($A_zM_sO_t$) oxide between the two extremes found for $M_x - O_y$ and $A_m - O_n$ such that

$$\overleftarrow{M_x - O_y > A_zM_sO_t > A_m - O_n}$$
Increasing O(1s) binding energy or
decreasing ionic nature of oxygen

One should also note that the described results are exemplified in Figures 1 and 2 of Reference 46.

4. All of this can be readily expressed in terms of a relatively simple covalency/ionicity argument for the chemical bonding exhibited between M and O, and A and O. Thus, in this new concept the relative degree of ionicity of a metal (metalloid)-oxygen bond can be made to vary substantially during complex oxide formation depending upon the type and extent of ionicity of the other metal (metalloid)-oxygen system that is *chemically* introduced into the "oxide" lattice. In this regard, we have also noted that it is much more appropriate not to consider the oxide ion itself as a spherical species of constant "size" and charge, but rather to consider it to be a variably sized entity that may be, in fact, polarized when effectively flanked by an M on one side and an A on the other, particularly where these two species are of distinctly different electronegativities.[45b,46]

5. It is also possible for these shifts in binding energies to occur for cases in which one simply varies the ratio of [s]/[z] in the general formulation $A_zM_sO_t$, e.g., in the case of sodium zeolites where s designates the number of silicons and z the number of aluminums and sodiums in a unit cell.[34,40] Thus, cases may arise in which several "independent" complex oxides differ only in their ratios of the relatively electropositive elements (or groups), but even in those cases the ESCA binding energy shifts suggest that *these systems are chemically quite distinct compounds*.[34,46]

3. ESCA Verification of the Concept

The experimental confirmations of these claims based on the sodium zeolites have been described in Chapter 2 and presented elsewhere in more detailed publications,[22,34,46] particularly a recent extensive survey and analysis of the oxide data obtained from several independent sources.[34] As further justification herein of these propositions, we present below four specific nonzeolitic examples of a figurative rendition of the data that relates the different oxides involved.

(a) Two select examples where interjection of more ionic, A type, units into select M_xO_y forms complex oxides with increased covalency (decreased binding energy) of the M in the resulting $A_zM_sO_t$ systems (all values in electronvolts $\pm \sim 0.3$ eV):

	M	**M_xO_y**	**$A_zM_sO_t$**	**Ref.**
(1)		Al_2O_3	$Na_2Al_2O_4$	
	Al(2p)	73.8	72.9	34, 40
(2)		H_2SeO_4	$(NH_4)_2SeO_4$	
	$Se(3d_{5/2})$	61.2	59.2	53

(b) Two select examples where interjection of more covalent M (metalloid) units into A_mO_n forms complex oxides with increased ionicity (increased binding energy) of the A in the resulting $A_zM_sO_t$ systems:

	A	**A_mO_n**	**$A_zM_sO_t$**	**Ref.**
(1)		MgO	$MgSO_4$	
	Mg(2p)	50.6	51.6	53
(2)		BaO	$BaSO_4$	
	$Ba(3d_{5/2})$	777.9	780.8	52, 53

Further examples of this type of shifting pattern were presented in Table 2-3 in Chapter 2.

4. Extent and Limitation of the Mixed Oxide Bonding Concept

Obviously, the diverse mixed systems that fall under the mantle of the aforementioned guidelines are too numerous and complex not to provide some exceptions to these criteria. We list below several of the most obvious *reasons* for these possible exceptions:

1. As indicated above, it is often the case that mixed simple oxides or mixtures of various complex oxides are just that, *mixtures* without the formation of obvious chemical bonds between the subunits involved. In some cases these mixtures are quasi alloys without compound or even solid solution formation.[47] This means that there is no theoretical restriction on the [p/q] ratio for oxide $C_p - B_q$, where C and B refer to less

complex oxides, such as our previous A_mO_n. In addition, cases exist where the range of p to q is restricted by the bounds of solid solubility.[47] In fact, it should be noted that all units C (or B) are always, at least, somewhat soluble in any other unit, B (or C). The degree of solubility, however, may in many cases be so slight (only a fraction of a percent) that for many considerations of these cases solubility may be ignored. In either case, however, few examples exist wherein ESCA shifts have been detected for situations in which no recognizable bond formation (change) occurs. Therefore, we specify that ESCA is immune to the aforementioned mixtures and, also, generally impervious to solution formation.

2. It also seems apparent that the registration of bond change (true compound formation) by ESCA becomes less and less specific as the relative differences in electronegativity $(i_C - i_B)$ of the metal (metalloid) species involved become smaller and smaller.[33] Thus, the ESCA can readily detect changes when MgO and SiO_2 are combined to form Mg_2SiO_4, but very little (ESCA registered) effect occurs when, for example, one combines SiO_2 and GeO_2 to *attempt* to form a germanium silicate or K_2O and Na_2O are mixed in an attempt to form a mixed alkali oxide. It should be obvious why this is so, and it also should be apparent that, whereas complex compound formation is often relatively straightforward when $i_C - i_B$ is large, it is difficult and may in some cases be impossible to achieve true compound formation when $i_C \approx i_B$ (*in a direct reaction from the simple oxides*).[20,33,47] Everyone should be aware, however, that the Hume-Rothery rules point out that for the ready formation of a *solid solution* the arguments concerning the electronegativities (i_x) are actually the reverse of these complex compound formation "rules."[43,54]

3. In many instances the differences between two unique complex compounds may be simply a realignment of structure.

   A. Thus, for example, the zeolite mordenite may exhibit, on the one hand, a composition identical to a silicon-rich Y zeolite, whereas on the other hand a zeolite of the mordenite composition may have the same composition as a low silicon ZSM-5. Detailed ESCA studies in our laboratories[34,40,55] of these types of cases suggest that there may be finite (but small) *structural shifts* that may, for example, distinguish these three different zeolite cases.[46,55] The shifts determined in actual cases studied have always proven to be very small, however, and, in view of the other potential interfering effects noted in this discussion, we still reserve some judgment regarding the existence of such structural shifts. In general, we note that the relatively large experimental shifts detected for many different zeolites are too sizable to be attributed to structural changes, and are, therefore, attributed (by us) to be (group) chemical shifts.[34,46] The fact that structural differences should generally create only small ESCA shifts should be apparent. Most ESCA studies of oxides and complex oxides have involved the detection of core level spectra, e.g., see the results described above.[53] The latter are generally far too specific to the atom being detected and its immediate atomic environment to be influenced by structural features that generally require the involvement of, at least, several full lattices of atoms. It is possible that ESCA detection of more delocalized features, such as the valence bands or loss lines, may bear more readily discerned structural "fruit". Zeolite valence bands are now being investigated by our group[40,46] with structural considerations a matter of major concern, see Chapter 7.

   B. Another feature, which also suggests structural shifts, centers around our recent ESCA examinations of clay systems.[52] In these cases we have discovered reproducible ESCA core level and valence band binding energy patterns for the sheet like (clay) silicates kaolinite (Si/Al $\approx$ 1) and montmorillonite (Si/Al $\approx$ 2) that not only differ from their simple oxide progenitors, but also differ from the corresponding (framework) zeolites of essentially identical silicon to aluminum ratios, see Section VI.A of this chapter.

4. As mentioned above, it should also be noted that it is possible for the process of ESCA measurement to interfere with (and even produce) the resulting shifts. Thus, as described in detail in Chapters 2 and 3, ESCA investigations of oxides are often encumbered with shifts due to charging (and the floating Fermi edge[21,56]) or final-state relaxation effects[21,57,58] or both.[21]

5. The Acid-Base Connection

Those interested in a type of "complex" oxide that exhibits the behavior described above and that also spans the periodic table need only investigate the hydroxides. Consider first those complex A–(O–H) bonding units that are realized when simple metallic oxides (anhydrous bases) from the left side of the periodic table are hydrated. In every case in which an H unit is substituted for an A unit in –A–O–A–, the ionicity of the A–O bond is increased while the resulting O–H bond in A–O–H is made more covalent than in H–O–H. Thus, in each of these cases the binding energies of the A unit increase (to reflect their enhanced ionicity), see Table 8-7. The binding energies of the O unit, on the other hand, *increase* to a value between that for A_xO_y (low side) and solid H_2O (high side), indicative of the (partial) increase in covalency induced by the hydrogen replacing A in the oxide.[22]

As the A unit under consideration gets closer to the top of the middle groups of the periodic table (upper 3A and/or 4A), the essential features of amphoteric behavior begin to arise. These metal (metalloid) oxides/hydroxides are apparently characterized by decreasing differences in ionicity of the bonding of the M–O and O–H units that gradually fades to zero. Thus, amphoteric behavior is the evolving feature. The XPS core level (and valence band) spectra readily reflect this feature, e.g., see Table 8-7. Thus, the Al(2p) binding energies for various Al(OH)$_3$s are apparently only slightly larger than those for various Al$_2$O$_3$s,[53] while this characteristic is even more apparent for silicon oxide (hydroxide) systems, where no chemical shift of Si(2p) or O(1s) is found for such species as silica-oxalic acid and silicic acid compared to α-quartz.

It is anticipated that this process will "flip over" as consideration moves to the VA through VIIA groups, particularly near the top of the periodic table. Very little of this has yet been documented, however, because of the inherent corrosive nature of the corresponding anhydrous acids and their hydrated forms for stainless steel, copper gasketry, and other facets of UHV technology. The few measurements already achieved for these oxides, however, support the continuity of our argument. A number of the pertinent results are summarized in Table 8-7. For much more detailed information, interested parties are directed to the appropriate tables in Reference 53.

It should be apparent that the present arguments are nothing more than a surface chemical alternative for conventional considerations of Bronsted and Lewis acidity. Although at present primarily qualitative in nature, the use of ESCA chemical shifts for this basis also seems to provide, at least, the semblance of a semiquantitative acidity/basicity scale that in these cases is applicable to the surfaces of these solid materials.[46]

In Table 8-8 we see a further rendition of one of our fundamental arguments. The Na–O bond is generally more ionic than the H–O bond. This should be particularly true for the phosphates. Thus, in Table 8-8 we find the P–O bond driven to its most covalent state (lowest) binding energy for P ($2p_{3/2}$) for the Na$_3$PO$_4$. As hydrogens are substituted for sodiums in these systems, the P–O bond becomes increasingly ionic, and the $P_{3/2}$ binding energy progressively grows. However, on the other hand, the H–O bond attached to phosphorous is more ionic than the P–O bond attached to phosphorous. Thus, the P–O binding energy in P$_2$O$_5$ is shifted to higher binding energy than it is in NaH$_2$PO$_4$. All of this reflects on the relative acidity of the H–O bond in the case of the phosphate system, a feature that is reversed for the relatively basic metals in Table 8-7.

Table 8-7 **Oxide to hydroxide binding energy (BE) shifts in eV ± 0.2 eV**

Elements	State	Oxide BE	Source	Hydroxide BE	Source
Y	$3d_{5/2}$	157.1	a	158.5 (YOOH)	a
Co		780.4 (CoO)	b	781.3 $Co(OH)_2$	b'
Cu	$2p_{3/2}$	933.7 (CuO)	a'	934.7 $[Cu(OH)_2]$	a'
	$2p_{3/2}$	933.8 (CuO)	c	934.5 $[Cu(OH)_2]$	d
Cd		404.2	e	405.1	f
Al	2p	73.85	g	74.3 $Al(OH)_3$	g
Sn	$3d_{5/2}$	486.1	a'	486.4	a'
Pb	$4f_{7/2}$	137.25 (PbO)	h	137.95 $Pb(OH)_2$	h
Se	$3d_{5/2}$	58.8	i	59.0	i

a. Barr, T. L., *Quantitative Surface Analysis of Materials*, McIntyre, N. S., ASTM STP 643, 1978, 83.

a'. Barr, T. L., *J. Phys. Chem.*, 82, 1801, 1978.

b. Nefedov, V. I., Firsov, M. N., and Shaplygin, I. S., *J. Electron Spectrosc. Rel. Phenomena*, 26, 65, 1982.

b'. Schenck, C. V., Dillard, J. G., and Murray, J. W., *J. Colloid Interface Sci.*, 95, 398, 1983.

c. Wagner, C. D., Riggs, W. M., Davis, L. E., Moulder, J. F., and Mullenberg, G. E., *Handbook of X-Ray Photoelectron Spectroscopy,* Perkin-Elmer Co., Eden Prairie, MN, 1979.

d. Barr, T. L., recent reexamination, different ESCA.

e. Gaarenstroom, S. W. and Winograd, N., *J. Chem. Phys.*, 67, 3500, 1977.

f. Wagner, C. D., Gale, L. H., and Raymond, R. H., *Anal. Chem.*, 51, 466, 1979.

g. Wagner, C. D., Six, H. A., Jansen, W. T., and Taylor, J. A., *Appl. Surf. Sci.*, 9, 203, 1981.

g'. Taylor, J. A., *J. Vac. Sci. Technol.*, 20, 751, 1982.

h. Pederson, L. R., *J. Electron Spectrosc. Rel. Phenomena*, 28, 203, 1982.

i. Bahl, M. K., Watson, R. L., and Irgolic, K. J., included in Reference f.

Table 8-8 **Changes in phosphorous binding energies (BE) for select phosphate systems, in eV**

Compound	P2p Value	Source
P(red)	130.2	j
P_2O_5	135.2	k
Na_3PO_4	132.3	l
Na_2HPO_4	133.1	m
NaH_2PO_4	134.2	l

j. Schaerli, M. and Brunner, J., *Z. Physik,* B42, 285, 1981.

k. Clark, D. T., Fok, T., Roberts, G. C., and Sykes, R. W., *Thin Solid Films,* 70, 261, 1980.

l. Swift, P., *Surface Interface Anal.*, 4, 47, 1982.

m. Wagner, C. D. and Taylor, J. A., *J. Electron Spectrosc. Rel. Phenomena*, 20, 83, 1980.

IV. XPS AND CORROSION SCIENCE

The early efforts to combine ESCA and corrosion science were reasonably success-ful,[59] but it was not until the last decade that the traditional corrosion scientists and engineers truly recognized the value of these contributions. The scope of these ESCA efforts was indicated by the chapter on the "Uses of Auger and Photoelectron Spectroscopies in Corrosion Sciences" provided by McIntyre in the first edition of the widely used text on *Practical Surface Analysis* by Briggs and Seah.[60] The growth of this interaction is further reflected in the substantial expansion of that chapter by McIntyre and Chan in the 1990 second edition of that book.[61] In the latter, the number of references almost doubled compared to the former. The emphasis in these reviews was on oxidative corrosion, and McIntyre devoted substantial time to careful discussions of oxide onset, growth and spread, and also spatial and depth profiling forms of analysis. McIntyre then moved throughout the periodic table, metallic element by element, to categorize the realization of their oxidative surface chemistries. Consideration was also given to the corrosion of the most prominent alloys, with detailed analysis of the ESCA studies of steels. McIntyre also provided some interesting details of the existing XPS analyses of a wide spectrum of protective coatings, paints, and inhibitors. The general importance of this section rests, in part, on the detailed mixing of XPS studies of polymeric surfaces interacting with metallic ones. The similarity of these results with those achieved in catalysis and biosurfaces should be obvious.

It should also be very apparent that a certain amount of license must be exercised in order to differentiate between the ambient oxidation layers formed at room temperature on nearly all metals, described in Section II of this chapter, as *natural passivation* (NP) and the onset of corrosion. Careful consideration indicates that NP is actually a state that interposes between the virgin material and corrosion, and thus the formation and retention of NP is a crucial (and generally neglected) part of the total process.

Generally ignored in the McIntyre reviews is the fact that many of these processes of natural protection and corrosion are features that affect all materials, not just elemental metals and alloys. The author of this book became acutely aware of these extensions during some ESCA studies of the extent and breakdown of NP.[16] Our early investigations, however, began on the traditional course and included studies of the air induced corrosion of Cu_{98}-Be_2,[62] and efforts to investigate the dezincification of α-brass.[63] We therefore begin by reviewing these studies before generalizing.

A. ALLOYS: NATURAL PASSIVATION AND CORROSION

Obviously the previously described NP process will play an active role during the "conditioning" and use of alloys.[14] It should be apparent, however, that the added complications of alloy formation must substantially increase the numbers of previously mentioned exceptional cases. Surprisingly, for many common alloys the only major novel feature introduced into the low temperature oxidation (LTO) mechanism by the presence of a second (or more) metal is the occurance of selective (more favorable) oxidations and preferential migrations (particularly segregations).[14-16]

In this regard, the author has studied the preferential migration of $Be°$ to the surface of a $Cu_{98}Be_2$ alloy, induced by only modest increases in temperature.[14,62] The substantial preponderance of Cu in the alloy produces the expected ready presence of copper oxides during room temperature oxidation. The $Be°$ released from the heated copper-beryllium lattice, however, exhibits a dramatic preference for oxygen. Thus, the released $Be°$ literally getters oxygen in high vacuum, producing growing islands of BeO that eventually swamp all of the aforementioned copper oxides,[16,62] Figure 8-8.

α-Brass ($Cu_{75}Zn_{25}$) proved to be an even more revealing case. In this regard XPS analyses were employed to examine the progressive growth of oxidation products formed by a variety of processes on top of coupons of the initially clean $Cu°/Zn°$ alloy. In addition, several representative systems already subjected to oxidation were examined with XPS following progressive applications of a moderate energy (~1 keV), Ar^+ ion sputter etching. (One must be careful not to over-interpret the latter results, as sputter etching may alter the newly exposed surface.) These results were then tied together in ~30-Å segments to provide the chemically detailed, but structurally somewhat idealized, layered structures suggested in Figure 8-9b. Thus, we found, for example, that room temperature oxidation[15,16] of this alloy produces an initial layer of ZnO that grows to a

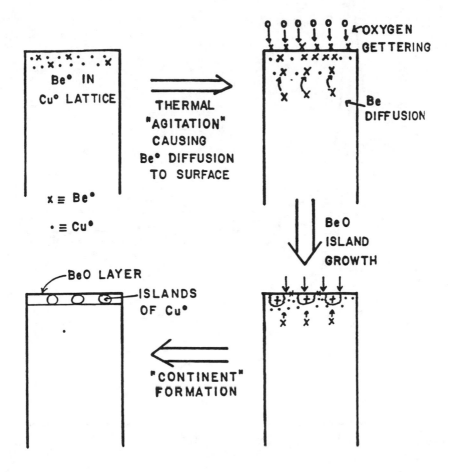

Figure 8-8 Stagewise development of the selective segregations and oxidations induced in $Cu_{98}Be_2$ by heat and (limited) oxygen exposure.[17]

thickness of ~50 Å, and then appears to stop. On top of this layer a series of copper oxides are then produced, oxidizing the Cu° that is apparently trapped in the early forming ZnO lattice. The air-induced chemistry of the α-brass-formed copper oxide lattice closely resembles that of the oxides produced on elemental copper,[4,5] i.e., Cu(I) oxides are first formed, then covered by Cu(II) oxides and finally the expected terminal hydroxides, see Figures 8-9b and 9-1. However, unlike the case of Cu alone, α-brass has the additional large subsurface ZnO barrier preventing corrosion. For these reasons, we find that air-exposed α-brass exhibits a somewhat brighter gold-like color, which retains its sheen for a much longer period, that is, common air-exposed α-brass exhibits more successful NP than elemental copper.[14–16]

Certain processes, however, easily overcome the natural oxide coating on α-brass. Perhaps the most interesting of these is the dezincification induced in α-brass by common ocean water. The onset of this process was investigated in detail by our group[63] and a mechanism was developed. It was found that ~3% salinated (simulated sea) water was able to dissolve some of the protective copper oxide layers exposing the subsurface of more porous and reactive zinc oxide. The extent of growth of zinc oxide was then

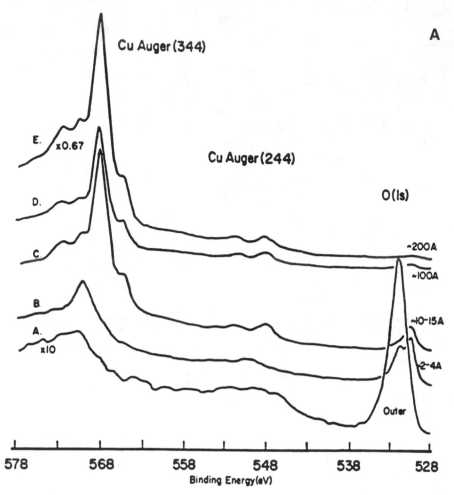

Figure 8-9 Evolution of the (~50 to 75 Å) oxidative layers formed on α-brass (Cu$_{75}$Zn$_{25}$), during natural passivation. (A) ESCA spectra depicting growth development of copper and oxygen. (B) ESCA developed rendition of layer structure showing chemistry and approximate quantification following natural passivation of α-brass.[14,15]

magnified many times producing a thick, porous sublayer. The latter eventually poured through [cracks(?)] in the thinner and noticeably weaker copper oxides, forming a surface film of zinc oxide, Figure 8-10. The latter caused the disappearance of the original copper-gold sheen of the brass.[15] Subsequent development suggested that the two layers of ZnO eventually unite into a series of "plugs" that are structurally at odds with the balance of the brass surface. Any pressure on these plugs may therefore easily induce their ejection, as occurred in the case of the steamship boilers of the British fleet immediately preceding World War I.[64,65]

Similar alloy passivation and corrosion studies, using XPS, have been conducted by our group[16] and others.[66] As suggested, in all cases the processes were substantially influenced by preferential segregations.[67] Studies have confirmed that these may be thermally induced simply as a result of the lack of structural and chemical match between the metallic components. As a result of this mismatch a surface tension-induced (Gibbsian) segregation often occurs. These effects have been theoretically described by Medima[68] and Williams and Nason,[69] and have been the topic of numerous surface analysis investigations.[67] In

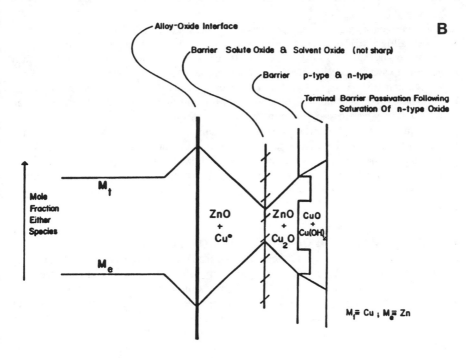

Figure 8-9B

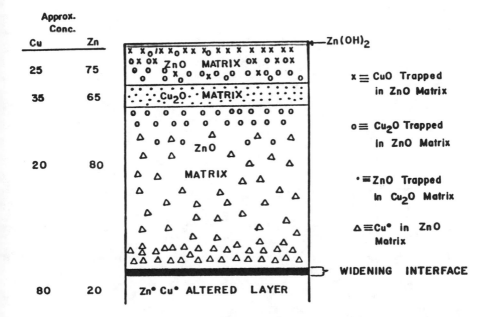

Figure 8-10 ESCA revealed (>1000 Å) layer structure formed on α-brass following completion of the "onset" stage during the dezincification of α-brass.[19,63]

addition, surface segregation may be induced by the preferential tendency of one component in an alloy to oxidize. An interesting specific example of these processes occurs when one considers the ternary alloy $Cu_{97}Be_2Au_1$. It has been shown that gold will preferentially segregate in vacuum to the surface of a room temperature gold-copper alloy by Medima-type segregation.[68] This process thus produces a surface with enhanced resistivity to oxidation/corrosion, and thus enhanced integrity for controlled surface conductivity. Simultaneously, it may be necessary to promote other regions in the same material with dramatically curtailed surface electrical conductivity. This can be induced by heating specific regions of the ternary alloy to release beryllium from the lattice. If the latter is exposed to even small quantities of air, it will preferentially oxidize forming the previously described, nonconductive BeO. The latter, of course, also has unique optical properties, so that a contiguous material may be produced with an assortment of very selective, regional properties.

B. THE SPECIAL CASE OF MIXED HYDROXIDES

Reference was made above to the obvious existence of exceptions to our previously proposed arguments. Many of these exceptions are primarily caused by unique structural factors[8] and are therefore outside of the specific thesis of this book. One, however, is of particular chemical nature, and should be considered herein. It concerns the obvious cases of metal (e.g., Co, Al, etc.) and metalloid elements (e.g., Si, Ge, etc.) that readily form some type of inner lattice hydroxide or, at least, oxyhydroxide system, i.e., elements for which hydroxide formation is not primarily an oxidation termination process.[8]

There must be a number of particular contributing factors to this bulk, or polymer, hydroxide situation, but one chemical feature may have semi-generality. It concerns the ability of some metals (metalloids) to form bridging hydroxide (β) units (as well as bridging oxides). These features must compete with cases where hydroxide species form primarily top site (α)-type bonding. Thus, at least three cases seem to arise:

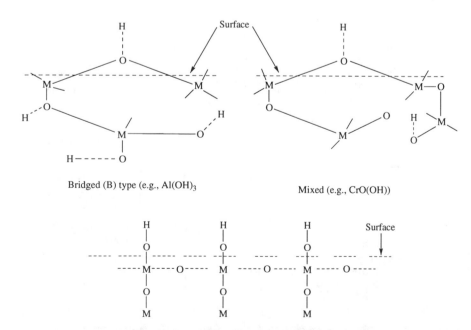

Bridged (B) type (e.g., $Al(OH)_3$)

Mixed (e.g., $CrO(OH)$)

Top (a) site only (e.g., $Cu(OH)_2$ on top of passirating copper oxides).[2]

The exact causes of these various cases and their relative stability are unknown, although there is some reason to believe that an increased tendency to form relatively stable hydrogen bonds, e.g.,

must play a significant role.[11]

The significance of these "polymeric" hydroxide units to the subsequent stability of the NP (and corresponding resistance to corrosion) is readily apparent. Thus, as numerous researchers have demonstrated,[9] hydroxides provide the bond weakening ingredients that initiate the corrosive effects of water on the oxide lattice. This may be argued in terms of the substantial polarization apparently experienced by the oxygen atom in an M–O–H unit or in terms of an ionic model similar to that induced by aquated chloride ion and other ionic impurities.[9] In any case, these highly polar (ionic) units seem to wreck havoc with the stabilizing oxide polymeric lattice structure, providing encroachment channels that permit corrosive attack.[9]

It should be noted that on the positive side, the presence of bridged hydroxides (with some hydrogen bonding) in mixed metal (metalloid) systems is also instrumental in providing variably strengthened, mixed sited, solid Bronsted acidity in such systems as zeolites,[46] e.g.,

These variable acid sites, combined with the unique, different, structural patterns of these systems permit the use of different zeolites as the key ingredients in the selective, acid-catalyzed cracking reactions that are one of the backbones of the petroleum industry.[1b]

C. NATURAL PASSIVATION OF SEMICONDUCTORS

While developing the concept of NP for metals and alloys our group also conducted XPS studies of the surface stability of certain compound semiconductors.[16] During these studies it became readily apparent that exposure of the latter materials to air invariably causes a variety of surface modifications that could only be considered as slightly modified forms of NP. Similar XPS studies have been conducted by others,[70] particularly the group of Wilmsen,[71] and therefore we will not present a detailed discussion here. We will, however, outline some of the generalities and unique points of disagreement.

In general, one should note that all semiconductor materials (even the most pristinely grown thin films) following air exposure are rapidly subjected to selective oxidation. In some cases, the resulting products are generated on the semiconductor surfaces to sufficient thicknesses to induce a unique chemistry, and to alter the predisposed interfacial characteristics and surface functionality of these materials. It should be noted that these effects are much more sophisticated than the crude ideas of a simple variance in the form of a surface "tarnish" as proposed only a few years before by Many et al. in their purportedly detailed book on the surface properties of semiconductors.[72]

Thus, when the compound semiconductor GaAs is exposed to air, it is rapidly subjected to the formation of oxide products that tend to passivate its surface from further (often corrosive) attack. This process may, of course, be a desired effort on the part of the device producer, i.e., by overlaying on the GaAs some other, relatively nonreactive, species, but generally this effect is unexpected, and may even be deleterious. In any case it is readily apparent that the air-induced oxidized products formed on GaAs exhibit both the selective layering of the oxidation products found on metals, such as $Cu°$, and also the preferential layering of components found in the case of alloys. Thus, regions containing oxidized forms of both Ga and As are detected by the XPS. An almost identical result has been found following air oxidation of GaSb, Figure 5-7a. The latter analysis was performed in a manner similar to that described above for the Cu/Zn alloy except that for the semiconductors even more use was made of angular resolution. The primary difference in interpretation between Wilmsen[71] and our group[16] concerns the degree to which complex mixed oxides of the group III and V elements are actually formed. Wilmsen suggests that mixed oxides are the principal products in the layered structures, whereas our interpretation suggests that the primary species formed in the oxide layers are simple oxides, with the possibility for the formation of more complex oxides, primarily at the moderately regular interfaces between sublayers. Thus, our interpretation of the oxide (and hydroxide) layers produced by air exposure of GaSb and related systems are presented in Figure 8-11. One of the reasons for considering the formation of complex oxides as the oxidation products of these III-V materials is the relative ease of production of such mixed oxides as metal nitrates and phosphates. However, additional results suggest progressive reduction in stability of these complex oxides going down the group V column, i.e., as one goes down the group V column the relative stability of simple oxides in solid form grows compared to the corresponding complex forms.

Whatever the products form, several implications are important related to the potential use of III-V systems. Generally III-V's are expected to function selectively as thin film, intrinsic or extrinsic (n or p) semiconductors, with their interfacial characteristics (when sandwiched with other components) of utmost importance. It should be noted, however, that the oxides detected by the XPS on the surface of any air-exposed III-V are, in themselves, wide band-gapped semiconductors, i.e., Ga_2O_3 is n type, whereas Sb_2O_3 is p type. Thus, regular surface arrays of the latter could under certain circumstances critically alter the behavior of the III-V or any other species involved. This is a particularly important problem since all semiconductors are subject to varying degrees of this same passivation "problem".

Another interesting effect concerns the air-exposed by-products formed by various III-nitrides, e.g., InN.[73] In these cases, we also detect an outer surface of nitrogen oxide species,[74] but because of the lack of stability of solid forms of the latter we suggest that our results display the presence of such compounds as In-(ON) and In-(NO$_2$). In these cases, it appears that one of the most interesting results that may be realized during angular resolved XPS (AR-XPS) studies is the condition sensitive preference of either the N or O to bond directly to the In.[75]

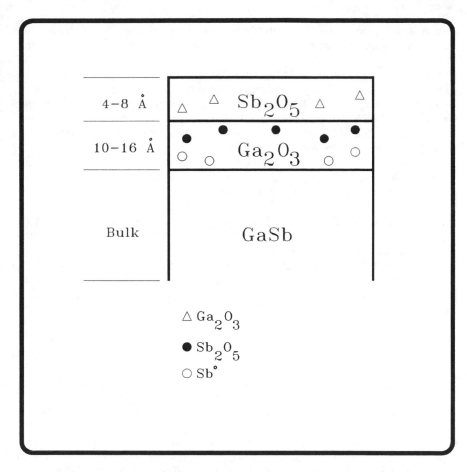

Figure 8-11 Natural passivation results for a typical III-V semiconductor: GaSb. Approximate rendition of the layer structure produced by GaSb during natural passivation. (Note that very similar results were produced by InSb.)

D. NATURAL PASSIVATION AND CORROSION OF "INERT" POLYMERS — BIOLOGY AND THE WORLD AT LARGE

The previous results suggested to the author that the process of NP has no bounds. In other words, nearly all solid materials, no matter what their origin, must be subject to the growth of either passivating oxides or the terminal hydroxide species. Thus, we contend that only those few systems whose Cabrera-Mott field results in a combined diffusion potential above $KT_{300°K}$ will avoid some form of room temperature, air-induced oxide growth. We have amplified above in Table 8-2 that even the latter materials are all subject to terminal hydroxide and/or carbonate formation. All of this suggested to the author that biological surfaces should also be subject to NP[16] — an interesting concept in view of the assumption generally promoted by biologists that the outer surfaces of biological materials (e.g., skin and leaf tissue, etc.) are represented by outer epidermal-type cells that chemically resemble those found in the subsurface.[76] In order to challenge this hypothesis the author decided to investigate the oxidative experiences, under various environmental conditions, of a polymer that was purported to be "nonreactive". Polypropylene was chosen for this purpose based on the assumption that whatever happened to relatively unreactive polypropylene would probably be a moderate version of the surface effects experienced by the outer surface of potentially more reactive biological entities, such as skin or leaf tissue.

For these purposes, therefore, a film of high-purity catalytically prepared polypropylene was subjected to a variety of oxidative treatments.[77] These included the equivalent of NP and a variety of controlled corrosion-type studies that simulate various environmental states. The details of the subsequent AR-XPS study were presented in Chapter 5.

In addition to the AR results, it was important that a careful spectral pattern analysis occur at this point in order to determine the purity of the polypropylene polymer and the characteristics of its carbon-only spectrum. In particular, it was important to discern those features, if any, that differentiate the XPS characteristics of polypropylene from other (C–H) only systems. For this purpose the C(1s) principal peak binding energies and the important loss splittings (Chapter 6)[78] from that peak were also recorded.

In addition to oxidation of the ambient polymer, evidence was also provided that suggests that treatments such as stretching the polymer with or against its grain significantly influences the oxidative behavior in all of the common environments. There was also an indication that controlling the degree of exposure to solar radiation and the amount of oxygen dissolved in the various fluids influences the extent and type of oxidation. The latter parts of this study are, however, still too preliminary and crude to reach any detailed judgments.

The most important feature detected in all these results was the general nature of the oxidation chemistry formed in the semidiscrete layer structures during exposure to NP air and the fluids and, in particular, the obvious interrelationship that exists between these results for this polymer[16] and those described earlier for the metals and alloys.[4,5,14-16]

A comparative summary of the key results for exposure of the polymer to NP air, fresh water, and simulated sea water is presented in Figure 5-6. The key features produced are obvious. Thus, fresh water significantly increases the amount of oxidized products in the polymer with a dramatic, and perhaps not surprising, increase in the presence of alcohol (C–OH)-type species. Simulated sea water, on the other hand, treats the polypropylene much the same way that it does α-brass. Thus, there is a substantial (near total) production of oxidized products that extend well below the 60 to ~70-Å cutoff of AR-XPS visibility. Also, the interfaces of the product layers are effectively destroyed, and those species that are almost entirely subsurface following NP are now pulled out toward the outer surface.[63]

The interesting aspect of these results (Chapter 5) may be realized by noting that the carbonyl species formed are oxidized versions of the ethers, that is, the carbon in our supposedly inert polymer is first oxidized in a relatively thick layer of species in its lowest common oxidation state, followed by a relatively thin, but contiguous, layer of carbon-oxide with carbon in its highest common oxidation state. The latter layer seems to be terminated by a layer of hydroxide. It should be *apparent that one may replace carbon in these arguments with any of the group A metals in Table 8-1, such as Cu, and the general layered structural (NP) pattern would be exactly the same.*

1. Generality

The previously described polymer results thus suggest a generality for NP (and perhaps even some of the features of subsequent corrosion that may be induced by harsher environments) that extends over most, if not all, types of solid surfaces. Thus, the nature of the NP process is that air reacts with nearly every surface through low temperature (Cabrera-Mott) oxidation (LTO) forming a series of thin, relatively contiguous oxide-layered structures, often of progressively growing, oxidation state. When conducted in air, with a certain amount of humidity, these layers are capped by either hydrated oxides or hydroxides.

It should be noted that the proposed layers are, in themselves, relatively contiguous, thin films of wide band-gapped semiconductors, that should be extremely sensitive to select interfacial changes in conditions, such as the relative O_2 or H_2O content. Thus, besides being first the inhibitor of, and finally the initiating point for, corrosion, these NP films should be extremely sensitive oxygen sensors, a feature that has intriguing possibilities, particularly in biological areas.

E. ADDITIONAL IMPLICATIONS OF NATURAL PASSIVATION

These observations concerning oxidation, NP and corrosion have been marked by their generality. However, at the same time, we have found several systems that for one reason or another are exceptions to the stated patterns. Several interesting points deserve special mention regarding these features.

1. Termination of Passivation

One of the more striking aspects of the previously described arguments is their apparent inevitability. Thus, (1) almost everything exposed to STP air forms oxides, but also (2) these oxides do not form natural terminal bonds, that is, if available all oxidized systems will terminate with either hydroxides (or hydrated oxides) or carbonates or both. With this in mind it should not be difficult to realize how many XPS users have recently been guilty of some serious errors in interpretation. The question concerns the XPS analysis of the recently discovered, very exciting, high T_c superconducting oxides. Most of these XPS analyses were conducted on these materials after they were transported through air. However, it should be apparent that all oxides are subject to reactions when air exposed. Thus, for example, even a lower valent oxide species, such as FeO, will rapidly form an oxidized top layer of Fe_3O_4 and Fe_2O_3.[79] In addition, oxides that have completed this oxidation process will all still form varying amounts of the previously mentioned capping hydroxides and carbonates. Thus, almost all of the early XPS studies of purportedly pure superconducting oxides such as the $YBa_2Cu_3O_{7-x}$ system were actually primarily detecting a mixture of hydroxide and carbonate by-products. The details of this problem were described by Miller et al.[80] and most recently summarized by Barr and Brundle, see Figure 8-12.[23,81] The carbonate situation is particularly important here. It should be noted that nearly a decade ago this author demonstrated that the outer surfaces of reagent grade $BaCO_3$ and BaO were identical, i.e., both are entirely converted to a mixture of $BaCO_3$ and $Ba(OH)_2$, see Figure 8-13.[35] The drive toward carbonation by the alkaline earth oxides is substantial.[20] The problem inherent in surface analyses of the superconducting oxides is, of course, a major one, as conventional sputter etching (to remove the terminal by-products) will also generally destroy the tenuous coupling to the superconducting, mixed oxide.[11] More will be said about this later in this chapter.

2. Metals at or Near the LTO Limit — The Catalytic Connection

Those who are familiar with some of the criteria for chemical catalysis are aware that proponents often refer to the interconnections between the fields of oxidation, corrosion, and catalysis. It seems appropriate to ask, therefore, if there are any obvious connections between our present arguments, and what is often labeled as molecular or basic catalysis. Generally, the latter area seems to be dominated by cases in which key structural features are investigated through selective chemisorptions of CO, CO_2, H_2O, or O_2 on specific crystallographic faces of metals such as Pt, with subsequent "catalytic" conversion of these adsorbents, without any apparent chemical alteration of the metal.[82] An obvious question at this point concerns why Pt is picked for this process and not Fe or Mo or even Ni.

One of the major reasons for this preference can be discerned from considering a combination of certain features in Table 8-2 and Figure 8-2. Thus, we find that, if we are trying to preferentially adsorb and react at the elemental metal surface certain structurally sensitive molecular units such as CO, H_2O, and OH, which we have previously labeled as terminal oxidizers, we need to concentrate on metals that exhibit only weak Cabrera-Mott fields, i.e., those metals whose LTO limit is at or above room temperature. Thus, metals of this type, such as Pt, Au, and, to some extent, Pd, do not form room temperature oxides, but are readily covered with near monolayer chemisorptions of terminal-type species. In addition, interjection of small amounts of O_2 will not disturb the important, direct involvement of these elemental metal surfaces and the terminal-type groups. Thus, the proper synergistic reactions that depend so intimately upon the structural features of the metal will be preserved. If, however, a metal with a relatively large, Cabrera-Mott field is chosen for these processes (e.g., Fe or Ni, etc.), then a tiny interjection or creation of oxygen in the system will result in the rapid, and not easily reversible, interjection of oxides into the key bonding positions adjacent to the elemental metal surface, and such features as the metallic structure will exert little or no influence over any of the subsequent reactions experienced by the –OH and CO– groups. In fact, the latter will lose their special status as catalytic structural signposts for elemental metals, and become merely terminators in a typical NP oxide layer process.

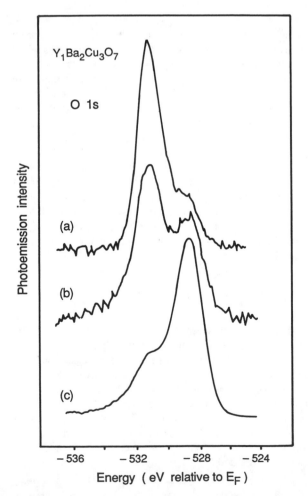

Figure 8-12 Evolution of the ESCA O(1s) for a $YBa_2Cu_3O_{7-x}$ (bulk) system during drastic (but necessary) cleaning to remove surface oriented degradation products (apparently primarily $BaCO_3$ and Ba $(OH)_2$ (a) Outer (dirty) surface. (b) Moderate cleaning. (c) Nearly clean.[80]

V. ESCA STUDIES OF THE CARBONACEOUS SYSTEMS INVOLVED IN POLYMERS, COMPOSITES, AND RELATED CHEMISTRIES

A. FUNDAMENTAL PROBLEMS

Analyses of carbonaceous systems have produced both XPS successes and failures. Many of the successes have occurred during investigations of alterations of functional groups on polymers or carbonaceous composites.[1d] Failures have often occurred when fragile, nonconductive polymers have been investigated during processes in which only the type and/or degree of various carbon-carbon (or carbon-hydrogen) bonds are changed. For many years this problem was accentuated by the (reputed) lack of chemical shift found in the C(1s) binding energies for these various bondings.[1d] Recently it has been reported that the C(1s) binding energy of graphite is different from that for most hydrocarbons.[83] Even more recently it has been demonstrated that there may be small, but regular shifts between the C(1s) values for different hydrocarbons.[56] The ready registration of these shifts, however, still may be difficult. Fortunately, several auxiliary, XPS-based

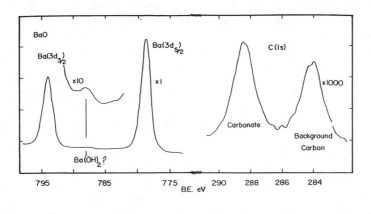

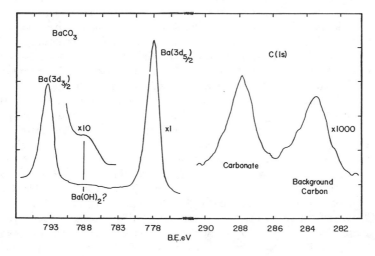

Figure 8-13 ESCA analysis of outer surfaces of reagent grade BaO and BaCO$_3$. Note that independent *bulk* analyses found both systems to be 99.7% pure as labeled.[35,37]

methods have been demonstrated that may be used to attack these problems. In general, these studies have involved examining less substantial XPS features than the C(1s) peak. For example, a variety of research groups have exploited the unique nature of the $\pi \to \pi^*$ loss peaks.[84] Beginning (as usual) with the groups of Siegbahn[85] and Shirley,[86] it has been noted that there are often substantial differences in the valence bands of different carbonaceous systems.[87] In addition, as described in Chapter 6, the author has also demonstrated the use of the different (plasmon-dominated) loss splittings as a means to study many of these systems.[21b,78]

It is now apparent that these auxiliary forms of XPS may, in fact, be employed as the primary methods for analysis of select carbonaceous problems, and Sherwood, in particular, has employed experimentally and theoretically (X-α) realized valence band results to produce a detailed map to follow select carbonaceous systems in both fresh and altered environments.[88]

No one it seems has yet tried to examine a select group of carbonaceous materials (containing only carbon-carbon and carbon-hydrogen bonds) employing all of these techniques together to see which ones seem to describe more appropriately such features as degree of aromaticity, structural integrity, etc. We have, therefore, initiated this type of analysis and report herein the preliminary results. In addition, we have followed these same features (and some of the additional binding energy evolutions) during a common dramatic change in chemistry — the oxidation induced when these materials are treated for varying periods with conventional ozonation.

During the latter treatment the author has also detected a set of binding energy shift results for the peaks C(1s), O(1s), and O(2s) that provided a unique signpost of the resulting chemistry. In other studies reported in Chapter 2, we have employed XPS to investigate the oxidation and oxides of numerous metal and metalloid systems including the IVA group (Si $\to$ Pb).[89] Among other things these studies have revealed certain consistent patterns of binding energy shifts.[21,22] As we describe below, carbon breaks these patterns in a manner that graphically reflects its unique chemistry.

Due to a variety of problems many of these XPS studies are much more difficult than may be apparent. In particular, some of these systems, primarily those with carbon-oxygen bonds may be destroyed by the combination of vacuum, plus high-intensity X-ray beam effects (heat?). Thus, the ratio of some types of carbon-oxygen bonds seem to be susceptible to these effects for certain materials — but, on the positive side, we found that the effect could be sufficiently controlled to ensure the general results to be described.

Listed in Table 8-9 is a representative selection of the carbonaceous materials employed in these studies. It should be noted that the group includes systems ranging from very pure polymeric hydrocarbons to very pure (but nonoriented) graphite. It should also be apparent, of course that this analysis will not detect impurities in concentrations below ~0.1 to 0.2 mol %, structural anomalies, or significant bulk* features. All of the materials examined were subjected to exposure to laboratory air before insertion into the vacuum chamber of the ESCA. Since no surface so exposed is totally impervious to all forms of oxygen chemisorption and/or reaction, we must certainly expect the same even for the relatively nonreactive hydrocarbons and pure graphite. These ambient processes, however, resulted in extremely sparse involvement. [Note the % (0) for air-exposed graphite, etc. in Table 8-9.] The ozonation, on the other hand, always produced a substantial increase in surface oxygen — often producing more than a magnitude increase in relative amount. The types of oxides induced during these processes varied as will be described below. The ozonation was rendered using a Griffin Techniques ozone generator. The system was operated in the prescribed manner, employing 8 psi of gas flow for 30 min in each treatment.

---

* The bulk purity of most of these materials were assured by their manufacturers. Generally, we found no reason to disagree with these specifications.

Table 8-9 Carbonaceous materials and their ESCA characterization

Carbon-containing material	Character of valence band	C(1s) loss eV	[O] % oxygen content	O(1s)eV	O(2s)eV	II→II*	C(1s)eV	Conductivity
Graphite	100% graphite	30.6	~1.0	~533	None	Large	284.7	Yes
O_3-treated graphite	100% graphite	29.0	4.2	532.7	28.2 26.2 [S]	Medium	284.9	Yes
Baker microfiber	100% graphite	27.8	7.0	—	27.0	Small	284.7	Yes
O_3-treated microfiber	100% graphite	~26.0	14.7	534.2	28.0	No	286.1	No
Carbon black	100% graphite	26.0	7.8	533.1 ⟨531.9 534	26.5 ⟨26.0 27.5	No	285.1	Yes
O_3-treated carbon black	1/4 graphite 3/4 polymer	25.5 [L] 22.0 [L]	8.9	532.9	28.0 [L] 26 [S]	No	285.1	Yes
Extensive O_3 carbon black	100% graphite	25.5	13.4	535.3	28.5 29.5 ⟨30.5	Medium	286.8	No
Carbon paper	1/2 graphite 1/2 polymer	26.0	8.2	532.9 ⟨531.8 534	28 [L]	Small	285.5	Yes
O_3-treated carbon paper	90% polymer	25.5 [L] 22 [S]	13.6	533.1	26.5 [S] 28 [S] 26.8 [S] 25 [M]	No	285.5	No
High-density polyethylene (HDPE)	100% polymer	20.5 ⟨21.6 22.0	1.3	534.7	None	No	286.6	No
O_3-treated HDPE	100% polymer	21.0	7.0	535.3	30.4 29.2	No	287.4	No
Low-density LDPE	100% polymer	22.0	1.5	535.1	None	No	286.8	No

Note: Relative amounts: [S] = small; [M] = moderate; [L] = large.

1. Analysis of Select ESCA-Based Methods for C–C and C–H Bond Identification

As mentioned above, one of the motivations for the present study is to try to examine collectively several of the techniques that have been proposed to supplement and, in some cases, substitute for the conventional method for study of carbonaceous surfaces employing C(1s) spectra. The results achieved in the present study for the several materials listed are presented in Table 8-9. As we proceed with this comparative analysis, it should also prove useful to employ the results of Table 8-9 as a means to review the various individual methods.

2. C(1s) Binding Energy

It has been argued that the C(1s) binding energy of graphite materials is noticeably lower than that of polymeric hydrocarbons with a value of 284.2 eV predicted by some for 100% graphite.[83] It also has recently been argued that the C(1s) values for polymeric hydrocarbons exhibit regular shifting progressions from values as low as ~284.5 eV to values near 285.5 eV.[56] Although the progressions appear to be complex, they may be loosely described as increasing in binding energy with increasing "complexity" of the carbon chains.[56]

In view of these suggestions, therefore, the C(1s) values for all of the systems were analyzed. Unfortunately, charging (see below) tends to complicate the implementation of the aforementioned arguments. The C(1s) of the graphite-dominated systems are, however, unaffected by charging and peaked near 284.7 eV rather than 284.2 eV suggested by some other researchers. Following removal of charging, the C(1s) values of the polymeric hydrocarbons are generally larger than that of the graphites, but only marginally. The one feature of the C(1s) that does seem to distinguish hydrocarbons from graphite is the substantial reduction in line width exhibited by the latter.

3. Charging and Conductivity

A detailed discussion of the charging effect is presented elsewhere in this text.[21] In the present case, it is most important to note that a charging shift generally results in ESCA when the material being examined is a sufficiently poor conductor so that it does not maintain Fermi edge coupling with the spectrometer probe during photoelectron ejection. As a result, a positive charge builds on (or around) the surface of the material. This effect generally can be verified by applying an electron flood gun and looking for movement of the C(1s) peak. Such an effect obviously occurs for the hydrocarbon polymers in the present study, but similarly does not occur for graphite. In between these two extremes it is difficult to gauge how sensitive this parameter is to alteration. Thus, we see in Table 8-9 that all of the predominantly graphitic materials retain sufficient conductivity not to produce charging shifts, despite the fact that their surfaces may vary significantly in graphite content.

After a certain point, ozone treatment apparently has a definite effect on charging. Thus, ozonation of pure graphite results in ~4% retention of oxygen and no charging, as does the initial ozone treatment of carbon black, [O] = 8.9%, but ozonation of the latter to [O] = 13.4% results in a definite charging shift.

4. $\pi \leftrightarrow \pi*$ Transitions

A number of years ago Clark and Dilks[84] and others[1d,90] noted that many carbonaceous systems containing conjugated π bonds produced obvious satellite peaks approximately 8 eV upfield from the principal line in their C(1s) spectra. These particular satellites are apparently caused by the consumption of part of the kinetic energy of the process by a transition from the ground (π) state to the corresponding ($\pi*$) excited state. These effects are thus equivalent to certain band to band transitions for metals. The ready presence of these satellites seems for the present situation to be dependent upon the degree of aromaticity in the materials in question. It should be noted that previous studies have tried to correlate the size of the satellite shift with type of π conjugation. In the present case, our observations are restricted to simple presence or lack thereof. In this regard, one finds an indication of a very sensitive and exacting measurement, see Table 8-9. Thus, only the purported true graphite exhibits a substantial $\pi \rightarrow \pi^*$ transition. The microfiber and carbon paper that are supposed to have almost totally graphitic surfaces exhibit only modest $\pi \rightarrow \pi^*$, whereas carbon black does not exhibit any semblance of the satellite. Surprisingly, after extensive ozone treatment, this same carbon black exhibited a definite $\pi \rightarrow \pi^*$ satellite. The exact cause of the latter effect is unknown, but the satellite is more distinct for ozonized carbon black than for ozonized graphite.

5. Character of the Valence Bands

It has also been well documented that most carbonaceous systems produce distinct signatures in their valence bands.[1d] The valence band for graphite was first produced by the Siegbahn group[85] and subsequently described in detail by Shirley and colleagues.[86] In the latter study care was employed to produce not only a detailed graphite spectrum (from approximately the Fermi edge to 30 eV) but also to explain the various parts of this spectrum and compare it to similar materials (e.g., crystalline and amorphous graphite, diamond, etc). One of the critical features of this analysis centered on what was not detected. Thus, it was shown that the π density adjacent to the Fermi edge produced virtually no detectable peak structure compared to the various types of σ density further downfield.[86] Thus, an XPS-generated valence band structure similar to Figure 8-14 was produced. It is interesting that, in the McFeeley et al.[86] study, the character of this band structure changes very little between the crystalline and amorphous forms of graphite.[86,87] In the present case, the pure graphite, microfiber, and carbon black yield essentially the same (graphitic) valence band. For some reason the carbon paper that seemed to exhibit a detectable $\pi \rightarrow \pi^*$ loss transition produced a singular valence band spectrum (Figure 8-15) that suggests the extensive presence of hydrocarbon.

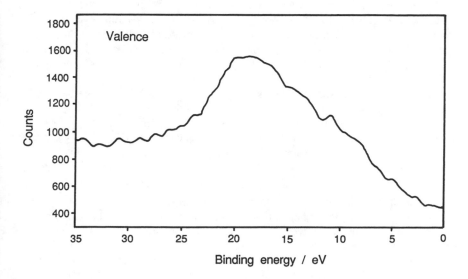

Figure 8-14 Typical XPS valence band pattern for (apparently) pure, polycrystalline graphite.

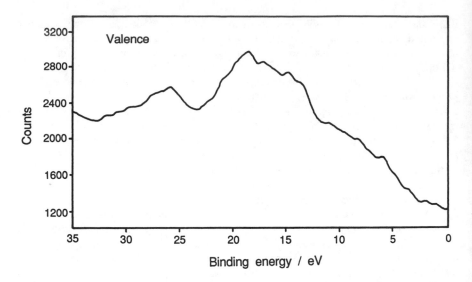

Figure 8-15 XPS valence band pattern for moderately oxidized (graphitic) carbon paper.

Polymeric hydrocarbons have also been shown to produce a set of valence bands distinctly different from graphite, and also distinguishable from one another.[1d] Pireaux et al.[91] have reviewed this subject in detail, and it has been extensively exploited by Sherwood in his numerous studies of different carbonaceous systems.[88] Once again the present study reproduces most of these features (see, for example, Figure 8-16). It is interesting to note that the conglomerate peak structures realized in the valence band of the carbon paper seem to suggest a mixture of graphite and one of these hydrocarbon polymers.

Ozone treatment of these systems results in changes in the XPS-detected valence bands, but the patterns do not demonstrate any consistency. Thus, only the "moderately disturbed" carbon black, [8.9% O], exhibits any significant degradation from the valence band of graphite, producing characteristics of a mixture of hydrocarbon polymer and graphite. Inexplicably, with further addition of oxygen, this alteration is substantially reversed and thus the carbon black with 13.4% O exhibits a near 100% graphite character in its valence band.

There are also some subtle and consistent changes in the XPS valence bands of the hydrocarbon polymers that are described later in this discussion.

All in all, one may say that, although the XPS-generated valence bands for carbonaceous systems seem to be good signposts for the degree and type of $C(2s^22p^2)$ hybridization exhibited by the dominant species present, at the resolution achieved in this study this feature is apparently not a very sensitive indicator of modest changes in these materials.

6. Carbon(1s) (Plasmon Dominated) Loss Spectra

As described in detail in Chapter 6, it has recently been determined that one may use the position of the relatively weak, broad, loss peak that arises between 20 and 30 eV upfield from the principal C(1s) line as a means to determine various changes in chemistry for a carbonaceous system. The approach centers on the concept that for reasonably good conductors the splitting between the principal line and its largest companion loss peak, E_L, is almost exclusively due to valence band, plasmon-type transitions, and even when the suggested conductivity is not present, that these plasmon effects still *dominate* the position of these loss lines for most oxides and similar systems.[21b]

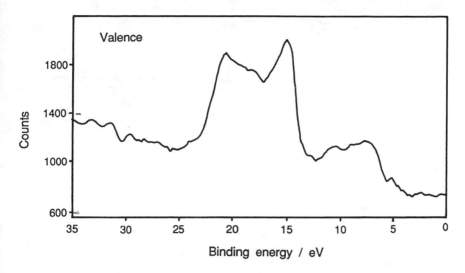

Figure 8-16 XPS valence band pattern for low-density polyethylene.

In the case of carbonaceous systems, we have determined that the loss splitting for true graphite should essentially duplicate the free electron plasmon calculated result of ~31.0 eV. Variations from this may be indications of a reduction in *graphitic character*.[78] As can be seen in Table 8-9 only our various samples designed as graphites approached 31 eV. The hydrocarbon polymers, on the other hand, exhibit dramatic reductions in their E_L values. Thus, the reduction in ΔE_L from graphite to HDPE (~9.0 eV) is expected, since a large portion of the valence density in the latter case is tied up in localized covalent bonding. Although all of these loss peaks are broad and difficult to fix (see, for example, Figures 6-26 through -28 in Chapter 6), the ΔE_L shifts are generally dramatic and readily suggest the suspected changes in bonding character. Perhaps the most expressive indicator of the latter effect occurs following ozone treatment. Interjection of oxygen into the graphite-dominated systems is found always to reduce the E_L value, suggesting a reduction in the degree of total delocalized conjugation, see Figure 8-17. On the other hand, interjection of $-\overset{\overset{\text{O}}{\|}}{\text{C}}-$ groups into hydrocarbon polymers is often found to increase the E_L value, suggesting an increasing in conjugation for these previously localized systems.

7. Oxidation of Select Carbonaceous Systems

Perhaps the most important aspect of the detected involvement of oxygen with these systems is that it provides two quite diverse signposts,[16] one about surfaces in general and the other about these carbonaceous systems in particular. First, it should be noted that the behavior of these materials reinforces the contention described in Chapters 2 and 3 that no air-exposed surface of any kind will be entirely devoid of some involvement with oxygen, at least of a chemisorbed nature. Second, it should be apparent that these particular carbonaceous surfaces are extremely reluctant to oxidize. Thus, even under the relatively harsh oxidizing conditions of ozonation, all of these surfaces (and subsurfaces) experience, at best, only modest oxidation.

This reluctance is readily suggested in the total % oxygen content, [O], detected for the individual systems before and following forced oxidation. Thus, those air-exposed materials that are prepared as pure, low-surface area, poor adsorbents, i.e., the graphite and hydrocarbon polymers, exhibit only slight chemical involvement with oxygen, whereas those systems that are prepared as adsorbents with large surface areas exhibit significant [O], even before oxidation, i.e., the case with the microfiber, carbon black, and carbon paper.

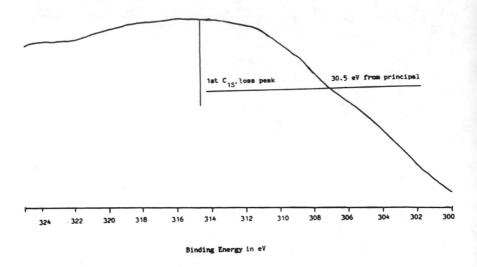

Figure 8-17 (First) plasmon-induced loss for clean graphite.

8. O(1s) Results

All of the O(1s) peaks realized in the present study were broad and appeared to be many faceted. This is, in part, a result of the limited resolution of the measurements, but in view of the relative narrow, singular structure of the C(1s) one can be sure that this breadth is primarily due to the creation of a chemical variety of oxygen-carbon bond units. In fact, in some cases it was possible roughly to partition the O(1s) through peak fitting and generate a manifold of subpeaks. In all of these cases the manifold suggested substantial presence of C–O–C (B.E. $\approx$ 532.4±0.4 eV), $-\overset{\text{O}}{\overset{\|}{\text{C}}}-$ (B.E. $\approx$ 531.8±0.4 eV), $-\overset{\text{O}}{\overset{\|}{\text{C}}}-$O (B.E. $\approx$ 533.3 ± 0.4 eV), and C–OH (B.E. $\approx$ 532.8 ± 0.4 eV) bonding units.[16,92] In view of the poor resolution, and problems with charging, it is impossible to suggest any quantification of these results, but the general evidence does suggest that the carboxyl and hydroxide (alcohol) units are formed somewhat superficially on the outer surface. On the other hand, the ozone treatment tends to introduce a significant presence of single-bonded oxygen, either of the ether or epoxide type.

It should be noted that the previous arguments are supported by a corresponding but even poorer-resolved analysis of the oxidized shoulder peaks on the C(1s).

9. Analysis of the O(2s) Region

Perhaps the single most important aspect of this study arises in an investigation of the corresponding O(2s) peak region. As with the case of the O(1s), the (much weaker) O(2s) system provides definite evidence of a manifold of peaks. When the charging is subtracted there is generally substantial proof of the presence of a principal peak structure near 27.5 eV which may, in turn, be a conglomerate of at least two lines of slightly higher and lower binding energy, see Figure 8-18. There is also some evidence of another peak structure at ~26 eV, and perhaps one even lower in binding energy, although the latter group is substantially masked by the leading edge of the valence band of the system. (The latter extends up to ~24 to 25 eV for both the graphitic and polymeric materials.)

The significance of these results is not the presence of this O(2s) peak system, but rather its position and relative size. As has also been demonstrated by Galuska et al.,[87] and later by Sherwood,[88] one finds that this peak manifold obviously increases in size with increased [O]. It has been found, however, that the relative size of the peak structure in the 27.5-eV area does not necessarily increase in proportion with the O(1s) peak. The

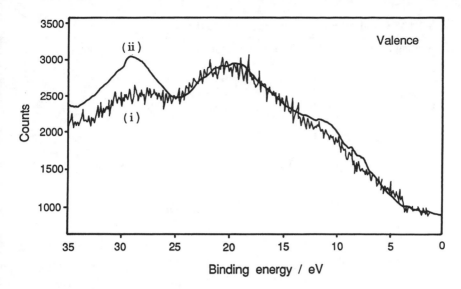

Figure 8-18 Changes of valence band of graphite (i) before and (ii) after ozonation. Note O(2s) dominated region.

latter is not entirely surprising in view of the possibility that some of the "O(2s)" may be "hiding" under the lead part of the valence band, but even considering this, one still finds a "short fall of density".

The reason for concentrating on these O(2s)-dominated features is that based upon comparisons with other oxide materials these peaks for carbonaceous systems are substantially out of position. By this we mean that, based upon the presence of O(1s) peaks for carbon-oxygen systems between ~533.5 and 531.5, one would not expect to find an O(2s) at ~27.5 eV. Based upon analogy with all other oxides, the latter would correspond to an O(1s) of ~535 eV! This suggests that the manifold of peaks maximizing at 27.5 eV must result from a mixture of atomic states, and do not reflect simply a core-like O(2s) (see below).

10. Comparisons with Other Oxides

For most of the oxide systems described in Chapter 2, it turns out that there is a fairly direct correspondence between the O(2s) binding energies and the above-mentioned O(1s)[12] as suggested in the results in Table 8-10. Note that, just as the binding energies of the O(1s) for these systems range from ~528.5 to ~533.5 eV, those for the corresponding O(2s) range from ~20 to ~25.5 eV. The reason for this is that these O(2s) electrons are essentially as much a part of the oxide *core* electron system as the O(1s) and as a result the useful (but approximate) "rule of chemical shifts", i.e., *with changes in chemistry all core photoelectron peaks of the same element shift essentially the same amount*, applies.[1]

In the case of C_xO_y *units, however, this rule generally does not seem to apply.* Thus, although it may be true that the peak that seems to be forming near 25.5 eV results from a –C–O unit with O(1s) of ~533.3 eV, one must be unsure of this. Thus, the latter species may correspond to SiO_2, but the presence of double bonds in the former make it unique, and even this case seems to be outside the previously stated "rule".

In view of this, it is our determination that the peak structure around 27.5 eV is totally unique to carbon-oxygen bonding. This seems to occur because of the extensively unique covalent bonding that exists in select carbon-oxygen systems. This bonding produces the dramatic overlap of carbon, valence, hybrid states with both atomic oxygen 2s and 2p

Table 8-10   **Comparison of key group IVA oxides**

A.   Binding energies (in eV $\pm$ 0.2) for select metal and metalloid oxides

	O(1s)	O(2s)
SiO_2	533.0	25.4
GeO_2	531.3	23.8
SnO_2	530.1	22.4
BaO	528.4	20.4

B.   Binding energies (in eV $\pm$ 0.4 except as indicated) for select carbon-oxygen systems

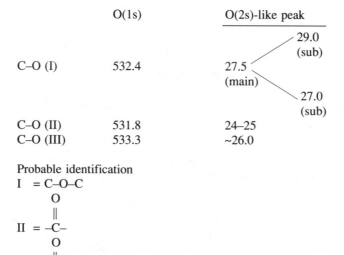

	O(1s)	O(2s)-like peak
C–O (I)	532.4	29.0 (sub) / 27.5 (main) \ 27.0 (sub)
C–O (II)	531.8	24–25
C–O (III)	533.3	~26.0

Probable identification

I   = C–O–C

II =

$$-\overset{\displaystyle O}{\underset{\displaystyle \|}{C}}-$$

III =

$$-\overset{\displaystyle O}{\underset{\displaystyle \|}{C}}-O \longleftrightarrow \text{related to } SiO_2$$

(valence) orbitals. Thus, the chemical interaction is primarily with the O(2p), but in the case of these systems the carbonaceous (hydrocarbon polymer or graphite) valence band has finite density above 20 eV in the region occupied by the O(2s). These two densities are of proper symmetry to interact with one another creating a "molecular orbital" mix. Based upon arguments from simple perturbation theory, a mixture of two such atomic states will create two new molecular states that generally will be pushed away from one another by the square of the amount of the perturbative interaction, divided by the separation of the two original states. Thus, the size of this interaction depends very much upon how close the original O(2s) and carbon valence bands are to one another.

The new partial valence state(s) (above 25 eV) will in cases of "maximum mix" apparently be pushed up to ~27.5 eV. Therefore, this peak should not be referred to as an O(2s) structure, *but rather a C–O valence band state that is substantially O(2s)*.[88]

Since the bonding chemistry is reflected in the scope of this perturbation, one may also suggest that the state of maximum shift should be indicative of the carbon-oxygen bond type of maximum covalency.[10] Therefore, it is argued that *the peaks in or around 27.5 eV are those due to carbon-oxygen single bond formation from either an ether or epoxy-type situation*. The fact that these peaks become most significant during ozone treatment is consistent with this assumption since, for example, that treatment of graphite is known to form graphite oxide in which the oxygen is found to bond between the graphite sheets

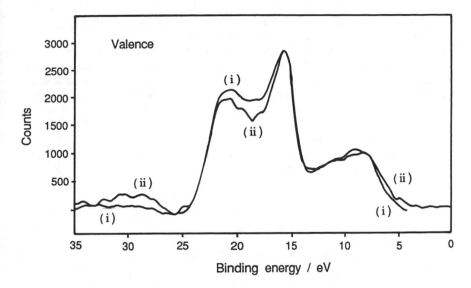

Figure 8-19 Valence band pattern of HDPE, (i) before and (ii) after oxidation. Note oxidation induced growth in O(2s) region and decrease in density near top of band.

in an epoxide-type bridge.

A more complete analysis should, of course, indicate the obvious differences between the ozonation induced –C–O–C units formed with graphite and those realized by the hydrocarbon polymers. This has not been done, but will be pursued in the future. Perhaps one of the most interesting select results to date relates to the question of what happens to the carbonaceous valence band density during this "mixing", see the papers of Sherwood et al. for an analysis of this and similar problems from the point of view of true oxides.[88] Obviously, one should expect the introduction of some finite band density due primarily to O(2p), but based upon analogy and preliminary observations this occurs in the region between ~6 and 12 eV, see Figure 8-19. In addition, one may ask if any change occurs in the leading part of the band, due to the mixing of part of it into the previous O(2s) region. An apparent realization of this effect is seen in Figure 8-19, where during the ozonation of HDPE as the density grows in the region about 27 eV the density near the top of the carbon valence band decreases. This may reflect the breaking of C–C σ bonds, and the insertion of single-bonded oxygen.

11. Comparative Summary of the Different Methods

The collective results suggest that it is impossible to ascribe to any one of these approaches the label of "best" method for analysis of the previously defined carbonaceous characteristics. The $\pi \rightarrow \pi^*$ approach may be too sensitive to be employed to try to spot partial decreases in aromaticity. The valence band method, on the other hand, may not be sensitive enough (note the poor detection of π bands). The plasmon loss technique seems to be the most sensitive registrar of the aromatic feature, but its resolution is very poor.

The fact that the valence band spectra of hydrocarbon polymers seems to be so sensitive to changes in their environments, including modest oxidation, may make this approach the method of choice for XPS analyses of polymer surface modification.

All things considered, at this point the recommended practice is to always include several of these methods during any study of closely related or modestly altered carbonaceous systems where the feature under investigation is the degree and type of change induced in the carbon-carbon and carbon-hydrogen bonds.

B. ESCA CHARACTERIZATION OF CARBON MICRO-FIBERS AS REINFORCEMENTS FOR VARIOUS EPOXY RESINS[93]

Graphite fibers, which possess high degrees of preferred orientation, consequently, high strength and high modulus, are widely used in the fabrication of high-performance resin matrix composites. However, large conventional fibers complicate the continuous fabrication of the strength and stiffness needed for reinforced-polymer composites. In contrast, shorter, micro-fibers should offer ease of production of the above-mentioned polymer composites via extrusion, injection modeling, and forging techniques.

Catalytically grown carbon micro-fibers (CMF) are produced when hydrocarbons are passed over certain heated metal surfaces. The material consists of a duplex structure, a graphitic sheath surrounding a more disordered core and the ratio of these two components can be controlled by the choice of the catalytic entity and the hydrocarbon source. Furthermore, the stacking arrangement of the graphite platelets constituting the skin is also a function of the metal employed as the generating catalyst. Numerous studies[94] have shown that the ferro-magnetic metals, iron, cobalt, and nickel are the most active catalysts for the formation of this type of carbon.

A suggested model for the formation of CMF was proposed by Baker and co-workers[95] which depends on diffusion of carbon through the metal catalyst particle from the hotter leading face in which hydrocarbon decomposition occurs to the cooler trailing faces at which carbon is deposited from solution. Growth ceases when the leading face is covered with carbon built up as a consequence of overall rate control by the carbon diffusion process. In this model the catalyst particle is carried away from the surface and remains located at the tip of the growing fiber. It has been found that relatively large amounts of CMF can be produced from the decomposition of ethylene catalyzed by copper:nickel alloys, the amount and conformation in which they are produced being dependent on both the temperature and catalyst composition.[96]

Previous studies have tended to focus on inhibiting the growth of this form of carbon since it exerts a number of deleterious features when formed in hydrocarbon conversion processes. Characterization of such deposits, however, suggests that CMF may possess some unique attributes for use in a number of areas and this aspect has provided the impetus to examine properties of the material which are important for their potential application in reinforcements of various polymeric matrices. Recent discoveries with ESCA, using X-ray photoelectron induced loss spectra (ESCALOSS), have been shown to be particularly revealing in questions related to the graphitic character of carbon materials,[78] and therefore it was decided to utilize this and the C–O valence band spectrum approach in the present case, along with conventional ESCA,[1] to evaluate the CMF and related surfaces both before and following select reactions.

One of the eventual objectives of this study is to determine the influence that the various surface conditions of the finely divided carbon fillers, filler loading levels, and types of curing agents or catalysts may have upon the curing characteristics of epoxy resins.

Carbon black and CMF fillers seem to behave quite differently in the epoxy-curing reaction, suggesting different surface chemistries for these two carbon systems and a possible utility for ESCA analysis. As mentioned earlier, conventional ESCA has substantial difficulties in the analysis of many carbon compounds, because carbon groups such as C–H, C–C, C=C, C≡C, etc. cannot be easily differentiated by their principal [C(1s)] peak chemical shifts.[1d,56] As a result, the C(1s) loss method, labeled as ESCALOSS,[21b,78] was included in the present analysis.

C(1s) loss results, Chapter 6 suggests that the micro-fiber surfaces are more graphitic than the carbon black. On the other hand, following the ozone treatment, the principal C(1s) peak of carbon black exhibits a larger shoulder on its left side resulting from oxidized chemical states of carbon materials, while for CMF, although this effect is significant, it is less prominent, apparently due to the more stable nature of its graphitic

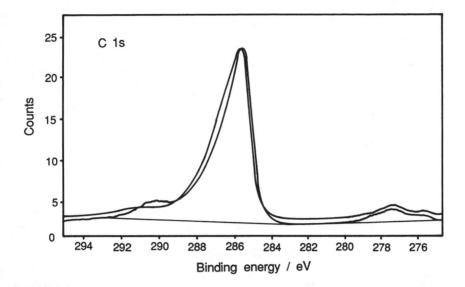

Figure 8-20 C(1s) spectra for several resin filler systems (microfibers) exhibiting noticeable C=O shoulder peaks ~4eV up field.

surface. The extent of this shoulder effect is thus related to the preferential formation of carbon-oxygen functionalities. This, we feel, is a reflection of the greater reactivity that generally exists on the surface of carbon black compared to CMF.

After surface treatment of carbon fillers with ammonia (e.g., immersion for 1 h), there is little change in the carbon and nitrogen ESCA spectra, indicating that very little ammonia was adsorbed on the surfaces of both carbon species. To test the extent of this property, pyridine was used as an alternative amine reactant. As an aromatic compound, pyridine is supposed to be adsorbed more easily on most graphitic carbon surfaces. Based on the N(1s) ESCA spectrum (not shown), we find a significant increase in nitrogen content compared to the background on the carbonaceous surfaces, suggesting substantial nitrogen adsorption on these surfaces. The chemistry indicated by the C(1s) and N(1s) binding energies for these species suggests the appearance of a new N(1s) peak after the pyridine treatment of both carbon fillers. This corresponds to an oxidized chemical state.

As mentioned above, ozone treatment of these materials produced a significant shoulder on the left side of the C(1s) principal peak for both carbon fillers (Figure 8-20). From the binding energies of the C(1s) and O(1s) principal peaks, one can determine that there were more C–O single bonds formed in the treatment than C=O double bonds. In the valence band region, a peak also emerges at 28 eV [apparently due to the previously described carbon valence band-O(2s) hybrid state] thus providing further evidence for the occurrence of oxidation.

There are also some interesting features in the development in the C(1s) loss spectra for these oxidized systems. For example, as mentioned above, oxidation of graphite surfaces generally shifts the C(1s) loss peaks to smaller splittings from their principal peak (see Figure 8-21), whereas oxidation of polymeric carbon surfaces shifts their C(1s) loss peak further away from the principal peak. This may suggest the disruption of the double bond conjugated electron system in the first case and an increased conjugation in the latter case. The loss splittings of the two extremes thus approach each other after significant oxidation. In the present case, we note that the C(1s) ESCALOSS of ozone-treated CMF exhibit decreases in their C(1s) loss splittings. This probably suggests some

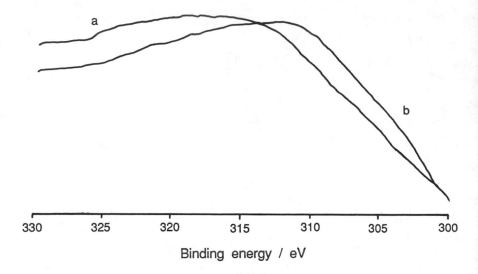

Binding energy / eV

Figure 8-21 Shift in the position of the C(1s) loss peak induced by the oxidation of graphitic-type carbons.

alteration of the large conjugating system of this material, induced through the creation of carbon-oxygen functionalities onto the graphite surface. This ozone-treated carbon micro-fiber-epoxy composite has a lower maximum curing temperature than the sample without treatment, suggesting that a relatively stable, delocalized graphite surface was replaced by a more active one after formation of some oxide functionalities, which may be helpful in initiating the curing reaction. These effects were reversed for the carbon black composites. The speculation is that the reactive carbon black surface, when attacked by O_3, may have lost some special kinds of functionalities which are relevant for the curing reactions, particularly when accompanied by the formation of certain new types of oxidized species. These speculations also seem to be supported by the O(1s) spectra of carbon black. An obvious narrowing of O(1s) principal peak can be observed after the ozone treatment of carbon black (relative to similarly treated CMF surfaces) indicating possible destruction of some of the diverse oxide species for the carbon black system, while the total oxygen content was increasing (Figure 8-22).

VI. ESCA AND CATALYSIS

A. ESCA STUDIES OF ZEOLITES AND CLAYS
1. Zeolites
a. Introduction

Continuing ESCA studies of zeolites and clay systems[1b,34,40,46,97] are aimed at a more complete understanding of the surface properties of zeolite catalysts and related techno-logical systems. The present program features detailed studies of the special influence of multivalent (vs. monovalent) cations, the special properties of the (III-V) nonzeolite molecular sieves, e.g., $AlPO^4$-δ[97b] and $GaPO_4$[97c], the states of any Bronsted and Lewis acidity induced during diammoniation (and subsequent dehydration), and also the eventual evolution of certain key zeolites to ultrastability.[98] All of these features have been extensively investigated for many years in the *bulk* regions of these materials using such methods as infrared (IR)[99] and Raman absorption spectroscopy,[100] X-ray diffraction,[101] dielectric properties and electrical conductivity,[102] thermal adsorption and desorption characteristics,[102] ESR,[103] and, more recently, reflective IR[104] and magic angle spinning

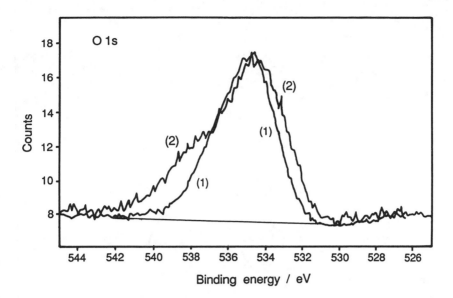

Figure 8-22 Change in the O(1s) spectrum due to ozone-induced formation of singular type of oxide.

(MAS)-NMR.[105] *Corresponding forms of analysis are now being reported of the outer surface regions (see below for definition) of these materials.*

A few years ago our research group discovered that, contrary to the prevailing view, the surfaces of zeolite systems with different framework structures and/or different Si/Al ratios (e.g., NaA, NaX, NaY, Mordenite, ZSM-5, etc.) exhibit progressive variations in the peak positions of their X-ray photoelectron (ESCA) core level spectra (see Figure 2-8 and 8-23 and Table 2-5).[34,55,106,107] Subsequently, we determined that related dissimilarities also exist in the valence band spectra of these materials, see Chapter 7.[40] Consistent explanations have now been provided for these unique spectra features,[46] and they have been utilized to monitor the selective surface presence of different zeolites, and impurities on zeolites,[55] and also to help describe the useful solid-state acidity of these materials, see Chapter 2.[46] These ESCA-based zeolite studies have also been integrated with those of numerous other simple (CaO) and complex ($CaSiO_4$) oxide systems to provide a novel hypothesis for the gradations in ionic/covalent bonding experienced by these materials, see Chapter 2 and earlier sections of this chapter.[22] Many of these discoveries have been confirmed by others[108] and they are now proving useful in numerous investigations of many applied zeolite problems,[109] *but unfortunately their utility has often proved limited because of the lack of surface information in certain key areas.*

One central feature permeates all of the present deficiencies in surface zeolite analysis. *No successful method has yet been developed to determine and control the state of the hydration (and/or hydroxylation) at the outer surface of zeolites.* This attribute has been shown to be critical to the bulk behavior of zeolites during changes in cations, acidity, and stability.[98-102] The effects of variations of the *surface* hydration of different zeolites, therefore, should play a central role in future research. Investigations of this problem are extremely difficult because the primary surface analysis method to be employed, ESCA, only indirectly registers the presence of hydrogen. In order to circumvent this problem, it will be necessary to be able to produce *controlled, unique modifications* in the surface regions of select zeolite systems. This means not only modifying zeolites in the conventional (bulk) sense, but also causing "selective" attenuations of the surfaces and then

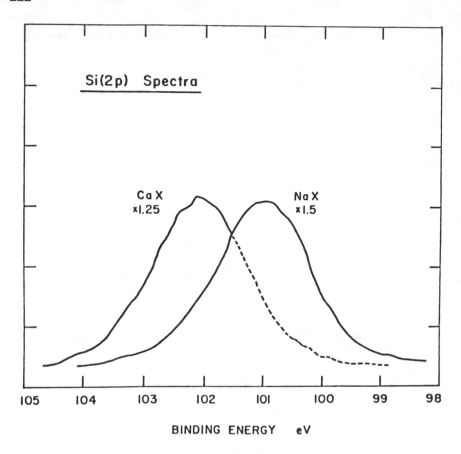

Figure 8-23 Select examples of core level shifts in zeolites created by substitution of Ca^{++} for Na^+ in the X system.

confining these materials in a manner to assure that their surface integrity is maintained.

Many of the features of zeolites have been successfully analyzed in their bulk regions, and some have suggested that zeolites are "all surface", thus suggesting that separate surface analyses are in some cases redundant. The answer to this challenge lies in the results obtained, particularly those achieved on various applied systems that demonstrate that these materials have important *distinct* surface and bulk regions. For example, it is not generally well known that in most cases zeolites are not employed in commercial processes by themselves. Thus, due to their somewhat fragile nature, in most instances zeolites employed in catalysis and separations are combined with other materials (e.g., clays, aluminas, etc.) that function as matrices or *binders*. When most of the resulting composite systems are produced by conventional means, and subsequently investigated by standard bulk analysis methods, e.g., IR or MAS-NMR, they appear to be configured as expected, i.e., with zeolite particles apparently bound together by the matrix material. When examined by ESCA, however, a quite different picture arises. Most of the zeolite particles are found to be entirely coated by a thin (often less than 100 Å) film of binder species.[110] The thickness of this film is found to vary markedly with the method of mixing and the type of binder employed, but it is always found to be present.[110] The detection of these surface-oriented morphological features was made possible by the development of our previously mentioned ESCA methods for differentiation of different zeolites and related materials.[1b,55] The importance of these observations, however, is that this coating often affects the absorptive and catalytic performance of the composite system in a manner not detected by any of the bulk analysis methods. (1) Thus, for example, in the case of a system employing CaY to separate different types of sugars, kaolinite clay employed as a binder was detected by ESCA to coat the zeolite, and it was determined that the poor initial (separation) performance of this system was due to this coating, rather than other suspected effects, such as possible residual Na$^+$ in the zeolite.[110] Further, following use, it was found by both bulk and surface analysis that some of the Ca^{++} was selectively removed from the zeolite, but the system continued to function. This (bulk) perplexing result was explained by ESCA analysis which showed that the extracted Ca^{++} ions were ion exchanged into the clay coating. Unfortunately, the detailed nature of the surface behavior of the Ca^{++} (and other multivalent cations) is not yet understood. (2) In another example involving binders, Al_2O_3, with and without added $AlPO_4$, was employed to bind Ga exchanged ZSM-5 zeolites, intended to catalyze select dehydrogenation reactions.[110] Once again, the binder was found to coat extensively the zeolite and as a result the selectivity and activity of the total system was found (by the ESCA and not by bulk methods) to depend very much on whether the Ga^{+3} was exchanged before, or after, the addition of the binder.

Many other examples of this "surface selectivity" exist. For example, using surface-oriented beams of neutral atomic emissions (FAB), Dwyer et al.[111] suggested that much of the aluminum removed from a Y zeolite during ultrastabilization was, in fact, "deposited" on the surface of the system. We have also begun analyzing this process with ESCA (see below) and detected not only the amount but some of the chemical status of the nonzeolitic, surface-oriented species.[1b,110] In all of the cases mentioned above, the degree of selective surface hydration should play a major, but, unfortunately, undetected role.

b. A Preliminary XPS Study of the Effect of Multivalent Cations

More than a decade ago we first noted that the ESCA peak positions of the Si, Al, and O spectra for a typical tetrahedrally oriented crystalline, $(SiO_2)_x \cdot (AlO_2^-)_y$, zeolite lattice seem to depend upon the ratio of x to y in that lattice, Figure 2-8 and Table 2-5.[43,55,106] This is apparently true because of the enhanced ionicity of the Al–O bond in the (tetrahedral) AlO_2^- subunits, compared to the more modest ionicity of the corresponding Si–O bond in the silica-like subunits.[22,46] Similar arguments have been formulated for related oxides.[24,36,89,112] In view of the progressive shifts in the ESCA peak positions detected as the Si/Al ratio of zeolites is increased from ≈ 1 (the A system) to several hundred (ZSM-5[113] and/or silicalite[114]), Table 2-5, it has been postulated that there is a correspondence between the x/y ratio and the types of bonds formed, and through this a direct relationship to the degree of Bronsted acidity.[115–117] In this manner, as x/y increases the dominant covalency of the Si–O bond causes the ionicity of the Al–O– bond system to increase (i.e., the Al binding energies grow).[22,89,118] As a result, the relative ionicity of the cations shared at the Si–O–Al bridge (that also neutralizes the AlO_2^- unit) are shown to increase with increasing x/y,[46] which, when the cation is H$^+$, corresponds to the progressions found in relative acidity.[46,119]

During the course of these studies a variety of preliminary XPS examinations were also conducted of several zeolite systems in which multivalent cations (e.g., Ca^{++}) were substituted for the typical monovalent forms (e.g., Na$^+$). As a result, core level peak distributions such as those presented in Table 2-5 and Figure 8-23 have been generated. It has become apparent that the change from Na$^+$ to Ca^{++} (and other multivalent cations) within a particular zeolite system also produces progressive, positive shifts in the resulting binding energies of all of the elemental peaks. Displayed in Table 8-11 is a select group of the line widths exhibited by many of the spectra listed in that table. In this case one can see that exchanging a multivalent cation (e.g., Ca^{++}) for a monovalent one (e.g., Na$^+$) not only produces a general positive shift in the positions of all the

224

Table 8-11 **Line widths for zeolite peaks listed in Table 2-5 in eV ± 0.15 eV**

	ReY	Ca(Na)Y	KY	NaY	CaX	NaX	Na(Ca)A	K(?)A	NaA
Si(2p)	1.84	1.84	1.71	1.68	1.82	1.77	2.00	1.84	1.72
Al(2p)	1.88	1.71	1.50	1.55	1.86	1.60	1.90	1.78	1.56
O(1s)	2.00	2.05	2.08	2.05	2.08	1.65	1.88	2.10	1.75
Na(2s)	—	—	—	1.65	—	1.67	1.67	1.9[a]	1.65
Na(1s)	—	—	—	1.95	—	1.95	1.90	—	1.92
C(1s)	2.0[a]	1.8[a]	1.50	1.50	—	1.52	1.66	—	1.44
Ca(2p$_{3/2}$)	—	1.92	—	—	1.94	—	1.90	—	—
Mg(122)	—	—	—	—	—	—	—	—	—
K(2p$_{3/2}$)	—	—	1.70	—	—	—	—	—	—

[a] Precision reduced to ± 0.3 eV.

elemental peaks, it also seems to broaden each peak manifold. Although previously presented,[34] little effort has been made to try to explain these shifting and broadening patterns.

In order to try to understand the effects induced in a zeolite system as a result of the aforementioned exchange (typified by Na$^+$ → Ca^{++}), it should be informative to compare the diverse features that seem to lead to the previously described general increase in binding energy of the elemental core level peaks. These effects are typified in Figure 8-24. Apparently, if one simply exchanges a multicovalent cation into a zeolite system (while keeping the x/y ratio and lattice structure the same), this will also induce an effect that simulates a related increase in ionicity of the Al–O–M unit. Insertion of trivalent cations (e.g., RE^{+3}) may be even more effective at this bond "alteration" than the divalent variety (e.g., Ca^{++}), see Table 2-5. Several possible explanations no doubt exist for these changes, but it is intriguing that the same type of ESCA shifts that are induced by increases in [Si/Al] also occur for a particular [Si/Al] when the corresponding cation is changed from monovalent to multivalent. Both of these types of compositional changes also generally result in increased catalytic (acidic) character.

c. Preliminary Study of Ultrastable Systems

There is some indication that the ultrastable Y system,[120] if properly constructed, may be utilized in place of REHY as a fluid cracking catalyst, particularly because of the very low coke formation often exhibited by the former. A preliminary ESCA study has therefore been accomplished on a few representative zeolite systems during the purported evolution to ultrastability.[110] Select results are reported below.

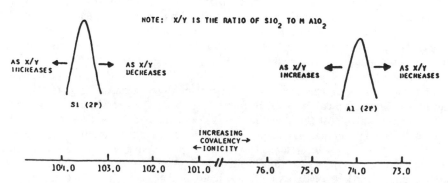

Figure 8-24 Generalized argument for the core level shifting pattern induced in zeolites (and other framework silicates) due to change in Si/Al.

Table 8-12 **Faujasite [Si/Al] ratios using various techniques**

	$[Si/Al]_S$	$[Si/Al]_B$
NaY	2.7[a]	—
NaNH$_4$Y	3.2	—
Rare earth exchanged Y	3.7	—
LA-1	1.7	4.5
LA-2	1.55	7.6
LZ-20S (12-h steam)	1.21	—

Note: B ≡ NMR determination, s ≡ ESCA determination of surface.

[a] Conventional formulation produces [Si/Al] = 2.5 for Y faujasite.

All of the zeolites were initially conventional NaY, faujasite systems. These were subsequently modified to produce the systems displayed in Figure 8-25.[10] Table 8-12 includes the detected surface Si/Al ratio, $[Si/Al]_s$ and, for comparison, bulk Si/Al ratios, $[Si/Al]_B$, obtained by NMR. The surface $[Si/Al]_s$ values reported in Table 8-12 pertain to the *total* surface aluminum present. Unlike the NMR, the ESCA is unable to quantitatively distinguish, with ease, between tetrahedral and octahedral types of aluminum in these systems.[46] It is apparent from the ESCA results, however, that the chemistry of the surface aluminum is changing with ultrastability treatment such that these surface "depositions" are primarily $Al(OH)_x^{+x}$, rather than alumina residue.[111,121]

Following the steps reputed to induce ultrastability, e.g., steam treatment, the binding energies of the Si and Al increase, Table 8-13, but in this case we attribute this to the extraction of the aluminum from the lattice, which then becomes stabilized in a high silicon (high binding energy) state. The extracted Al is, at least partially, converted to $Al(OH)_x^{+x}$ species which also have relatively large binding energies for Al(2p), see Table 8-13.[24,110]

As already mentioned, the valence band structures also provide a novel means for analysis of zeolite systems,[24,40] changing noticeably from system to system and with purity, see Figure 7-22. The NH$_4$Y system produces a valence band spectrum (Figure 8-26) that largely duplicates the principal features of NaY (Figure 7-22), except for the expected slight shift to lower binding energies.[24,40,118] In the case of the ultrastable systems [LA-1 and LA-2] the extraction of AlO_2^- is reflected in the valence band results by the significant central band indenture for LA-1, Figure 8-26, i.e., where the peak structure due to AlO_2^- arises in NaY.[24,40] LA-2, on the other hand, exhibits a new set of subband peaks attributable to Al-O bonding. The fact that the latter are better resolved for LA-2S, Figure 8-26, suggests that these new subbands represent the contribution from the deposited $Al(OH)_x^{+x}$ which apparently reaches its maximum for LA-2S.

A series of tentative hypotheses is presented in Table 8-14 for these ESCA studies of different ultrastable and cation-exchanged Y-zeolite systems. Once again the results must be viewed as preliminary, with far greater control needed to reach any firm conclusions.

2. ESCA Studies of Select Clays[97]
a. Preliminary Analysis

In order to test the feasibility of conducting meaningful ESCA-based studies of all types of silicate systems, we have also employed the previously described innovations in ESCA spectroscopy to examine several representative clay cases, including detailed investigations of several kaolinites and montmorillonites (formed from relatively pure smectite clay minerals). (Versions of these clays also have been analyzed with ESCA in mixed systems following their use as binders with several zeolites. The ESCA differences described below readily permit identification of the separate components.)

Flow scheme of faujasite modification

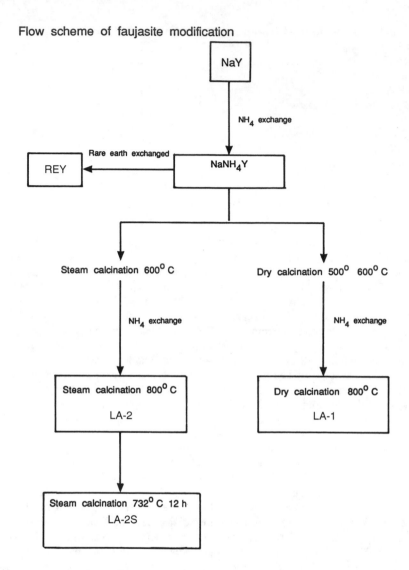

Figure 8-25 Flow chart for the production of several forms of ultrastabilized Y zeolites by different processes.

Any ESCA study of materials, such as these (purportedly pure) clays, must begin with careful qualitative analyses and (relative) quantification. With the exception of the modest impurity features mentioned below, the two clay systems being analyzed were found to be, as advertised, i.e., kaolinite $[Al_2(OH)_4Si_2O_5]$ and montmorillonite, $M_{x/n}^{n+}$ $(Al_{2-x}Mg_x) Si_4O_{10}(OH)_2$, where $M_{x/n}^{n+}$ represents the charge balancing cation, typified by Ca^{+2}, for many of the systems examined. The detected impurities were the expected surface presence of adventitious carbon, and between 0.1 and 0.5% of the following cations: kaolinite (Mg^{++}, Ca^{++}, and Na^+) and montmorillonite (Fe^{+3} and Na^+). Following ion etching, a modest, but persistent, amount of Ar (< 1.0%) was also found trapped in the clay lattices. It should be noted that in the case of montmorillonite we are hereafter more concerned with the behavior of the principal "sheet" constituents (e.g., Si and Al) than the less-specific counter cation(s), $M_{x/n}^{n+}$, or the rather sparse Mg.

Table 8-13 **Adjusted binding energies and line widths (in parentheses) for selected faujasite materials**

	NH4Y	NaY	ReY	LA-1	LA-2	LA-2S (12 h steamed)
Al(2p)	73.8	74.0	74.4	74.5	74.6	74.5
	(1.9)	(1.6)	(2.0)	(2.2)	(2.0)	(2.0)
Si(2p)	101.9	102.3	102.7	102.2	102.4	102.5
	(1.9)	(1.7)	(2.0)	(2.0)	(2.1)	(2.1)
O(1s)	530.9	531.5	531.9	532.0	532.1	532.1
	(2.3)	(2.0)	(2.2)	(2.7)	(2.4)	(2.4)
Na(1s)	1071.3	1071.8	1072.4	1073.0	1071.4	1073.5
	(2.0)	(2.0)	(2.3)	—	—	—
Valence band width	10.05	10.04	10.25	10.05	10.00	9.95

As often detected in the case of the many zeolites previously analyzed, both of these clay systems also exhibited *modest* excesses in their presputtered (surface) [Si/Al] ratios. These excesses seemed to dissipate with slight sputter etching.[1b] Further sputter etching, however, caused the expected selective removal of Al [creation of apparent $Al(OH)_x^{+x}$ species].[1b] In the case of the montmorillonite the [cation]/[Si] ratio also decreased with etching, whereas the [cation]/[Al] ratio remained fairly uniform.

b. Core Level Binding Energy Analysis

Because of their similar appearance, differentiation of the core level peaks for the clays is achieved using the resulting binding energies.[55,107] The first thing to note about

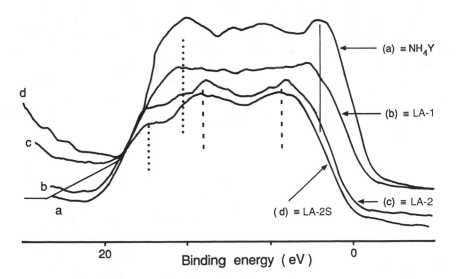

Figure 8-26 Valence band spectra resulting from ultrastabilization process, i.e., (a) NH₄Y, (b) dry treatment result, (c) steam calcined result, and (d) enhanced steam treatment. Note the shifts of the subband structure from those typical of Y-zeolite systems to new subband positions suggesting the formation of positively charged aluminum. Similar shifts are found in the core level spectra, e.g., Al(2p) shifts from 74.0 (Y zeolite) to 74.6 eV (Al⁺³).

Table 8-14 **Preliminary ESCA study of formation of ultrastability**

1. Ammonium treatment produces significant exchange of the NH_4^+ for Na^+, but retains some Na^+ in select sites that *may* be identified (in part) by the ESCA binding energy shifts.
2. A dry treatment to induce ultrastability (LA-1) seems to produce (at the surface) a system which at least mimics those features described in previous publications;[107] thus, the silicon part of the ESCA spectrum suggests (zeolitic) SiO_2, whereas the aluminum features depict mixed zeolitic Al and $Al(OH)_x^{+x}$.
3. Another technique to induce ultrastability, steam calcination, produces a system (LA-2) with a lower $[Si/Al]_s$ ratio than (LA-1); this suggests that the steam process is even more "successful" than the dry process at extracting aluminum from the zeolite lattice and depositing it on the surface [primarily as $Al(OH)_x^{+x}$ species]; detailed ESCA valence band analyses also suggest that all of these features seem to be enhanced by further steam treatment (LA-2S).

the binding energies realized on the outer (pre-ion-etched) surfaces of kaolinite and montmorillonite, Table 8-15, is that, just as is the case for the many zeolites examined, the binding energies of both the Si and Al core states increase with increasing [Si/Al] ratio. At this point, however, the similarity between the zeolites and clays ceases. The binding energies of the individual clays are all larger than those for zeolites with similar [Si/Al] ratios, i.e., compare the values in Tables 2-5 and 8-15 for montmorillonite with NaY and kaolinite with NaA. The reproducible nature of these "shifts" suggests that ESCA may be registering some type of chemical (?) difference between these systems. It should also be noted that the clays are also *obviously not* just mixtures of silica and aluminas.

In fact, the binding energies for each material reflects exactly the chemical nature of the subunits of which they are composed. Thus, both clays contain sheets of silica. The latter *alone* would have a Si(2p) binding energy of ~103.5eV, see Table 2-5. When united with layers of alumina and aluminate, a reduction in the binding energies of the individual silicate subunits ensues (see our numerous publications on this subject).[1b,40,46] As described extensively in one of the author's most recent publications, this occurs because the more ionic Al–O bond apparently forces the species on the other side of some of the oxygens (the silicon) to be bonded more covalently.[1b,22] Thus, the binding energy of the silicon must decrease (from silica) and will continue down as the [Si/Al] ratio decreases, i.e., Si(2p) kaolinite <Si(2p) montmorillonite, etc.[1b,22,97] However, because the aluminum containing units in the clays are in separate octahedral layers, rather than part of the tetrahedrally bonded silicate layer, *the relative effect felt by the silicons (in terms of*

Table 8-15 **Representative clay core level binding energies and line widths in electronvolts and selected valence band widths**

	Montmorillonite	Kaolinite
Si(2p)	102.75	102.45
	(2.05)	(2.28)
Al(2p)	74.8	74.3
	(1.98)	(2.21)
O(1s)	532.0	531.5
	(2.36)	(2.45)
Valence band width at half maximum	10.2	10.2

Note: C (1s) = 284.6 eV.

reduction in binding energy) is apparently less than that experienced in the various zeolites with similar [Si/Al] ratios!

On the other hand, the binding energy of the Al(2p) for the clays form a relatively unique set. (One must recall that we have already determined that the Al(2p) binding energy position of any aluminum is less susceptible to shifts due to the presence of silicon than is the reverse case.[34]) Thus, it appears that the Al(2p) for kaolinite reflects the presence of octahedrally bonded aluminum, but not in the form of Al_2O_3, [Al(2p) ~73.9 eV], rather the presence of near Gibbsite ($AlOH_3$) is detected, i.e., the Al(2p) is found above 74 eV.[53] On the other hand, the aluminums also reflect their increased ionicity due to their attachment to the sheets of covalent Si–O bonds. Thus, the resulting Al(2p) binding energy for kaolinite is slightly in excess of 74 eV, but not as positive as the Al(2p) for $Al(OH)_2{}^+$, see below. In the case of the montmorillonite the Al(2p) of 74.8 eV reflects the presence of more silicon (making each Al–O unit even more ionic[46]) and also the apparent conversion of some of the octahedral aluminum to $Al(OH)_2{}^+$-type cation units.[122] The latter constitutes one of the most ionic (positive) aluminum cases of all, and generally seems to exhibit Al(2p) above 74.7 eV.[22]

c. Valence Band Analyses

The valence bands for both kaolinite and montmorillonite, Figure 8-27, are quite different from any of the zeolite valence bands. In the case of the clays they are both very broad, see Table 8-15, almost the same width as silica. The definition of the subband features is generally poorer for the clays. Despite this, there are a number of interesting and potentially revealing sub-band features.

Two particular aspects seem to be reflected in these results. First, the general breadth of the clay bands (particularly that of the kaolinite) suggests that the amalgamation of the silica and aluminate systems found in the case of most zeolites and extensively described in Reference 40 [a result where two (or more) unique systems combine to form *one (amalgamated) tetrahedrally coordinated unit*[40]] is absent in the clays. Thus, the clay valence bands seem to reflect the structural presence of separate tetrahedral (silicate) and octahedral (aluminate) layers! Second, the individual subfeatures of the bands may, therefore, suggest the presence of the components in these substructures. In this regard the sharp peak (I) on the far side of the montmorillonite band seems to replicate a similar feature exhibited by silica, Figure 7-11, whereas the protrusions in the middle of the clay bands (II and II′) suggest (when compared to the band structure of pure aluminas, Figure 7-12) some type of octahedrally oriented Al–O system, perhaps in the case of montmorillonite (II) integrated with $Al(OH)_2{}^+$, which should be slightly higher in binding energy than the Gibbsite-only part in kaolinite (II′). Other features of these clay bands, such as the substantial breadth of the "N" parts [reflecting primarily the O(2p) density due to the nonbonding electrons], may be indicative of the sheet-like nature of the silicate layers.

d. Ion Beam Alteration of the Clays

Following treatment with beams of 1 KeV Ar+, both kaolinite and montmorillonite exhibited spectral alterations. As expected, all of the core level line widths broadened. Further the Al(2p) of all systems shift to ~74.6 eV reflecting the production of $Al(OH)_2{}^+$ species.[110] Just as in the case of the zeolites, the features of the valence bands of these clays were distorted by ion alteration to reflect the creation of possible Al(II)O.[24] (Note in Figure 8-28 the "condensing" of the density of states to a similar humped peak in the center of each band.) The balance of the post ion treatment valence band spectra should reflect, therefore, two features, i.e., the presence of the silica that should remain following extraction and alteration of the octahedral aluminate units plus the steady-state retention of some unaltered clay. A rough rendition of this scenario emerges if one compares an overlay of the valence band spectra for silica and ion altered alumina, Figures 7-13 and 7-14, with the ion altered clay bands in Figure 8-28.

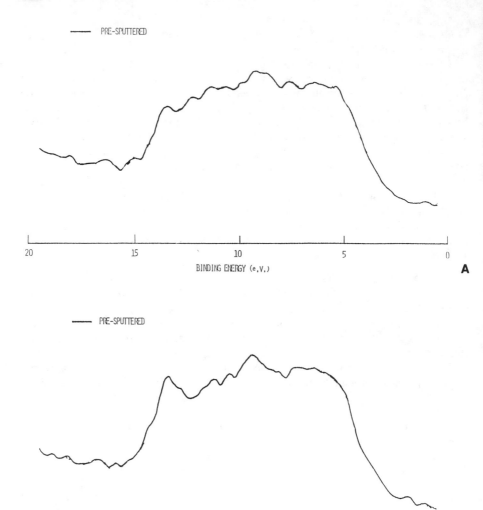

Figure 8-27 Valence band spectra for select (A) kaolinite and (B) montmorillonite clays.

The impact of the latter results comes into focus when one notes that similar conversions of ESCA core level and valence band features have been monitored for a variety of aluminas, zeolites, and clays in practical use situations following various forms of thermal abuse.[110]

e. Consideration of the Resulting Line Widths

A variety of suppositions are raised of the nature of these clay systems based upon the resulting core level and valence band shifts. It should be apparent that these same properties should be reflected in the resulting line widths at half maximum. In general, we are describing a situation in the case of most clays where species that begin as relatively pure (tetrahedrally bonded) silicas and (octahedrally bonded) aluminas are

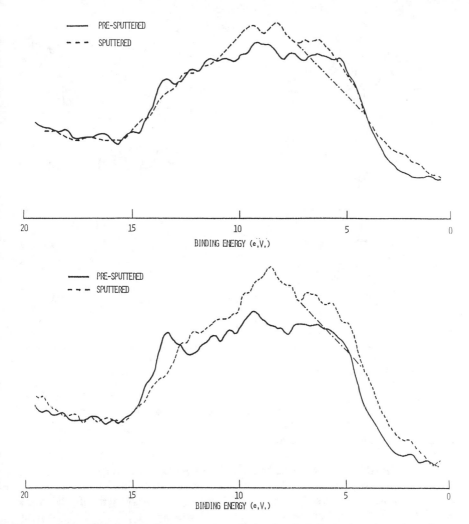

Figure 8-28 Representative examples of Ar⁺ ion-induced alterations in valence bands of select clay systems.

unified in sheets that essentially retain their integrity inside their respective sub-sheets, but lose that integrity across the (apical) interfacial region where the sheets unite. This suggests that the ESCA that detects these bonding features should reflect this combination of integrity-lack of integrity as a "perturbation" of the features of the pure subunits. This type of material behavior is generally reflected in ESCA spectra as a broadening of the resulting core lines.

This is indeed what we detect in the present situation. Thus, in the case of the clays under consideration the resulting Si(2p) and Al(2p) lines are relatively broad, Table 8-15, particularly compared to those produced by related zeolite systems, Table 8-11. The latter thus seem to reflect the aforementioned amalgamation into singular tetrahedral silica-aluminate lattices, whereas the clay results are indicative of interconnected sheets. This feature is particularly reflected in the O(1s) lines, where the diversity of oxygen cases for the clay compared to the zeolites is seen in the line broadening of the former.

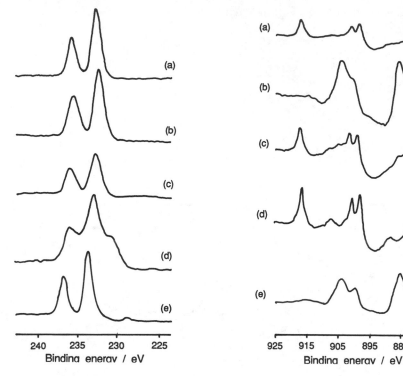

Figure 8-29 (A) $M_o(3d)$ photoemission spectra of β-$Ce_2Mo_4O_{15}$ (a), γ-$Ce_2Mo_3O_{13}$ (b), $Ce_8Mo_{12}O_{49}$ (c), α-$Ce_2Mo_4O_{15}$ after sputtering for 2 min (d), and MoO_3 on Mo foil (e). (B) $Ce(3d)$ photoemission spectra of β-$Ce_2Mo_3O_{13}$ (a), α-$Ce_2Mo_4O_{15}$ (b), $Ce_8Mo_{12}O_{49}$ (c), CeO_2 (d), and Ce_2O_3 on Ce foil (e).

B. ESCA STUDIES OF ACRYLONITRILE (ACN) CATALYSTS

This statement summarizes the details of an extensive ESCA study of a simulation of a possible acrylonitrile catalyst system (ACN). This is included to exemplify the type of detailed analysis that may be achieved of catalytic systems using ESCA. The most common ACN catalyst system is a Bi, U, molybdate first developed by SOHIO.[123] The present system (consisting of complex Mo, Ce, Te, and Ni oxides on SiO_2) involves significant differences in composition and use. Numerous samples were examined in an HP ESCA, after the present ACN system had experienced nearly every aspect of (simulated) use, degradation, and regeneration. The ESCA results were extremely detailed, with both binding energy/line width and quantitative data extracted for every constituent of the systems. As a result of this analysis, many of the most important chemical features of this potential catalyst system have been established. In addition, the evidence is sufficient to permit a postulation of the major chemical entities involved and their possible reactions during use and regeneration. The following conclusions were reached during this study:

1. The chemical status of *the fresh ACN system* can be mapped by ESCA. In this regard, a complex array of mixed [Ce(III), Mo(VI)] compounds was discovered along with [Te(IV), Mo(VI)] compounds[124] (see Figure 8-29). Also, several simple oxides seem to be present (see the postulated results in Table 8-16).

2. *Simulated use of the ACN system* is characterized by the reduction of some of the Mo(VI) to Mo(V), the combination of the Ni with some Te to form a complex compound, and the apparent change in composition of the suggested (Te, Mo) compound.

3. The *initial degradation* of the ACN system finds further conversion of Mo(VI) to Mo(V) and the splitting of the complex (Ce, Mo) compounds into simple oxides. Some of the Ni and Te seem to separate again with the production and loss by evaporation of some Te°. The most significant feature is that the metal dopants (in their simpler form) are effectively "rising" to the surface and there is a marked enhancement in (surface) dopant concentration, see for example, Figure 8-30.

4. Continued simulated use of the ACN system eventually results *in a spent catalyst*. This system is characterized by maximum Mo(V) [~4/1 Mo(V)/Mo(VI)] and total separation of (Ce, Mo) to simple oxides. The Ce during all the processes described is almost exclusively Ce(III). Complete degradation of the catalyst also seems to produce a significant amount of Te^{-2}, perhaps as NiTe. This result accompanies the maximum enhancement in visibility of Ce and Mo, while the Te and Ni concentrations fall.

5. *Regeneration of the ACN system is essentially impossible for a degenerated catalyst.* The regeneration process does effectuate the reoxidation of the metals: Mo(V) → Mo(VI), Te(O) → Te(IV), and even some Ce(III) → Ce(IV). However, at this point much of the Te is lost and the complex (Te, Mo) species do not form. Also, the Te^{-2} does not seem to readily oxidize. In addition, the oxidation of Mo and Ce does not regenerate the complex (Ce, Mo) compounds.

6. *Regeneration of a moderately used ACN system is possible.* The "magic" use limit beyond which regeneration is not effective appears to be ~2000 h (for the standard use conditions). At this point the major changes needed to reproduce a "fresh" ACN system appear to be oxidation of Mo_2O_5 to MoO_3, destruction of a complex (Ni, Te) compound, and modification of the complex (Te, Mo) compounds. All of these features would appear possible for the moderately used system.

7. *Other features of the ACN system* detected by ESCA to establish the above conclusions include effect of "refreshing"; effect of basic SiO_2 particle size (i.e., ball milling); and the effect of X-ray beam damage. In all cases these features were extensively tested and monitored and shown to not modify and/or produce the results summarized in 1 through 6 above; rather, in most cases these "effect" studies provided additional data in support of 1 through 6.

Table 8-16 includes the binding energies of a select group of ACN-related systems after their values are adjusted to a common C(1s) = 284.6 eV. This is done because such tabulations have been included in other studies of catalysts or catalyst/type systems.[1b]

Table 8-16 **List of possible species inherent to ACN**

Constituent (state)	(~) Critical binding energy	Oxidation state	Chemical species
Mo ($3d_{5/2}$)	232.6	VI	$Ce_2Mo_4O_{15}$
	232.2	VI	MoO_3
	231.7	V	Mo_2O_5
	231.4	V	$TeMo_4O_{13}$
Te ($3d_{5/2}$)	576.2	IV	Te_2MoO_7
	575.6	IV	$TeMo_4O_{13}$
	572.6	O	Te°
	571.5	II	Te^{-2}
Ce ($3d_{5/2}$)	882.8	IV	CeO_2
	882.0	III	$Ce(III)_2Mo_4O_{15}$ and Ce_2O_3
Ni ($2p_{3/2}$)	855.5	II	$NiTe_3O_7$
	854.5	II	NiO and NiO/SiO_2,[a]
	852.5	O	Ni°

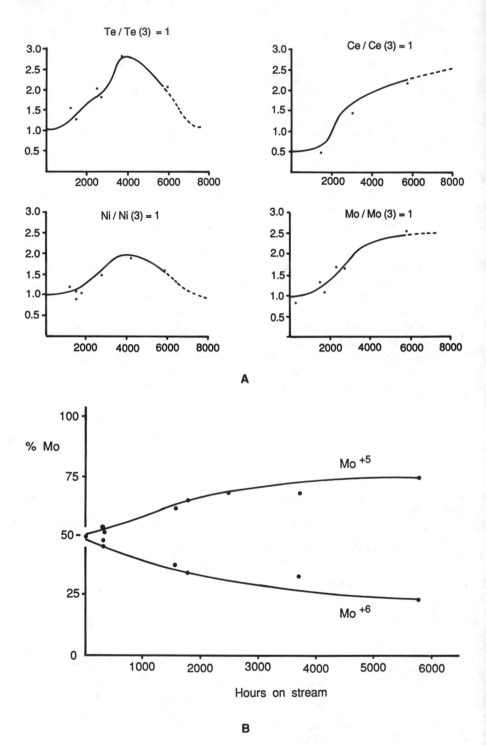

Figure 8-30 Temporal development of projected ACN catalysts with simulated use (A) Progressive ESCA "visibilty" of metal dopants. (B) Evolution of Mo^{+6} and Mo^{+5} vs. hours on "stream".

Table 8-17a **Suggested chemical interactions for hypothetical ACN system**

$Ce_4Mo_{12}Te_4Ni_1O_x$	Set up $\rightarrow$	$Ce(III)_2Mo(VI)_4O_{15}$ $+\ Ce_2O_3 + 2Mo_2O_5 + {}_2MoO_3$ $+\ 2Te(IV)_2Mo(VI)O_7$ $+\ NiO$

Use
↓ (catalyst still active)
typical of ~1000 h use

$2Ce_2O_3 + 4Mo_2O_5$		$Ce_2Mo_4O_{15}$
$+\ Te\ Mo_4O_{13}$	Degradation	$+\ Ce_2O_3 + 2Mo_2O_5$
$+\ Te^\circ \uparrow + NiO/SiO_2$	$\leftarrow$	$+\ Te(IV)Mo(5.5)_4O_{13}$
$+\qquad 2TeO_2$	typical of	$+\ Ni(II)Te(IV)_3O_7$
	~3000 h use	

↓ Spent
typical ~5500 h use

$2Ce_2O_3 + 5Mo_2O_5$
$+\ 2TeO_2 + 2MoO_3$
$+\ Te^\circ + NiTe$

Regeneration

Ce_2O_3		CeO_2
Mo_2O_5	O_2	$2MoO_3$
$TeMo_4O_{13}$	$\Rightarrow$	Te_2MoO_7
$NiTe_3O_7$	Δ	$NiO + TeO_2$
Te°		TeO_2
$NiTe$		?

This procedure removes some of the binding energy shifts discussed above (but does not affect the line widths) and because of this some doubt may be placed upon the previous analysis. However, this problem should not be considered too serious as the variations indicated in the previous tables *are* reproducible and the relative shifts appear significant. It is apparent that any carbonaceous impurity may be affected in a nonuniform fashion by these essentially different catalysts.[1b]

Based upon the data compiled and the chemical entities proposed in Table 8-17 it is possible to postulate a series of chemical steps covering the entire history of the ACN system; fresh — use — degradation — spent — regeneration. Some of these steps are fairly well founded; others are largely guesses. In general, the consistency of the results is surprisingly good.

The most important feature to note about these equations is that "use" immediately produces an increase in Mo(V); however, this production can be reversed by the regenerative process. Further use of the system, however, results in an irreversible degeneration of the (Ce, Mo) compounds and the reduction of some of the Te. The reduced Te *may* be vaporized in the form of $Te^\circ \uparrow$. Also, the (Ni, Te) compound seems to dissociate, perhaps leaving the Ni to associate with the support, as shown. Further use of the catalyst produces the maximum Mo(V) and may create a telluride (possibly NiTe).

C. STUDIES OF PT METAL CATALYSIS BY HIGH-RESOLUTION ESCA[125]

In the present examples we report ESCA examinations of composite systems realized when varying amounts of platinum metals (i.e., Pt or Pd) and other species are doped by some conventional means onto standard γ-alumina supports in a manner designed to simulate systems often employed in Pt metals catalysis.[1b,126] A unique feature of the present study is that the doping compositions are typical of those of real catalysts, e.g., 0.2 < Pt wt% < 0.75. Generally the Pt metals can be assumed to be the principal active metal species added, although the function of that dopant has been suspected to depend significantly on the other species present.[127] Examples of the latter are demonstrated along with the clarification of certain other questions for bimetallic systems.[126,128] It is not our intention to describe a catalytic process, but rather the materials alterations that seem to occur during common set-up and modification procedures, such as oxidation, reduction, and simulated use. We have also reported in Section I.D of Chapter 7 on a unique ESCA-based method that seems to determine the important feature of metals dispersion in the alumina matrix, a feature that may be realized for the present systems as a result of the "abuse" experienced by a catalyst.[1b]

As indicated, the range of metals doping employed herein is common to "real" reforming catalysts as reported in the published and patent literature.[126] It should be noted that these values are quite different from those generally reported in previous publications dealing with ESCA studies of Pt metal systems simulating the reforming process. In general, those studies examined materials with either unrealistically large metals compositions,[129] or inappropriate (but interesting) supports,[130] or ignored all together the Pt in their analyses.[131] This was done because these analyses were plagued by the problem of the blockage of the major ESCA signal for platinum, the Pt(4f) peaks, by the Al(2p). We have succeeded in circumventing this problem by employing extremely long scans of the much weaker (in cross section) Pt(4d) lines [compared to the aforementioned Pt(4f)] and exploiting the relatively high-resolution, acute sensitivity, and long-time stability of the HP-5950 ESCA located at the Allied Signal Engineered Materials Research Center and also monochromator-based ESCA systems at the University of Wisconsin at Milwaukee.

1. Analysis of the Alumina Support

Any analysis of an alumina-supported metals catalyst should begin with the alumina itself. Numerous ESCA researchers have examined various aluminas, but they have almost exclusively studied the prominent core level peaks, particularly the Al(2p) lines.[1d] We have done so as well, and find for all cases (including the former) nothing unusual, except that the Al(2p) binding energies detected in studies of conventional alumina powders (73.8 eV) are generally significantly smaller than many of the binding energies reported for Al_2O_3 grown as a film on $Al°$.[132] Generally, for example, little, if any, difference is detected between the ESCA core level results for different alumina phases (e.g., γ vs. α), and very little can be said about the nature of alumina (compared to other possible support matrices) from these core level results except to note that they suggest that neigher $Al(OH)_3$ nor $AlOOH$ is present.[1b] Recently we have extended analyses of oxides such as alumina to include high-resolution, XPS-generated, valence band spectra,[24,40] see Section I.D of Chapter 7. For details of these analyses one should consult that section.

2. Oxidized Pt on Al_2O_3 System

Most of the systems utilized for the purposes mentioned above were initially formed from oxidized versions of the platinum generally placed into the alumina matrix by quasi ion exchange of a particular platinum complex.[1b] One of the most popular procedures involves the use of the platinum (+4) acid, H_2PtCl_6. Generally, this system is then subjected to further oxidation, to ensure a consistency of materials states before subsequent reduction.

We have employed ESCA to examine a number of systems formed in this manner following oxidation, and have always detected essentially the same, somewhat surprising distribution of products. In all cases the principal platinum species found by the ESCA

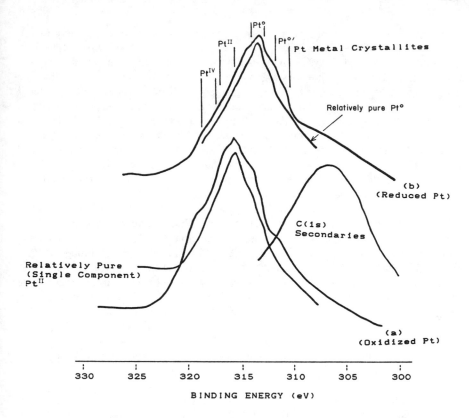

Figure 8-31 Pt(4d$_{3/2}$) spectra for select Pt/Cl/Al$_2$O$_3$ systems following oxidation and reduction.

is not Pt(+4), but rather Pt(+2). A representative case is displayed in Figure 8-31. It should be noted that we are not alone in detecting this "anomaly". Czaran et al.[133] have also determined that hexachloroplatinic acid yields a surface dominated by Pt(+2) species, with the Pt(+4) state retained at the surface only when the hexammine complex is employed. It should also be noted that the enumeration of the Pt(+2) state was achieved by determining the peak binding energies and comparing those to the values found for other platinum species in various oxidation states.[1b] This procedure and detailed analysis further suggests that modest amounts of the surface platinum in the oxidized system may be in the Pt(+4) and Pt° states, see Figure 8-31. The latter may, in fact, be distributed subsurface to the Pt(+2), and treatment of the resulting system with ion etching to remove the outer surface seems to suggest that some degree of this Pt(+2) "coating" is occurring.

3. Metals Reduction and Dispersion

A major question in metals catalysis, and many other fields of materials science, concerns the reduction of the metal(s), perhaps to their zero valent form, e.g., Pt°, and the distribution (or dispersion) of the metal throughout the oxide support matrix, e.g., Al$_2$O$_3$.[1b,126]

Since the Al$_2$O$_3$, in question, is a surface insulator the ESCA examination of these systems must consider the aforementioned charging problem. In addition, since there is some question herein as to the distribution of a conductive metal in this insulator, one must also be cognizant of the more difficult feature generally labeled as *differential* charging.

The important question of metals dispersion that arises for the reduced catalyst (well treated vs. abused) has been considered for Pt metals catalysis in Section I.D of Chapter 7. In our case there is 0.2 to 0.7 wt% Pt[A], generally well dispersed (essentially dissolved) in a matrix of Al_2O_3, and the previous argument would suggest a system that exhibits the charging properties of the alumina. Thus, despite the excellent conductivity of Pt° when it is a macroscopic continuum, in the present case it will, for all measurable purposes, charge along with the Al_2O_3 matrix, particularly if the Pt° is, as suggested, a uniformly dispersed (near solid solution) part of the mixture. This has been observed on numerous occasions during ESCA studies of platinum-doped alumina systems, where the Pt° has been confirmed to be well dispersed by those who (bulk) analyze and use these systems.[127,134]

If we attempt to compensate for the charging with the use of the electron flood gun, we find that we are able to essentially accomplish this for both the Al_2O_3 and Pt°. However, this observation does not readily facilitate ESCA detection of this system, since as noted above the relatively strong Pt(4f) lines remain hidden under the dominant Al(2p) peak. Fortunately, in the present case, one may exploit the high resolution and acute sensitivity of the monochromator-based ESCA's to realize the weak $Pt(4d_{5/2})$ peak, as displayed in Figure 8-31 for a representative case. This was the first presentation of a Pt ESCA spectrum with this wt% Pt in Al_2O_3, and several chemical states appear to be present. Although the spectrum is seemingly not singular, it is dominated by one line that peaks at ~314.0 eV. This value is almost identical to that $(4d_{5/2})$ produced by the Pt° in a continuous (sheet) of platinum metal.[5]

When the aforementioned system is thermally abused (in a manner that simulates certain cases of catalyst abuse) the Pt ESCA spectra are found to shift to a new set of peaks that suggest metal aglomeration and differential changing. This is described in detail in Chapter 7.

4. Bimetallic Cases

In addition to those Pt-doped situations, we have discovered that a result, perhaps related to the aforementioned differential charging effect, may in some cases be chemically induced for other systems. In this regard, we have examined a series of Pd-doped alumina systems that are also in some cases infused with different alkali cations. Even without physical abuses, these systems produce the aforementioned negative shift (with respect to "normal" Pd°), and one also finds that the size of the shift seems to depend, in a regular fashion, on which alkali cation is employed (i.e., the relative ionicity of the latter?), see Section I.D in Chapter 7. This result seems to suggest two possible explanations. Either the different alkali cations induce the presence of different-sized and -shaped Pd clusters, and the effect is as previously explained, or these alkali cations are able to force differing amounts of excess electrons into the partially vacant $Pd°d^8$ bands. At this time we are uncertain as to the origins of this effect, but based upon additional as-yet unreported work we suspect a mixture of both explanations.[1b]

Numerous other bimetallic situations arise in catalysis that produce results that are perhaps related to the addition of alkali cations. For example, Burch[127] has examined cases where group 4A species such as Sn are added to the $Pt/Cl/Al_2O_3$ reforming system in order to enhance its use stability, and also favorably affect related features. Sexton et al.[131] have examined similar systems with ESCA and, as noted above, have found that in the common catalytic doping range, e.g., Pt(0.25 to 0.75 wt%) and Sn(0.4 to 0.75 wt%), they were able to monitor the behavior of the tin and found that in the oxidized version Sn(+4) species were present, whereas following reduction surface Sn(+2) entities persisted.[131] We have also used ESCA to examine simulated Pt/Sn catalysts of similar composition and found that reduction yields Sn(II). This was consistent with the arguments of Burch[127] and seemed to be contrary to the alloy formation concept of Sachtler et al., as described by Clarke.[128]

Studies of the Pt(4d) spectra for reduced forms of these simulated $Pt/Sn/Cl/Al_2O_3$ catalysts with all constituents in typical reforming concentrations are even more revealing.

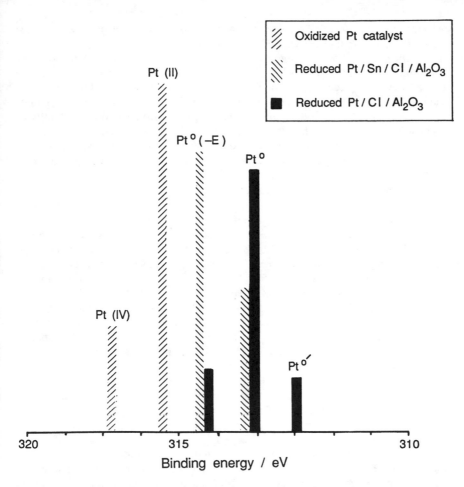

Figure 8-32 ESCA-detected evolution of Pt chemistry for oxidized and reduced versions of simulated reforming systems: Pt/Cl/Al$_2$O$_3$ and Pt/Sn/Cl/Al$_2$O$_3$.

Now, instead of large (almost exclusive) concentrations of elemental platinum, Pt° (as detected on the platinum-only catalyst), a new dominant Pt species is found with binding energy suggesting an oxidation state between Pt(II) and Pt°, see Figure 8-32. ESCA examination of the other constituents, i.e., the Sn(II) and Cl⁻, suggests that these new units consist of electron deficient Pt° configured to form Pt°(-e)-Sn(II)-Cl clusters on the reduced catalyst. Preliminary evidence suggests that it is the retention of these units that promotes the longevity of this bimetallic-bifunctional catalyst during lengthy reforming use under adverse conditions (as needed for "crude" crudes). It is possible that simple coke burn may not regenerate this cluster, whereas reconditioning of spent versions of these catalysts in acid chloride may reconstruct the aforementioned clusters, as well as creating Al-Cl sites to "acidify" the alumina.

VII. ESCA IN BIOLOGY AND MEDICINE

A. BIOMEDICAL ESCA STUDIES

ESCA has not been extensively employed in biomedical areas, but where it has been used the results have been quite rewarding. Several research groups have been primarily associated with the use of photoelectron spectroscopy in medical and dental areas, including those of Andrade,[135] Gardella,[136] Cooper,[137] Gassman,[138] and, in particular, Ratner.[139] Once again, however, significant leadership in this area has sprung from the point of origin of ESCA in Uppsala.[140]

In the latter case, "second-generation" photoelectron spectroscopist B. Lindberg has developed programs that employ ESCA to monitor the key cleanliness of potential inter-occular lenses, and also methods to investigate the nature of the blood compatibility of different surfaces. In work with Maripu[140] and others, Lindberg has been able to investigate the relative stability of heparinized blood with different substrates, e.g., various steels. The relative stacking of the sulfate vs. disulfide groups has been determined using AR-XPS.

The rapid growth of interest in the use of XPS in biomedical areas is indicated in the two well-attended symposia recently held under the sponsorship of the Physical Electronics Division of the Perkin-Elmer Co. in Minneapolis.[141] In most cases, the use of ESCA in support of these studies has concentrated on investigations of the interfacial features of the point of interaction between various biological species and a number of different inorganic substrates. Thus, in one case, the analyses included investigations of the oxidative species detected on modified FEP teflon *before* interaction with such biological species as albumin, fibronectin, or fetal calf serum.[142] In another study, Healy and Ducheyne[143] employed ESCA to catalog the surface interaction between titanium implants and the extracellular fluid (primarily the phosphates) from the unknown protein provided by surrounding bone tissue. Through comparative ESCA analysis the adsorbed species are presumed to be $Ti-H_2PO_4$ and $Ti-HPO_4$. The lack of involvement of Ca at this interface was also cataloged by ESCA.

Beebe and associates[144] used this same forum to describe some initial efforts to combine ESCA with either scanning tunneling microscopy (STM) or atomic force microscopy (AFM). (This combination has also recently been suggested by the present author as a means to examine the micromorphology and chemistry of cryotreated *in vitro* tissue samples.[16]) Beebe et al.,[144] however, suggest that an "intermediate" form of microscopy (SEM or TEM) may be needed to complete such studies.

Linton et al.[145] have suggested that AR-ESCA may be of value in attempts to investigate the layering structure of the Langmuir-Blodgett-type thin films apparently formed during certain types of cell membrane alteration.

Gassman and colleagues[146] have employed ESCA to investigate certain complex bioligands when the latter form regular complexes with such transition metal ions as ruthenium (II) or (III). The studies have concentrated so far on detecting features of the central metal ion. The latter are modified by the different bioligands.

Cooper[137,147] has employed XPS to study one of the now-classic problems in this area "the chemical status of plasma-treated polyurethane surfaces". The plasma is often supplemented with SO_2 treatment and Cooper has demonstrated how the various states of sulfur may be monitored using ESCA.

As mentioned above, the realization of the importance of surface analysis in biomaterials is new. In 1989, Ratner provided for the biomedical community the type of panoramic view of the *raison-d-etra* for surfaces that have existed for more than a decade in catalysis, corrosion, and ceramics.

Thus, for example, Ratner has demonstrated that the surface composition and structure of polyurethanes are generally different from the bulk.[139,148] Examples have been found of surface migration of low molecular weight ingredients (e.g., silicones). ESCA analyses have also been employed to investigate the ratios of soft to hard regions in polyetherurethanes.[149] Ratner and associates have also determined that certain chemical features at the surface may respond to the external environment. They have also investigated the hydrogel surfaces that constitute the interface between solid and liquid phases. ESCA studies related to these areas are still limited. The Ratner group has also employed ESCA in studies of titanium implant surfaces (also see Reference 143) and their compatibility with bone-bonding properties. Extensive studies have also been made by the Ratner group of plasma-deposited surfaces that are critical to certain biological interactions, particularly blood cell adhesion.[139,150]

B. UNCONVENTIONAL USE OF ESCA IN BIOLOGICAL RESEARCH

An unusual biological area for ESCA study concerned attempts to investigate the effect of the treatment of human hair generally labeled as "conditioning". The resulting examinations represent particularly expressive examples of the use of ESCA research as a means of making a direct impact in product evaluation. Surfactant hair conditioners are often cellulosic or related polymers containing polyquaternion cations, usually formed from phosphorous or nitrogen. The treatment process is designed to return the hair fibers to their original, prewash condition, with the proper balance of O, N, and S in the correct ratios of oxidation states, i.e., although shampooing cleans hair, it also removes and/or modifies the chemistry of some of its natural constituents. Conditioning is designed to restore these components.

ESCA analysis by Goddard and Harris[151] demonstrated that some types of (hydrophobic modified) conditioners are more effective than others in uniformly coating the entire fiber. Studies carried out in the present author's research group (D. Twu,[152] unpublished) have demonstrated that the most effective conditioners are those that create the largest ratio of sulfate to disulfide groups, and that this advantage seems to arise during treatments with certain quaternary phosphorous systems over comparable quaternary nitrogen species. Care must be employed in identifying the various species involved, as, for example, both $R–SO_3$ systems and $R–S–SO_3$ units are found on various fiber surfaces. High-resolution ESCA suggests a 0.5-eV chemical shift between these two species. Detailed analysis is possible only when the variable charging shifts of these systems are properly handled.

C. INVESTIGATIONS IN BOTANY

In addition to the ESCA studies reported above that relate to the biomedical and polymeric areas, through the years a variety of studies have been initiated in areas related to botany, but few, if any, have examined a singular problem with a detailed arsenal of many of the various ESCA submethods described in this book. Recently, we removed part of this deficiency by examining a problem in botany closely related to our general interests. In this regard a detailed examination has been made of the behavior of the surfaces of select tree leaves, before, during, and after the process of senescence. (Senescence is the process by which many leaf systems change from green to other fall colors, e.g., yellow then brown.) One of the types of leaves examined in detail (top and bottom, on and away from the large veins, before and after leaf fall, etc.) was the common American or red maple. The color change involved is due to a fairly well-understood phenomenon in which the chlorophyll recedes from a leaf down into the stem and root system. Before this happens the green color of the chlorophyll has been able to mask the yellow, orange, and brown colors of the always present caroten, and the reds of the anthocyanins. Thus, during senescence, gradual removal of the color blocking chlorophyll releases first one color, then another, as the plant or tree evolves through the fall season.

1. ESCA of Leaves
a. Introduction

To many observers the surfaces of leaves are an obvious feature that is continuously being examined by the human eye. To many others that surface is not very interesting anyway, since it is purported to feature many nonliving [dead(?)] cells that at most provide a barrier to the damaging ingestion of poisons and pollutants and/or a block to the harmful, premature ejection of water. Few people, however, seem to be aware that the outer surface of leaves (like that of all living matter) is in a state of continual evolution during its life cycle, and almost no scientific effort has been applied to study this evolution, particularly its chemistry. The first contention mentioned above is, of course, ludicrous, since, as we have mentioned elsewhere, the eye detects optical reflections and ejections from well below the outermost surface, and is essentially "blind" to the latter region of a leaf.

b. The Known Features of Leaf Surfaces

The outermost *tissue* area of a leaf is formed from its upper epidermis, or skin. This growth generally is found in the form of relatively flat structural cells. These, in turn, seem to be coated in many areas with a secreted, waxy film, or epicutical, layer. The latter seems to be emitted primarily to form a blockage to water emission. This outer, cutin material is described by biologists as "not living", and indeed some of the outermost cells of the epidermal layers are also assumed to be "dead" scales, in the act of being "sluffed". In fact, it seems natural to assume that most biological processes have ceased [are dead (?)] at the surface of a leaf and that since it is already a "dead zone", it should experience little or no chemical change as the leaf itself evolves from life (green) to death (brown). From another perspective, the chemical composition of this outer region of a leaf should consist of certain sugars, amino acids, and hydrocarbons (epidermal) and waxes and lipids (cuticle). In addition to little information about the general evolving chemistry, almost no information exists concerning differences in surfaces for different types of leaves.

2. Vacuum and X-ray Damage

The noticeable rise in pressure in the UHV system following leaf introduction and its continued elevation (from $\sim 1 \times 10^{-10}$ to $\sim 4 \times 10^{-9}$) suggests that the vacuum is forcing the leaf to "bleed". The nature of this modest extraction is not entirely understood, but the limited evidence suggests the extraction of primarily water vapor. This process seems to be accentuated in the soft ($Mg_{k\alpha} = 1256\,eV$) X-ray beam, such that with extensive (several hour) exposure to the X-rays the pressure increases to $\sim 7 \times 10^{-9}$ torr and all of the leaves exhibit a slight brown "stain". Comparison of the ESCA results achieved early in the analysis (after a matter of seconds in the X-ray) and those realized following hours in the beam reveal only modest changes in composition and chemistry. In general there seems to be a slight time-induced reduction in the [O/C] ratio, and probably an even more pronounced falloff in {[OH]/[C]}, but these effects seem to be well within the largely qualitative, relative boundaries imposed in the present study.

3. Leaves in Various Stages of Senescence
a. Green-Top

The top of a green maple leaf exhibited large relative oxygen, e.g., [O]/[C], and nitrogen ESCA spectral ratios and also ESCA peak positions that suggest the detection of epidermal amino acids and substantial air-induced passivation, Figure 8-33 and Table 8-18, as well as the expected long-chained hydrocarbons (probably from the cuticle).

The top of a leaf that had retained its green color but fallen to the ground was also examined. The [O]/[C] intensity suggested a substantial reduction in oxygen compared to the top of the green leaf just picked from a maple tree. The C(1s) loss spectrum, Figure 8-34, and valence band, Figure 8-35, for this fallen leaf system indicate the presence of primarily long-chained hydrocarbons. The O(2s) region of the latter is significantly retarded. Tilting this sample to enhance the surface detection (~ 10 Å vs. ~ 50 Å) produced an obvious enhancement of oxygen. The small N(1s) spectrum for the latter indicated that almost no epidermal tissue is present on the outermost surface.

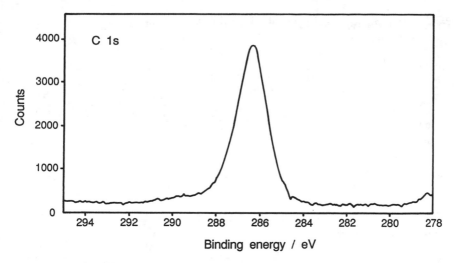

Figure 8-33 C(1s) for the top of a green maple leaf.

b. Green Back

The surface chemistry of the bottom of a green leaf is partially demonstrated in a C(1s) spectrum which confirms that the carbonaceous material detected in the first ~50 Å is primarily long-chained hydrocarbons, with carbonyls [esters (?)], i.e., the presence of substantial overlayers of cuticle in the form of paraffinic hydrocarbons and fatty acid ester (lipid) waxes. The ratio of [O]/[C] ~ 5.7 suggests that these materials are also subject to moderate air-induced oxidation (NP, see the discussion earlier in this chapter) in amounts that are also present, but significantly retarded, on the top of this leaf. The N(1s) spectrum, however, indicates that, although this cuticle-type layer is dominant, the ESCA spectrometer is able to detect (in the first 50 Å of this surface region) some amino acid-type material apparently indicative of the surface presence of ESCA-visible epidermal tissue.

The valence band spectrum also confirms the presence of predominantly long-chained hydrocarbons, with the oxygen in an (ESCA detected) mixture of C–O–C and C = 0. The latter is indicative of significant air-induced oxidation.

c. Green Leaf —Three Weeks After Picking

This system exhibits a valence band for both top and bottom that suggests significant oxygen presence that is predominantly of the C–O–C type. Almost the same relative amount of oxygen was detected on the bottom and top side of this green leaf, i.e., in the ratios of [O]/[C] ~4.1% top and ~4.6% bottom. Interestingly, the relative amount of oxygen goes down during examination with grazing incidence to ~3.7%. At the same time the relative amount of nitrogen found on the surface at conventional incidence suggests that the bottom of this leaf has significant amount of exposed epidermal skin (as well as cuticle). Grazing incidence suggests a reduction in this outer surface-exposed epidermal tissue, indicating that most of the exposed "skin" is still subsurface to the cuticle and passivation species. The top of this leaf, on the other hand, exhibits only a small amino acid exposure. This suggests that this top has either a more substantial passivation layer or a thicker retained cuticle layer than the bottom of the leaf.

d. Second Leaf Picked Off Tree — Yellow-Green

One of the most informative ESCA studies was rendered by examining a leaf picked from a maple tree after it had begun to turn partially yellow. This leaf was characterized herein by analyzing *in detail* both its top and bottom.

Table 8-18a **Relative concentrations of oxygen [O] and nitrogen [N] on various leaves**

Tree[a]	Top		Bottom	
	O	N	O	N
Green	L	L	ML	S
Green-yellow	VL	L	L	ML(1)
Yellow	L	ML	M	M(1)
Ground brown	S	~0	ML	~0

Ground[a]	Top		Bottom	
	O	N	O	N
Green	MS	~0	—	—
Green (3 weeks)	ML	S	ML	M
Yellow-brown (3 weeks)	M	L(1)	MS	M
Brown	S	~0	ML	~0

Note: VL ≡ very large, L ≡ large, ML ≡ moderately large, M ≡ moderate, MS ≡ moderately small, S ≡ small.

[a] Origin of leaf.

Table 8-18b **Comparative semi-quantifications: top to bottom of leaves**

Top Ground: Always Less O Green or Brown Almost No N

Top	O	N
Green	Large	Large
3 weeks after pick	Moderately varied	Small
Yellow-green	Very large	Large
Yellow	Large	Moderately large
3 weeks after	Moderate	Large
Brown	Small	Very small

Bottom	O	N
Green	Moderately small	Small
Green (3 weeks later)	Moderately large	Moderate
Yellow-green	Moderate	Large
Yellow	Moderate	Moderately large
Yellow-brown (3 weeks)	Moderately small	Moderately large

Top — The C(1s) of this side of the leaf indicates the expected predominance of long-chained hydrocarbons, with an unusually large amount of oxides. The O(1s) spectrum reveals not only a significant presence of oxide, [O]/[C] ~ 14%, but also that most of these oxides are in the form of ether-type (C–O–C) linkages. Since the latter make only a modest contribution to the possible cuticle waxes it is suspected that they arise largely from air-induced NP of the leaf surface. Both the C(1s) and O(1s) spectra also exhibit substantial $-\overset{O}{\overset{\|}{C}}-$, but the amount of carbonyls is less than the detected C–O–C.

Further analysis of these features requires the use of some of the less well known, but often more informative, forms of ESCA analysis described in Chapters 2, 6, and 7. For example, examination of the resulting C(1s) loss spectrum demonstrates the presence of hydrocarbon with little, if any, indication of conjugated unsaturation. Perhaps the most informative spectrum is the 0- to 30-eV valence band, Figure 8-36. In this result one sees that the valence band region is dominated by long-chain hydrocarbons.

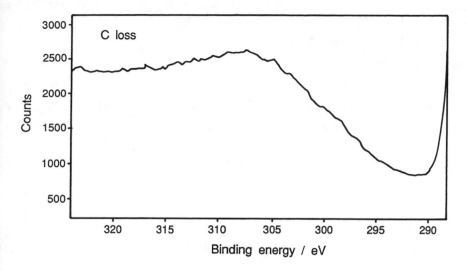

Figure 8-34 C(1s) loss spectrum for the top surface of a green maple leaf.

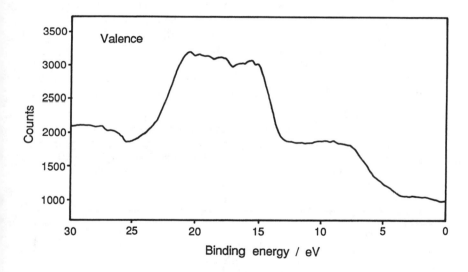

Figure 8-35 ESCA valence band spectrum for top surface of green maple leaf.

The nature of the oxygen species are reflected in the peak structures from 25 to 30 eV. The peak at ~29 eV (27 eV after subtracting the 2-eV charging) suggests the presence of C–O–C, whereas the C = 0 is indicated by the peak at 27.5 eV (25.5 eV without charging). It should be noted that there is obviously more of the former than the latter, particularly when one realizes that most of the –C = 0 lies on top of the C–O–C, see Section IV.D of this chapter.

Also detected in this analysis was a significant surface presence of nitrogen (see the N(1s) spectrum in Figure 8-37), [N]/[C] ~ 0.024 (2.4%). The detailed spectrum appears to have some structure, suggesting the presence of three or more chemical types of nitrogen. The weak structure at ~401.5 eV (398.5 eV without charging) suggests the presence of small amounts of some type of nitride. The other two possibilities occur at 403 eV and 402.5 eV

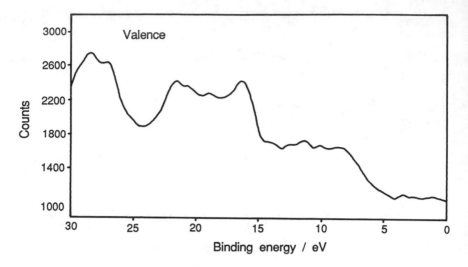

Figure 8-36 ESCA valence band for top surface of yellow-green maple leaf.

(400.0 and 399.5 eV without charging). These two peaks are both in the range of organic amine systems, apparently from the amino acids of the epidermal species.

Bottom — The bottom of this green-yellow leaf is generally similar to the top, but exhibits several, perhaps significant, differences. First there is less total oxygen, i.e., [O]/[C] ~ 11.4%. Once again most of the detected carbon is in the form of long-chain hydrocarbons. The oxygen distribution is reflected in the valence band spectrum between 25 and 30 eV in Figure 8-38. We see that there is now (at conventional incidence) much more C–O–C (B.E. $\cong$ 28.5 eV, 26.0 eV at zero charging) than $-\overset{\overset{\text{O}}{\|}}{\text{C}}-$ (B.E. $\cong$ 27.0 eV, 24.5 eV at zero charging). The nitrogen revealed on the bottom is also substantially different with only one type of amine nitrogen indicated in the N(1s), see Figure 8-39. This may reflect itself in the presence of a different distribution of epidermal structures, perhaps one that represents dead or "sluffed-off" leaf "skin", whereas the other may be indicative of live tissue.

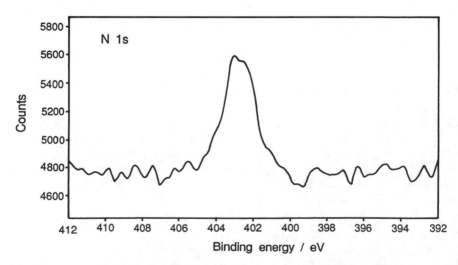

Figure 8-37 N(1s) spectrum from top surface of yellow-green maple leaf.

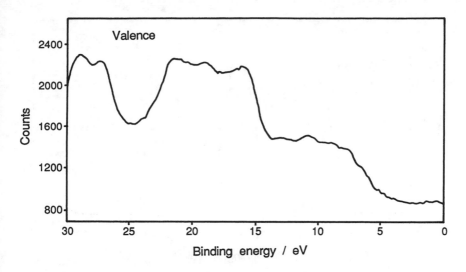

Figure 8-38 ESCA valence band from bottom surface of yellow-green maple leaf.

Grazing Incidence — An extensive grazing ESCA study was rendered of the outer-most surface (~10 Å) of the bottom of this green-yellow leaf. This reveals about the same [O]/[C] ratio as detected in the 0- to 50-Å conventional incidence survey into the subsurfaces. The distribution of oxygen species, however, is quite different. In this case the outer (grazing incidence) ESCA view, Figure 8-40, of the leaf bottom reveals a substantial enhancement in $-\overset{\text{O}}{\underset{\|}{\text{C}}}-$ relative to C–O–C, particularly compared to the 50-Å view. This suggests that much of the total $-\overset{\text{O}}{\underset{\|}{\text{C}}}-$ detected on the surface of these materials results from air-induced oxidation (NP) that we have shown to indicate the evolution of increasing degrees of oxidation as the oxidized products build, that is, in the case of these leaves air first produces C–O–C-type species that are subsequently covered with $-\overset{\text{C}}{\underset{\|}{\text{C}}}-$. (Note, in addition, the study of the films of polypropylene described in Section IV.D.)

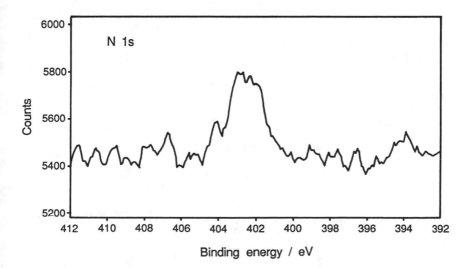

Figure 8-39 N(1s) spectrum from bottom surface of yellow-green maple leaf.

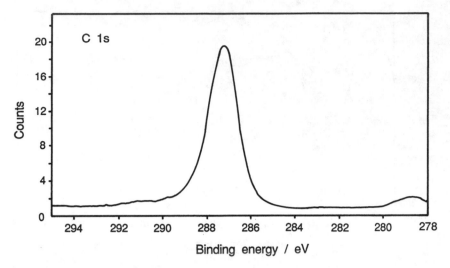

Figure 8-40 Grazing incidence (outer surface) study of a yellow-green maple leaf.

It should also be noted, Figure 8-39, that the presence of nitrogen found on the bottom of this leaf is substantially retarded on the outer surface compared to that detected at conventional incidence. This magnifies the previous observation that these nitrogens are of epidermal origin and generally found subsurface to the cuticle wax depositions and the air-oxidized by-products.

e. Yellow Leaf on Tree

A leaf that had entirely turned yellow but was still attached to the maple tree was also analyzed in detail on both its top and bottom surfaces.

Top — The top of this leaf exhibits a significant presence of oxygen on its surface, almost the same amount as detected on the top surface of the green leaf from the same tree. Once again this oxygen was determined to be primarily due to C–O–C-type oxygen, see Figure 8-41.

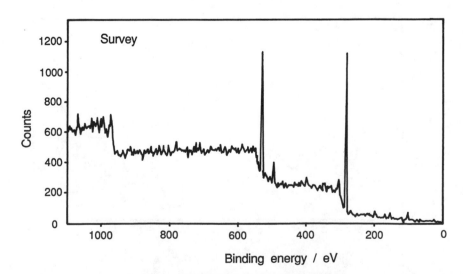

Figure 8-41 Survey scan for surface of yellow maple leaf.

The nitrogen revealed in the top-side N(1s) analysis seemed to reflect a single type of nitrogen that was similar to one of the two amine species found on the top of the yellow-green leaf, but this may be the species not found on the bottom of that leaf. This surface also exhibits a small amount of silicon, as does the green leaf from the same tree. All other features are the same as found on other leaves.

Bottom — The bottom of this leaf exhibits no silicon. (This is similar to the situation for the aforementioned green leaf.) Most of the carbon was again found to be in the form of long-chained hydrocarbons. As found with some surfaces (e.g., the yellow-green leaf) there is less oxygen than on the top surface, but in this case the drop off in [O]/[C] appears even more dramatic, since there is more oxygen on the top of the yellow leaf. In this case there is actually almost the same [O]/[C] as on the bottom of the yellow-green leaf. The amount of nitrogen is also less than on the top surface, but almost the same as found on the surface of the yellow-green leaf. The nature of these nitrogen species seem to be formed from the same amine groups as found on the yellow-green leaf.

This surface was also examined at grazing incidence. Once again the relative amount of oxygen on the outer ~10 Å increased compared to the subsurface detected at conventional incidence. The increase seems to be primarily due to enhanced $-\overset{\text{C}}{\underset{\|}{\text{C}}}-$. The relative presence of nitrogen, on the other hand, is found to drop precipitously, once again indicating that the source of these amine nitrogens is largely subsurface to much of the hydrocarbons and air-induced oxidation.

f. Yellow Leaf — Picked From Tree Three Weeks Before Analysis — Turned Partially Brown

This material was different from the brown leaf taken from the ground under the maple tree (see below) in that the present material completed its change in color from yellow to brown in a plastic bag in a relatively clean environment.

Top — The surface of this material revealed a small presence of silicon that was, on the other hand, the largest found on any of the leaf surfaces. The carbon was found to be primarily long-chain hydrocarbons, exhibiting a very singular C(1s), i.e., the surface organic species. The top of this leaf exhibited significantly more [O] than the brown leaf taken off the ground, with almost all of the oxygen due to either C–O–C- or C–OH-type species, Figure 8-42. There is also much more nitrogen detected on the top of this brown leaf than on the top of a fallen brown one. In fact, there may be more surface visible nitrogen than on any of the leaf surfaces analyzed. As indicated in Figure 8-43, this nitrogen also appears to be of a rather singular amine type, apparently expressing the ready presence of a small range of (surface-oriented) epidermal species.

Bottom — The bottom of this leaf detected by ESCA (to ~50 Å) revealed essentially no Si. The oxygen presence was also substantially reduced, from that on the top of the leaf. In this case an [O]/[C] of ~3.8 was detected. This is approximately one half the oxygen detected on the top of this leaf. The type of oxygen detected suggests a growth of $-\overset{\text{O}}{\underset{\|}{\text{C}}}-$ species relative to the C–O–C type. The amount of visible nitrogen is also dramatically lower compared to the top surface, Figure 8-44. In fact, the relative amount of nitrogen is now similar to that detected on some of the other "off tree" surfaces. As in the case of the top of the green leaf picked off the tree and held for 3 weeks, the present situation reveals a mixture of at least two distinct types of amine species, one being the species that seems to predominate on the top of this leaf.

Grazing Incidence — In this case grazing incidence exhibits almost no nitrogen, suggesting as with the other leaves a bulk origin of the nitrogen. Once again the total outer surface oxygen is also dramatically reduced in amount, whereas the ratio of $-\overset{\text{O}}{\underset{\|}{\text{C}}}-$ to C–O–C is up. It should be noted that the relative amount of top to bottom oxygen differs from that for the brown leaf found on the ground (see following pages).

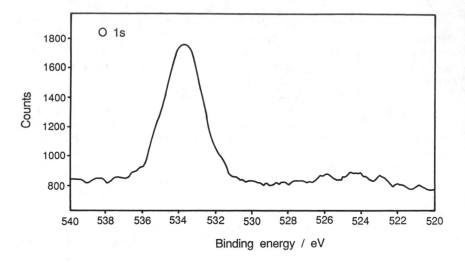

Figure 8-42 O(1s) spectrum of the top surface of a yellow maple leaf.

g. Brown Maple Leaf
a. Valence Band (VB) Analysis

In this case the VB demonstrates that the leaf surface is composed predominantly of polymeric hydrocarbons or at least long alkane chains. At conventional incidence, the relatively "deep" surface (~50 Å) shows almost no oxygen of any type, as indicated in the peak structure between 25 and 30 eV and the characteristic of the valence band pattern between 5 and 25 eV. The latter should be compared to that for polyethylene displayed in Figure 5-5 and analyzed in References 78 and 93. Interestingly, if one goes to grazing incidence (~10 to 15 Å), this brown leaf surface demonstrates a significant increase in the percentage of oxygen detected and, in comparison with the deeper C(1s) and O(1s) analyses, these outer surface oxide systems seem to be predominantly C–O–C-type species, with outer surface C = 0 in a distribution similar to that suggested above for the

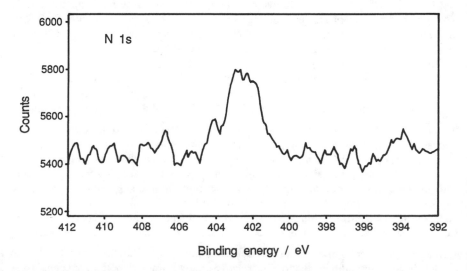

Figure 8-43 N(1s) for top of yellow maple leaf.

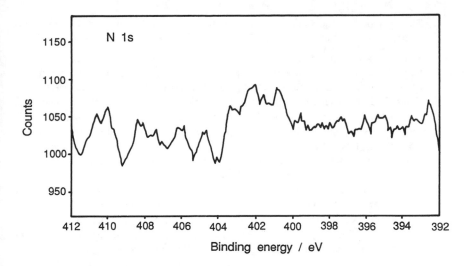

Figure 8-44 N(1s) for the bottom of yellow maple leaf.

air-induced oxidation of polyethylene, that is, our brown leaf is a layer of dormant cuticle that has been "naturally passivated" through air exposure.

All of the analyses put together suggest that our brown leaf has a surface dominated by naturally passivated hydrocarbon deposits, with more passivation on the bottom of the leaf than on its top. All of this is, of course, interesting, but very limited and should only be considered as a signpost to total reality.

h. Summary Statement

It should be noted that the limitations and omissions in this study preclude anything other than generalized speculations, but within the context of the very limited survey (only maple leaves were analyzed in detail, "harvested" during one fall season, from a single group of trees, etc.) several highly qualified statements may be made.

First, we consider those features common to all examined leaf surfaces:

1. The primary species detected on the surface of all leaves suggests the principal presence of long-chained hydrocarbons. This signifies that, as expected, all the surfaces are heavily coated by cuticle type waxy materials, dominated by long-chain, paraffin hydrocarbons. The layering of this material is, however, never found to be so thick and/or so uniform as to cover entirely any suggestion of epidermal material, to a thickness in excess of ~50 Å (see below).

2. All leaf surfaces, including the waxy cuticle, are also extensively covered by several oxygen-containing species. Some of these may, in fact, be part of exposed epidermal material, and still others may originate from various cuticle units, such as the long-chained esters that are common in certain vegetable waxes, but most of the oxygen is shown, however, to arise from the involvement of air with the surface species, that is, it has been found that these leaf surfaces readily suffer from the property we have labeled as NP.[4,5] Thus, angular resolution has confirmed that most of the resulting oxygen-containing species are layered on the outermost surface with first C–O–C units, topped by $-\overset{\overset{\text{O}}{\|}}{\text{C}}-$ -containing species, in a manner related to those previously documented on in exposed hydrocarbon polymers[78] (e.g., polypropylene[16]). This process is suggested to produce a form of a heretofore undocumented, protection or passivation for the leaf surface. *Versions of this same process have been postulated for all air-exposed surfaces.*[16]

3. The principal means employed to identify the ESCA-visible presence of some type of epidermal species is the detection of a series of –C–N–H-type species indicative of the amino acid-units inherent to the epidermis region. Angular resolution confirms that, although these amine species are generally detected in the top 50+Å, they are primarily located in the lower (25- to 50-Å) levels of the detected species (see the figures below). Thus, although several possible sources of nitrogen exist in those regions (including the atmosphere), the chemistry and location of most of the detected species suggest epidermal or "skin"-type amine and amino acid origins. *It must be remembered that although we feel reasonably certain that the organic amine units detected* originate from epidermal-type species, we cannot differentiate between live (attached) epidermis and dead (or "sluffed-off") epidermis.*

4. Although not discussed in detail, the following are also characteristic of these surfaces: (i) the presence of some alcohol (C–OH) or retained aquation; (ii) the creation of a small amount of uniform outer surface damage produced by the "action"** of the X-ray beam. The latter effect produces a modest, but noticeable, time dependent reduction in [O]/[C] ratio, and corresponding lowering of the relative amounts of C–OH and

    $-\overset{\overset{O}{\parallel}}{C}-$ detected on each system. The effect appears to be nonselective, and could be retarded, but not entirely curtailed, by switching to a monochromatic X-ray source. Except for the uniform reduction in [O]/[C], no adverse effect seems to occur during measurements that influence the present analyses. The effects reported, therefore, were primarily realized following achievement of steady-state conditions; (iii) no significant "impurities" were detected on these surfaces, with the exception of modest amounts of silicon found on the outer surface of the tops of those leaves still attached to their respective maple trees.

In addition to the above-mentioned "uniform" features each leaf surface exhibited certain unique differences and many of these have been found to vary in ways that seem to pattern select orientations and processes, such as (i) whether the leaf is attached to the tree or detached when obtained for study, (ii) top vs. bottom of the leaf, (iii) various stages of senescence, etc.

(i) The first difference to note is that larger amounts of both oxygen and nitrogen are detected on all types of surfaces of leaves that are freshly picked from the maple tree, whereas leaves of similar appearance that were found on the ground or removed from the tree a comparatively long time before XPS examination (~3 weeks) all exhibited much less surface *oxygen* and nitrogen. If one assumes that only excess overlayers of cuticle-like hydrocarbon can replace the nitrogen and oxygen, one comes to the conclusion that the "demise" of a maple leaf is characterized by substantial surface retention, or outpouring, of cuticle material, as well as removal of chlorophyll from the bulk.

Conversely, this suggests that, if most of the oxygen comes from passivation reactions between the O_2 of the air and exposed surface species, then it suggests that the nitrogen containing epidermal surface material that is more prevalent on the green and yellow leaves still on the tree are also more readily oxidized. The hydrocarbon-dominated cuticle material is less prone to passivation.

---

* Recall that the ESCA can differentiate:

(1) R–NH |	(4) N_xO_y and
(2) NH$_3$	(5) N_xCl_y, etc.
(3) M–N–H |	

** Radiation and/or thermal effects.

(ii) It is also apparent that the bottom of a particular maple leaf is much more extensively covered by the hydrocarbon-dominate cuticle material than the top of that same leaf, that is, both the oxygen and nitrogen contents are more prevalent on the top side. The only significant exception to this point occurs for the brown leaf. This may indicate that the side of the leaves that has the most water exchanging stromatii is also the side that has the largest amount of cuticle.

The fact that a green leaf loses surface oxygen and nitrogen (presumably gaining surface cuticle material) demonstrates that, although some botonists may assume that removal of a leaf from its stem kills the leaf, surface chemical processes continue for some time.

It also should be noted that close examination of the details of the ESCA spectra suggest a changing distribution of the elemental chemistry, particularly involving at least two types of organic amines. The results are far too sparse to speculate about the significance of these differences.

(iii) The evolving features documented for these maple leaves [and the comparable alterations detected for other trees, including Swedish mulberries (details not described)[153]] do confirm, however, that the surfaces of the leaves are far from dormant during and after senescence. In fact, leaf surfaces are engaged in their own unique chemical "dance" during this seasonal process of life.

VIII. XPS STUDIES IN HIGH T_c SUPERCONDUCTORS

A. BONDING AND ELECTRONIC STRUCTURE[23,37]
1. Introduction

In the numerous discussions concerning the composition, structure, and bonding of the new high-T_c superconducting oxides (see, for example, References 80 and 81) the obvious importance of the omnipresent Cu, and in particular the significance of the two-dimensional sheets of Cu and O, have dominated attention. This emphasis has led many authors to refer collectively to these systems, by way of a simple classification, as "copper oxides". It is also apparent that numerous experimental and theoretical studies have been devoted to investigating the apparent direct relationship of the high-T_c superconductors to the various simple copper oxides.

Contrary to the above supposition, it is the opinion of this author and one of his colleagues, based on a comparison of the X-ray photoelectron spectra of the various compounds involved, that there is little electronic-structure justification for a description of the superconducting materials in terms of conventional copper oxides. Rather, a key factor in the compounds that are superconducting is the (necessary) presence of very ionic cations (e.g., Ba), which apparently induce extremely large ionic fields. This, plus low-lying core levels adjacent to the O(2p) valence band, modify the bonding between the Cu and O atoms in the Cu-O counterions (anions), which should thus be described as unique, very ionic, cuprates rather than copper oxide structures. In these cuprate anions the O(2p) band extends down to E_F, whereas in the copper oxides there is a several-electronvolt gap between the top of the O(2p) band and E_F. The rationale for these statements is the nature of the oxygen-derived features in the X-ray photoelectron spectrum of the superconduct-ing oxides, and a comparison to those of copper oxides (and other transition-metal oxides) on the one hand and to those of the very ionic oxides (e.g., BaO) on the other. In making these comparisons between photoelectron spectra it is essential to be sure that the inherently surface-sensitive spectrum of the superconducting material is, in fact, that of the intrinsic material and not of contaminant phases such as $BaCO_3$, $Ba(OH)_2$, etc. which are ubiquitously present at the surface of these materials unless they are prepared and maintained under very exacting, UHV, conditions. One of the author's colleagues, with co-workers,[80,154] has gone to some lengths to establish the veracity of the photoemission spectra of $Y_1Ba_2Cu_3O_7$ (sometimes labeled as 1-2-3), and to demonstrate that the vast majority of papers published on the photoelectron spectra of this material to date represent predominantly, and in some cases even entirely, contaminant phases thereby adding nothing (except confusion) to the subject of the electronic structure or the superconductivity mechanism. This fact has also been recognized by other authors,[155] whose own results on the X-ray photoemission spectroscopy (XPS) of $YBa_2Cu_3O_7$ confirm the salient features of the photoelectron spectra referred to here and reported in more detail elsewhere.[23]

2. "Simple" Copper Oxides

The XPS spectra of the common copper oxides Cu_2O and CuO have been well established for more than 15 years, and it is well known that Cu^{II} species are easily distinguishable from Cu^I species by the chemical shift and presence of a shake-up satellite in the $Cu(2p_{3/2})$ and $Cu(2p_{1/2})$ spectra for the Cu^{II} species.[156] It is also common knowledge that to differentiate Cu° from Cu^I requires the additional use of the Cu Auger spectra, where a chemical shift exists between the two species, which does not exist in the $Cu(2p_{3/2})$ XPS spectra. Perhaps not so well known, but certainly extensively discussed by several authors,[4,157] is the fact that, though there is the above-mentioned large difference in $Cu(2p)$ spectra for Cu_2O and CuO, the binding energy (BE) of the O(1s) level is essentially fixed at around 530.1 ± 0.4 eV, see Chapter 2, and does not vary significantly (by more than a few tenths of an electronvolt), even following alloy formation with other metals.[15] In fact, a BE of 530.0 ± 0.5 eV brackets the range of *many* oxides, particularly those of almost all the transition metals.[22,23] Values for some of these oxides are included in Table 2-3 in Chapter 2. Oxides with O(1s) values in this range have been characterized as "normal" ionic oxides (NIO), see Section III.A of this chapter. It is also critical to the arguments developed to note that these types of oxide systems have valence-level O(2p) bands that are characteristic of the group. These valence bands have been studied in detail by such authors as Wertheim and Hufner,[158] for copper oxides and phenomenologically described by Goodenough,[41] also see Section III.A of this chapter. In this regard, the Cu_2O and CuO Cu(3d) valence bands were found to be fairly narrow with distinct peaks at around 3 eV above E_F (Figure 7-8). The O(2p) bands are weaker and broader, being centered at 7 eV and spreading down to an onset point at approximately 3 to 4 eV (Figure 7-8) with little distinction between Cu_2O and CuO, as is the case for the O(1s) levels as described above. Many other transition metal and other oxides have very similar O(2p) behavior.[41] According to Wertheim and Hufner,[158] this relatively distinct separation of the metal d and oxygen p bands is typical of most transition metal oxides. This is also an important feature of our subsequent arguments.

3. Superconducting Copper-Containing Oxides

When prepared as a *cleaned, undamaged* surface, the best example of which is an *in situ* cleaved, *good quality* single crystal (good quality means high-T_c value, narrow transition width, and high Meissner signal), $YBa_2Cu_3O_7$ samples exhibit, at room temperature, the following features distinctive of the true superconductivity phase material above T_c. The Ba $(3d_{5/2})$, $(4d_{5/2})$, and $(5p_{3/2})$ levels produce *single-featured, sharp peaks*, at BE's much lower (by 2 to 3 eV) than the equivalent Ba features for contaminant phases such as $BaCO_3$ or for the oxygen-deficient, nonsuperconducting version of this material, $YBa_2Cu_3O_{\leq 6.5}$.[159] The Y(3d) and (4d) features are also single featured and characteristic of Y^{3+} with BE's very similar to those for Y_2O_3. The $Cu(2p_{3/2})$ spectrum is dominated by a Cu^{II} line shape, which is only slightly different from that of Cu^{II} in CuO.[1,4,157] When the features described above are obtained free of any peaks due to contaminants, the O(1s) level has a dominating feature located at ~528 eV (Figure 8-45). Thus, for the intrinsic superconducting material the O(1s) BE exhibits a unique, very low BE in no way

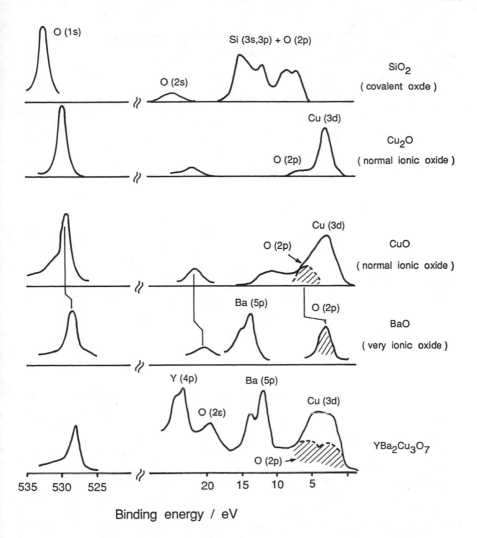

Figure 8-45 Schematic representation of the O(1s) core and the valence band and near valence-band XPS spectra of some representative oxides in the covalent (e.g., SiO_2 or Al_2O_3), normal ionic, where the BE falls in the range 530 ± 0.5 eV (e.g., Cu oxides) and very ionic (e.g., BaO) classes. For more details on these spectra, see References 21, 23, 41, and 158. The equivalent spectra for a *good* single crystal of $Y_1Ba_2Cu_3O_7$ (Reference 154) is also shown for comparison to the Cu oxides and the BaO spectra. Note the trend of *decreasing* O(1s), O(2s), and O(2p) BE's as one moves from covalent to normal ionic to very ionic classes.

indicative of a copper oxide. In fact, the O(1s) BE is very close to that in BaO, Figure 8-45 and Table 8-4. The same conclusions can also be drawn concerning the less-studied O(2s) level (Figure 8-45).

Other studies have revealed several interesting additional facts related to the above results. First, it should be noted that though the majority of literature reports represent contamination (see References 159 to 161), there are independent confirmations of the BE patterns described above for the $Y_1Ba_2Cu_3O_{7-x}$, materials, though they usually show some additional features in the Ba core and O(1s) regions, which were recognized as

Table 8-19 O(1s) XPS binding energies in eV for selected oxides

IA	IIA	IIIA	IIIB	IVB	VIIIB	IB	IIB
Na ~529.7	Mg ~530.9	Al ~531.5	Sc	Ti 529.7	Ni 530.0	Cu ~529.7	Zn 530.6 (ΔΔd)
K	Ca ~529.8	Ga 530.5 (Δd)	Y 529.3 (Δp)	Zr 529.9	Pd 530.1	Ag 528.3 (ΔΔΔd)	Cd 528.6 (ΔΔΔd)
Rb ~529.0 (ΔΔp)	Sr ~529.0 (Δp)	In 530.1 (Δd)	La 528.8 (ΔΔp)	Ce 529.1 (Δp)	Pt 530.0	Au 530.0? (ΔΔΔΔd)	Hg ? (ΔΔΔΔd)
Cs ~528.5 (ΔΔΔp)	Ba ~528.5 (ΔΔΔp)	Tl ? (ΔΔΔd)					

Note: Highest (common) valent oxide; BE values are ± 0.2 eV. Binding energies of insulators fixed by several methods, including C(1s) = 284.6 eV for adventitious carbon, Reference 21. Deltas (Δ) are the approximate separation of lowest-lying cation core peak from centroid of O(2p) dominated valence band: 11 to 13 eV Δ; 8 to 10 eV ΔΔ; 5 to 8 eV ΔΔΔ. (d) designates the d state; (p) designates the p state.

surface contamination. The conventional La_2CuO_4 system also has the same low BE for O(1s).[154] Finally, there have been a number of literature reports on the layered Bi superconductivity materials, for which clean surfaces are apparently much easier to obtain, which also show an O(1s) at a similar low BE position.

In order to understand the reasons for the results discussed above, it is useful to consider the general nature of oxides and how the different classes relate to one another. We have tried to systematize, using XPS, the bonding patterns exhibited by a variety of metallic oxides.[22] From this work we provide, in Table 8-19, a selection of O(1s) BE's for these oxides. A detailed discussion of all the factors affecting O(1s) BE's, including photoemission final-state effects, is given in Section V.A of Chapter 2 and Section III.A of Chapter 8.[22,23]

As noted in the latter discussions, many of the metallic oxides usually classified as ionic, including the Cu oxides, exhibit an O(1s) BE in the range 530.0 ± 0.5 eV, whereas more *covalent* oxides such as Al_2O_3 or SiO_2 have considerably higher O(1s) BE's (see Figure 8-45). It is the former class we refer to as normal ionic oxides, i.e., NIO.[22,23,37] The valence bands of group A oxides, which fall within this NIO class (e.g., Na_2O and CaO), are thus almost entirely dominated by electron density from the O(2p) orbitals, the only contributions from the metal valence orbitals being from the small residual covalency (11 to 24%). Within this group it has been found that this oxygen-dominated valence band shifts slightly closer to the pseudo-Fermi edge as the oxygen becomes more and more negative [i.e., as the percentage ionicity and the corresponding percentage O(2p) contribution to the valence band increases]. Thus, as we move to the *left* and *down* in the periodic table for NIO formed from group A metals, the progressive small increase in ionicity generally causes the leading edge of the valence band to move increasingly to lower binding energies.[22,23]

This "movement" of the O(2p) edge toward E_F with increasing ionicity is stopped if large metal density (d bands) are placed in the gap between the O(2p) band and E_F, e.g., for the group-B transition-metal oxides.[22] Thus, as was described earlier for the Cu oxides, we end up with O(2p) bands centered near 7 eV and reasonably separate metal 3d bands between this value and E_F.[23]

For those oxides formed by metals from the bottom of group A, and also for some oxides at the bottom of group B, the O(1s) values [and also the O(2p) band centers] move substantially *lower* than the NIO norm (see, for example, Table 8-19, and note, in particular, Rb, Cs, Sr, Ba, La, Ce, Ag, Cd, and probably Hg). Based on our arguments to date,[22,23] we would expect these to be the most ionic oxides of all, and indeed they are, producing a group we classify as very ionic oxides (VIO) with ionicities clustered around 94%. The increase in ionicity of this group compared to the NIO group is only about 7 to 9%, however, which is less than the spread of ionicities in the NIO group. We would therefore have expected only *slight* additional decreases in O(1s) BE and O(2p) band centers for this VIO group compared to the NIO group. Instead, a relatively *large* decrease is observed. For example, BaO (a typical VIO) produces an O(1s) of ~528.5 eV and a corresponding downfield shift in its O(2p) valence band to a band maximum near 4 eV and a leading edge near 1.5 eV. These are larger changes for a few percent ionicity increase than that observed, for instance, in going from GeO_2 to CaO, where the ionicity increase is 25 to 30%.[22,37]

Before discussing the possible significance of these shifts in the superconductors, it will be useful to reconsider why the shifts are so large, i.e., what effects there are in the VIO group, in addition to the relatively small increase in ionicity, which might affect the O(1s) and O(2p) BEs.

As noted earlier in this chapter, one effect stands out. All of the oxides in the VIO class have a relatively intense cation core-level state close to the valence band, and the gap between these two states decreases as the principal quantum number of the cation increases. Thus, in the case of BaO, the Ba5p level is only about 8 eV from the centroid of the O(2p)-dominated valence band. The *proximity* and *symmetry* of the cation level and the O(2p) band ensures their quantum-mechanical interaction. Based on simple perturbation arguments, we expect this interaction to increase in proportion to the square of the cross-sectional intensity of the core state and inversely with the separation between it and the valence band. The result of this interaction will be to push the two states further apart, the state with the weaker density, the valence band, moving most. It is estimated for BaO that the effect is about 1 eV, i.e., the O(2p) band is pushed closer to E_F by 1 eV by the presence of the Ba5p level. All other oxygen-dominated states, i.e., the O(2s) and O(1s) levels, of course, also move down by similar amounts in response to the O(2p) valence level changes. Thus, it would seem that this cation core-level-anion O(2p) band interaction may be a principal reason for the downshifted BEs for oxygen-related levels in the VIO group, *including the superconducting oxides*.[22,23] Examination of Table 8-19 also shows two elements, Ag and Cd, which one would not have expected to be highly ionic, and yet have lower than normal O(1s) values. Again we ascribe this to the effect of the interactions of the low-lying Ag and Cd(4d) levels [compared with the Ba(5p) levels] with the O(2p) band. It would be expected that this process grows for HgO, perhaps explaining its recent appearance in a superconductor with very high T_c.

4. Connections Between the General Oxide Arguments and the Superconductors

The critical features that we believe connect the arguments above to the nature of the bonding in the high-T_c oxides can be presented in a series of statements as follows.

1. It is, by now, well established that the controlling feature for many of the properties of the superconductivity oxides, including the superconductivity itself, manifests itself in the nature of the O(2p) band (e.g., that band crosses E_F and the superconductivity is transported by holes in the oxygen band at E_F).[160]

2. We believe that a *critical electronic behavior* of the oxide anions in these systems is primarily controlled by the presence of VIO cations, such as Ba^{2+} (or Sr^{2+}).

3. In all cases there must be present, as a prerequisite to the possibility of superconductivity, a substantial percentage of the total cation contribution in this VIO form, i.e., there must be present a significant number of cations from near the bottom of the periodic table, e.g., Ba, which have near-valence core states capable of interaction with the O(2p) band.

4. The superconducting oxides are, in effect, constructed into *complex oxides* in which the *normal* ionic cation (e.g., Cu) becomes part of a complex anion, (e.g., $Cu_3O_7^{-7}$) balanced by a *very* ionic cation (e.g., Ba^{2+}). Though the bonding between the complex anion and the VIO cation is very ionic (i.e., essentially full charge is donated to the complex anion), the internal bonding between Cu and O in the complex anion is rather covalent, that is, as discussed in detail in Section III.B of this chapter and elsewhere,[22,23] the introduction of a VIO oxide (e.g., BaO) into the lattice of an NIO oxide (e.g., CuO) will enhance the covalency of the metal-oxygen bonds in the latter. The net result of these effects is an indication of the ability of an oxide ion to "polarize"[46] (Section V.C of Chapter 2).

5. Though one can substitute one type of VIO cation for another and maintain the superconductivity capability, e.g., Sr^{2+} for Ba^{2+}, our argument that there is a special effect of these ions on the O(2p) BE implies that, for instance, swapping Mg^{2+} or even Ca^{2+} for all Ba^{2+} will damage the possibility of superconductivity, since neither Mg^{2+} nor Ca^{2+} are VIO cations.

6. It is important to reiterate at this point that the bonding between the Cu and O in these superconductors is substantially altered by the imposed effect of the very large ionic field of the VIO cation, such that there is *no* resemblance, electronically, to copper oxides. The proof of this is, in our opinion, contained in the observation that the O(1s) and O(2s) BEs in XPS of the superconductors are close to those of BaO, and far from those of the copper oxides (Figure 8-45), or other normal ionic oxides.[22,23] It is a prerequisite for superconductivity in these systems for this VIO interaction to be present.

5. Mechanistic Suggestions

There is nothing in the above list of statements that directly relates to the *mechanism* of the superconductivity process. In fact, we should emphasize that we have not examined the XPS of these systems below T_c, although others have been successful in observing small changes at E_F, on passing through T_c, which can be interpreted within the framework of the BCS theory for superconductivity (see the next section).[23] Despite our lack of data under superconducting conditions, the obvious interdependence of our VIO concept and the existence of superconductivity in these complex oxides seems to suggest a possible scenario, which we present here.

In the case of the copper oxides it has been demonstrated phenomenologically by Goodenough,[41] and confirmed experimentally by Wertheim and Hufner,[158] that the valence band consists of two components, designated as the Cu(3d) and O(2p) bands, which, though in close proximity, are relatively distinct. The O(2p) band lies at higher BE than the Cu(3d) and is centered at about 7 eV for both Cu^{II} and Cu^{I} oxides. BaO and the other VIO oxides, on the other hand, are typified by oxygen states pushed to low BE, e.g., the O(2p) band for BaO is centered around 4 eV. The O(2p) band in the high T_c superconductivity oxide systems is thus also pushed by the dominant VIO interaction (i.e., extreme ionicity, plus the cation near-valence core states) several electronvolts closer to E_F than is the case for the copper oxides. As a result of this, the O(2p) band in $YBa_2Cu_3O_7$ (and related superconducting systems) is pushed directly under the Cu(3d) band.[23] This should create the ingredients needed for an extensive ligand field-type interaction between the two bands. It would be expected that this interaction should affect the lower-density O(2p) band much more than the Cu(3d). We postulate that, owing to this interaction, the part of the O(2p) band (predominantly ionic in nature) originally *below* Cu(3d) is shifted to lower binding energy by the additional 1 eV necessary to couple to the Fermi edge, while the part originally *above* Cu(3d) is shifted up enhancing the covalent aspect of this part of the band (see Figure 8-45). Thus, one has the O(2p) character at E_F necessary for the production of holes, plus the possibility of resonance covalency in the Cu-O planes, which may correspond to the resonant valence bond models of Pauling,[162] and of Anderson et al.[163]

Another way of expressing the arguments above is to relate them to the formal band models that have been used for oxides,[164] and recently for superconducting oxides.[165] In order to try to explain the electronic behavior of oxides, Hubbard introduced the one-band (metal d) Hamiltonian, with the inclusion of dd' correlation. Recently Sawatsky and colleagues[165] have borrowed an idea of Anderson[163] and provided an extension to the Hubbard model to include two-band (metal d and oxygen 2p) effects. As a result, (dp) mixing terms were introduced, such as hopping or hybridization (t_{ij}) and mixed Coulomb repulsion (U_{pd}) integrals. Shen et al.[166] have extended those arguments to the high-T_c superconductors. They have, however, largely just expanded upon the same model that confirms that CuO is an insulator. Our model, on the other hand, would result in a novel three band, Hubbard-Hamiltonian in which (for $Y_1Ba_2Cu_3O_7$)Ba(5p) states are coupled with the O(2p) band and, through this combination, with the Cu(3d). Thus, much more involved t and U terms occur, and the initial O(2p) band is more like that of BaO than CuO. Perturbation solutions with this three-state Hamiltonian shifts part of the O(2p) band to make contact with the Fermi edge resulting in the equally unusual metallic behavior above T_c (see below).

B. STUDIES OF THE OXYGEN INTENSITY IN 1-2-3 SYSTEMS

The short history of the use of X-ray photoelectron spectroscopy to investigate the surface characteristics of the new high T_c superconducting oxides has been filled with errors and frustrations. In the case of the most extensively studied 1-2-3 system, the errors have primarily centered around the mistaken identification of the surface oriented, air-induced by-products, such as $Ba(OH)_2$ and $BaCO_3$, as the true high T_c oxide. Corrections of these mistakes were finally achieved by several research groups, particularly those of Brundle[154] and Weaver,[155] through the examination, at room temperature, of carefully cleaved single crystals of 1-2-3. These successes were frustrated, however, by the lack of detection for these materials of any certain (finite) density at the Fermi edge, although Brundle et al.[154] did suggest a possible protrusion in some of their measurements. A finite Fermi edge density was carefully documented by Veal et al.[167] at 20 K for cleaved 1-2-3, when the material was in the superconducting state, but these researchers found that this density could be made to decrease, essentially to zero, by either decreasing the oxygen content, x, in $YBa_2Cu_3O_x$ below x = 6.5 or as a result of heating the material being examined above 70 K.

It has been the contention of our research group that the high T_c superconducting oxides are unique not only for their superconductivity below T_c, but also for their (normal) conductivity above T_c. We also felt, that although there are perhaps unusual reasons for the superconductivity of these materials, their conductivity should be due to conventional reasons, i.e., as a result of the presence of a finite density of s and/or p state electrons straddling the Fermi edge at room temperature that are, in turn, not covered by the density of highly localized d state electrons. This means that in studying this problem we are addressing two difficulties: (1) trying to detect the presence of a small, but definite, Fermi density for 1-2-3 at, or near, room temperature, similar to that detected by Veal et al.[167] near 20 K and (2) assuming we find this density, explaining why it was not detected by Veal,[167] Weaver,[154] or Brundle[155] above T_c.

It was our assumption that two previous results reported by our group play a key role in these investigations: (1) like Brundle[154] and Weaver[155] we found that all 1-2-3 systems exposed to air are subject to a certain (significant) degree of surface degradation, resulting in the production of $BaCO_3$, $Ba(OH)_2$,[35,36] (see Figure 8-13) and (2) exposure of 1-2-3 to vacuum removes from the surface environment some of the aforementioned oxygen, $x \rightarrow 6.5$, which apparently may be reinvested (in some cases all the way back to the original x value) simply by exposing the vacuum treated 1-2-3 to S-T-P air.[168] The latter observations were made during investigations of the resistivity vs. temperature behavior of 1-2-3 systems.

All of these previous results suggested to us that some set of circumstances surrounding the handling of the 1-2-3 materials and/or the method of XPS measurement were either removing or at least hiding the presence of a room temperature finite Fermi edge density from XPS detection.

The materials examined in this study included a number of conventionally prepared, bulk-ceramic 1-2-3 superconductors, and also several ion beam deposited (by A. Yen and L. M. Chen at the EIC Co.) 1-2-3 superconducting thin films. Initially, all of these systems exhibited good T_c behavior, i.e., 83 to 92 K. All of the materials were polycrystalline and exposed to STP air prior to ESCA examination. This means, of course, that the surfaces of all of the samples exhibit substantial amounts of the previously described air-induced degradation.

In order to investigate these problems more thoroughly, a colleague and the author employed the very high resolution, SCIENTA ESCA system (Chapter 3) at Lehigh University to analyze a conventionally prepared, ceramic 1-2-3 system ($T_c \sim 92$ K). The expected outer surface layer of air-induced "junk" [dominated by $Ba(OH)_2$ and $BaCO_3$] was detected,[37,154] but a good "view" of subsurface species suspected to be the 1-2-3 material [binding energy of O(1s) = 528.25 eV] was also produced, see Figure 8-46. Surprisingly, we also found distinct evidence of the aforementioned finite Fermi surface in the XPS valence band spectrum, Figure 8-47. This is to our knowledge the first XPS generation of a finite Fermi surface for this material at *room temperature*. The X-ray radiation used in this study was monochromatic Al Kα produced by the aforementioned high-speed rotating anode.[169] The same sample was also subsequently analyzed at the University of Wisconsin-Milwaukee Surface Analysis Facility (SAF), 1 month later, using a VG ESCALAB II system with a conventional Al Kα anode (no monochromator). In the latter case, although the core level results resembled those achieved on the SCIENTA, we did not detect any finite density at the Fermi surface, see Figure 8-48.

Subsequently, we analyzed one of the superconducting 1-2-3 system thin film made at the EIC Laboratory Company. These ESCA measurements were also produced on the (SAF) VG ESCALAB II system. In this case, however, we used an Al Kα monochromator as the radiation source, and evidence of a finite Fermi density protrusion was detected, Figure 8-49, in a manner similar to that produced on the SCIENTA ESCA. (Note that correction for charging has overshifted the Fermi edge.)

As expected, the surfaces of all the samples exhibited a certain amount of the previously described air induced degradation, i.e., primarily $BaCO_3$ and $Ba(OH)_2$ by-products. In the resulting O(1s) spectra, one finds displayed, for example in Figure 8-46, not only a predominant peak at ~531.25 eV (indicative of these various surface by-products), but also an additional well-resolved peak at ~528.25 eV, that we have previously indicated (based on the arguments of Miller et al.[170] and Meyer et al.[155] strongly suggests the detection of the 1-2-3 superconductor system.

The most interesting observations in these studies, however, were produced in the valence band region. In the case of the SCIENTA results (Figure 8-47) these high-resolution spectra were achieved at approximately the same resolution as obtained by Veal et al.[167] on the Stoughton Synchrotron Radiation (SR)-PES system. (The SCIENTA

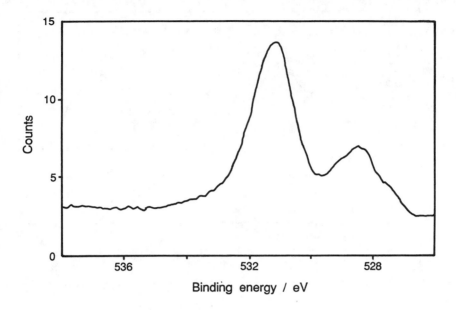

Figure 8-46 O(1s) spectrum obtained with a SCIENTA ESCA for a $YBa_2Cu_3O_7$ sample with a T_c = 92K. Notice the presence of the surface "junk" indicated by the peak at 531.2 eV and the subsurface 1-2-3 material characterized by the peak at 528.5 eV.

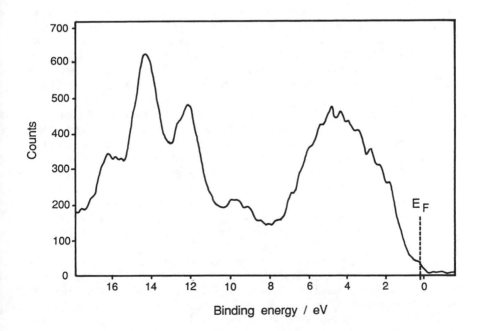

Figure 8-47 A valence-band spectra of a $YBa_2Cu_3O_{7-x}$ sample exposed to air, as obtained with a SCIENTA-ESCA system. The spectrum was obtained with a conventional (normal) incidence X-ray beam with a 0- to 20-eV scan.

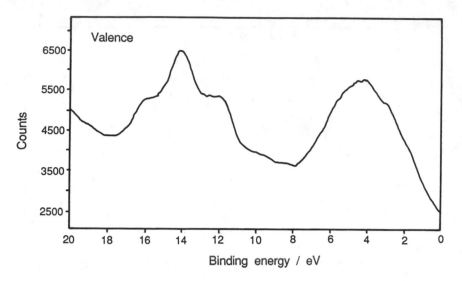

Figure 8-48 Valence-band spectrum of a $YBa_2Cu_3O_{7-x}$ bulk sample using VG ESCA with conventional $Al_{k\alpha}$ anode.

actually has a superior signal-to-background.) In many respects, our SCIENTA result resembles the valence band spectra reported by Veal et al.[167] for 1-2-3 at 20 K, with x ~ 6.7, but most important, since the SCIENTA result is obtained at room temperature, the position of this finite density is indicative of a conductor. Detailed analysis of all the spectra produced by this system eliminates all sources for this Fermi edge straddling density other than the (528.25 eV) oxide system, i.e., the 1-2-3 species. (Remember that a species producing a Fermi edge density of the size revealed in Figure 8-47 must also produce core level peaks that are several orders of magnitude larger, and no peaks of this size other than those due to the 1-2-3 are found.) Thus, all of these considerations put together suggest that we are detecting substantial amounts of the above — T_c $YBa_2Cu_3O_x$, with x > 6.5.[167]

In order to further analyze this supposition, we also recorded similar AR-XPS spectra with the SCIENTA at near grazing incidence. This produced (Figure 8-49) modest, but definite, differences when compared with the previous described results achieved at conventional (normal) incidence. For example, in the grazing incidence results, the relative intensities of the peaks derived from the surface impurities such as $BaCO_3$ and $Ba(OH)_2$ were noticeably increased relative to the subsurface 1-2-3 species. In addition, it was found that the small protrusion near the Fermi edge almost entirely disappeared (see Figure 8-49). A subsequent series of spectra (not shown) were then regenerated at "conventional incidence" reproducing all of the original features, including the critical indication of a finite density of states near E_F. This strongly suggests that the XPS-detected Fermi density protrusion is a very sensitive part of a subsurface conductor, i.e., the above T_c, "normal metallic" phase of the 1-2-3 material.

As expected, the aforementioned VG spectra produced with conventional anode exhibited a noticeable reduction in resolution compared to the SCIENTA results, but the most significant feature was the absence in the VG-generated results of anything other than a slight change of slope in the range of binding energy between 1.0 eV and E_F, Figure 8-48. A detailed examination of the VG valence band results suggests that these bear a closer resemblance to a "dirty" version of those obtained by Veal et al.[167] for 1-2-3 with x ≥ 6.3, than for the 1-2-3 system with x ≥ 6.7, as suggested by the SCIENTA results, Figure 8-47. All of this, of course, is very speculative, but three sources of this difference seem possible. First, it might be suggested that some change has occurred in the subsurface

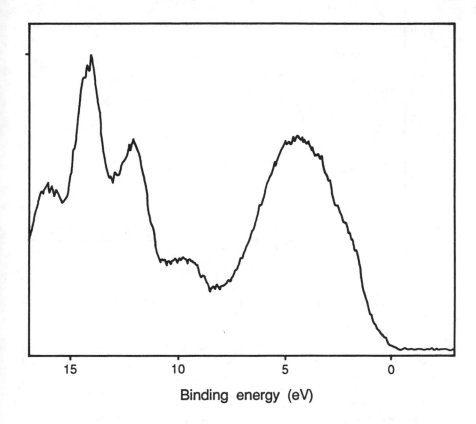

Figure 8-49 Valence-band spectrum of a $YBa_2Cu_3O_{7-x}$ sample obtained with the Scienta X-ray beam impacting at grazing incidence on the surface of the sample. Notice the near total disappearance of the Fermi edge peak (compare with Figure 8-47) and the larger size of the features attributed to the surface "junk".

material between the time of examination with the SCIENTA and the VG that may have significantly decreased the finite density of states just below the Fermi edge. Second, the lack of a detected Fermi density may be attributed to the difference in resolution of the two ESCAs involved. Third, much of the loss in intensity detected during the VG analysis (as well as its absence in numerous other investigations) may be due to the damage of the delicate 1-2-3 system caused by the use of an ESCA spectrometer with direct X-ray impingement.

In order to further explore this question, select XPS studies were also conducted of 1-2-3 thin films. For these investigations, the same VG ESCA system was employed, but an Al $K\alpha$ monochromator was used as the radiation source to "soften" the beam and improve the resolution. The core level results obtained from one of these films were similar to those achieved with the 1-2-3 ceramic. Thus, once again, the detected portion of this film revealed a mixture of air degradation and subsurface (apparently) 1-2-3 species. For example, in the resulting $O(1s)$ spectrum, the peak intensity at 528.5 eV seemed to correspond to that suggested for the 1-2-3 ceramic system [with a $C(1s)$ shift indicating charging of ~0.25 eV]. If the ESCA resolution and damage caused by direct focus XPS is truly a limitation to the detection of the previously noted density protrusion, then, in the present case, the Fermi density should be revealed. (This 1-2-3 film has a detected T_c of 83 K.) Figure 8-50 is the valence band spectrum obtained from the present

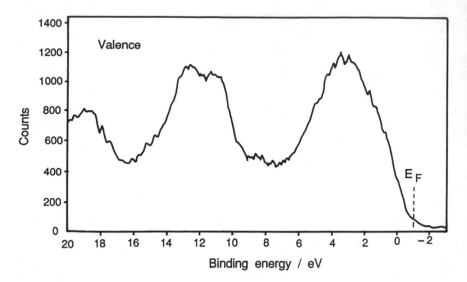

Figure 8-50 Valence-band spectrum of $YBa_2Cu_3O_{7-x}$ thin film using VG ESCA with $Al_{k\alpha}$ monochromator.

experiment. The moderately resolved finite Fermi density suggest that spectrometer resolution and sensitivity both play roles in the detection of the aforementioned Fermi density protrusion. However, it is also suggested that the conventional X-ray source employed to examine the previously described bulk ceramic in the VG system may be of sufficient energy intensity to penetrate the "junk" layer and damage the underlying 1-2-3 system, thus apparently removing most of the presence of the Fermi density protrusion. Comparing these thin film results with those obtained by Veal et al.[167] suggests that a substantial amount of $YBa_2Cu_3O_x$ with $x \geq 6.5$ is being detected in the VG ESCALAB system (using the monochromatic X-ray source).

Thus, apparently oxygen depletion causes the conductivity as well as the superconductivity (T_c) to decrease (where both features are revealed in XPS by the presence of a finite density at the Fermi edge). We are making a detailed study of this problem by XPS to try to understand the mechanism of the aforementioned oxygen depletion of these HTSC materials. In this regard, our present results indicate that, (1) placing a "surface clean" 1-2-3 sample in vacuum above ~70 K tends to "pull" oxygen out of the high T_c lattice. This may explain why Brundle,[80,154] Weaver,[155] and Veal et al.[167] do not detect a finite Fermi edge above T_c for their clean surface, single crystals. (2) The presence of a certain amount of surface oriented "junk" may, on the other hand, provide just the protection needed by the 1-2-3 system to block most of the depletion effect of the vacuum. Thus, our XPS examination of air-exposed 1-2-3 "benefits" from the air-induced degradation of the outer surface high T_c material. If, of course, the degradation is too extensive then the protection provided will be self defeating, as the ESCA will be unable to "see" the subsurface 1-2-3. (3) At the same time there is evidence (in the VG vs. SCIENTA results) that suggests that, if the intensity of the photoelectron process is too intense, then the 1-2-3 surface may be damaged, even when the 1-2-3 is covered by a "protective layer of junk". Studies performed on related Cu-containing systems suggest that the thermal effect of the X-ray beam is the principal cause of this damage.[14] Thus, our XPS investigations (and our allied R vs. T studies[168]) indicate that a combination of vacuum plus temperature may result in an oxygen depletion that reduces the x value for these high T_c $YBa_2Cu_3O_x$ superconducting materials from 6.9 to 6.5 or even less, and thus the finite Fermi edge density (that apparently requires an x value of 6.5 or above) disappears.

Perhaps of even greater significance is our discovery that the vacuum-induced degradation can apparently be reversed. Thus, we have found that exposure of the vacuum-damaged material to air seems to restore the original T_c, apparently by returning x to its original value.[168] Thus, in the present study, we have XPS evidence that shows that materials that have lost their Fermi edge density have that feature restored, simply by exposing the system to air and returning the material to vacuum and XPS analysis cycles the whole process through again. This process may apparently be repeated many times, although our T_c measurement results suggest that a material subjected to this type of "abuse" gradually deteriorates. We are thus left with the uneasy feeling that the 1-2-3 system may have a surface that is characterized by hypersensitivity. When exposed to air it suffers extensive degradation, yielding a layer of "junk" dominated by barium salts. On the other hand, when these species are exposed to vacuum they are subject to extensive oxygen depletion. Since this process is apparently restricted to the materials surface it may be of little consequence for the behavior of these systems in bulk-oriented processes, but in the case of such interfacially dependent features as Josephson Junctions the "problem" may be substantial.

REFERENCES

1. **Briggs, D. and Seah, M. P., Eds.,** *Practical Surface Analysis,* 2nd ed., John Wiley & Sons, Chichester, England, 1992. (b) **Barr, T. L.,** *Practical Surface Analysis,* 2nd ed., Briggs, D. and Seah, M. P., Eds., John Wiley & Sons, Chichester, England, 1992, chap. 8. (c) **McIntyre, N. S. and Chan, T. C.,** *Practical Surface Analysis,* 2nd ed., Briggs, D. and Seah, M. P., Eds., John Wiley & Sons, Chichester, England, 1992, chap. 10. (d) **Briggs, D.,** *Practical Surface Analysis,* 2nd ed., Briggs, D. and Seah, M. P., Eds., John Wiley & Sons, Chichester, England, 1992, chap. 9.
2. **Kay, E. and Brundle, C. R.,** private communication.
3. **Ohno, T. R., Kroll, G. H., Weaver, J. H., Chibante, L. P. R., and Smalley, R. E.,** *Reports in Surface Science,* ISISS Issue, Barr, T. L. and Saldin, D., Eds., 1993, in press.
4. **Barr, T. L.,** *J. Vac. Sci. Technol.,* 14, 650, 1977.
5. **Barr, T. L.,** *J. Phys. Chem.,* 82, 1801, 1978.
6. **Barr, T. L.,** Address, University of Uppsala, Sweden, October, 1985.
7. **Siegbahn, K., Nordling, C., Fahlman, A., Nordberg, R., Hamrin, K., Hedman, J., Johansson, G., Bergmark, T., Karlsson, S. E., Lindgren, I., and Lindberg, B.,** ESCA atomic molecular and solid state structure studied by means of electron spectroscopy, *Nova Acta Regiae Soc. Sci. Ups.,* Ser. 4, Vol. 20, 1967.
8. **Kubaschewski, O. and Hopkins, B. E.,** *Oxidation of Metals and Alloys,* Butterworths, London, 1962.
9. **Evans, U. R.,** *The Corrosion and Oxidation of Metals,* Arnold, London, 1960.
10. **Cabrera, N. and Mott, N. F.,** *Rep. Prog. Phys.,* 12, 163, 1948.
11. **Fehlner, F. P.,** *Low Temperature Oxidation,* Wiley-Interscience, New York, 1986.
12. **Wagner, C.,** *J. Electrochem. Soc.,* 103, 627, 1956.
13. **Kubaschewski, O. and Hopkins, B. E.,** *Oxidation of Metals and Alloys,* Butterworths, London, 1962, chap. 3; **Phillips, J. R.,** *Solid State Physics,* 37, 93, 1982.
14. **Barr, T. L.,** *Surface Interface Anal.,* 4, 185, 1982.
15. **Barr, T. L. and Hackenberg, J. J.,** *Appl. Surface Sci.,* 10, 523, 1982.
16. **Barr, T. L.,** Recent Advances of ESCA in Corrosion Research: Natural Passivation and its Consequences, Corrosion 50, NACE, Houston, 1990, 296.
17. **Barr, T. L.,** presentation at Int. Conf. Metal Coatings and Thin Films, ICMCTF-94, April 1994, San Diego, to be published, *Thin Solid Films.*
18. **Lebugle, A., Axelsson, U., Nyholm, R., and Martensson, N.,** *Phys. Scr.,* 23, 824, 1981.

19. (a) **Briggs, D.,** *Faraday Discuss. Chem. Soc.,* 60, 81, 1975; **Brundle, C. R.,** *Faraday Discuss. Chem. Soc.,* 60, 152, 1975. (b) **Brundle, C. R. and Carley, A. F.,** *Faraday Discuss. Chem. Soc.,* 60, 51, 1975; **Spicer, W. E., Lindau, I., Miller, J. N., Ling, D. T., Pianetta, P., Chye, P. W., and Garner, C. M.,** *Electron Spectroscopy,* 61, 169, 1977.

20. **Cotton, F. A. and Wilkinson, G.,** *Advanced Inorganic Chemistry*, John Wiley & Sons, New York, 1988.

21. (a) **Barr, T. L.,** *J. Vac. Sci. Technol. A,* 7, 1677, 1989. (b) **Barr, T. L.,** *Crit. Rev. Anal. Chem.,* 22, 113, 1991.

22. **Barr, T. L.,** *J. Vac. Sci. Technol. A,* 9, 1793, 1991.

23. **Barr, T. L. and Brundle, C. R.,** *Phys. Rev. B,* 46, 9199, 1992.

24. **Barr, T. L.,** *Crit. Rev. Anal. Chem.,* 22, 229, 1991.

25. **Fuggle, J. C., Hillebrecht, F. U., Zeller, R., Zolnierek, Z., and Bennet, P. A.,** *Phys. Rev. B,* 27, 2145, 1984.

26. **Johansson, B. and Martensson, N.,** *Helv. Phys. Acta,* 56, 405, 1983.

27. **Gaarenstroom, S. W. and Winograd, N.,** *J. Chem. Phys.,* 67, 3500, 1977.

28. This has been suggested to one of the authors (TLB) by several researchers as a result of detailed, but as of yet unpublished, studies (Reference 17 is an example). Several others have obtained negative shifts similar to those of Reference 27 and the authors; see for example, **Humbert, P. and Denville, J. P.,** *J. Electron Spectrosc. Rel. Phenomena,* 36, 131, 1985.

29. **Davis, D. W. and Shirley, D. A.,** *J. Electron. Spectrosc. Rel. Phenomena,* 3, 137, 1974.

30. **Barr, T. L., Yin, M. P., and Varma, S.,** *J. Vac. Sci. Tech. A,* 10, 2383, 1992.

31. **Pollack, R. A., Kowalczyk, S. P., Ley, L., and Shirley, D. A.,** *Phys. Rev. Lett. A,* 29, 274, 1972.

32. **Jorgensson, C. K.,** Photoelectron spectra showing relaxation effects in the continuum and electrostatic and chemical influences of the surrounding atoms, in *Advances in Quantum Chemistry,* Löwdin, P. O., Ed., Vol. 8, Academic Press, New York, 1974, 137.

33. **Pauling, L.,** *Nature of the Chemical Bond*, 3rd ed., Cornell University Press, Ithaca, NY, 1960.

34. **Barr, T. L.,** *Appl. Surface Sci.,* 15, 1, 1983.

35. **Barr, T. L.,** *Am. Lab.,* 10, 65, 1978.

36. **Barr, T. L. and Liu, Y. L.,** *J. Phys. Chem. Solids,* 50, 657, 1989.

37. **Barr, T. L.,** Electron spectroscopic analysis of multicomponent thin films with particular emphasis on oxides, in *Multicomponent and Multilayered Thin Films for Advanced Microtechnologies: Techniques, Fundamentals and Devices*, Anciello, O. and Engemann, J., Eds., (NATO/ASI Ser. E, Vol. 234), Kluwer Academic, Amsterdam, 1993, 283.

38. **Auciello, O. and Krauss, A.,** AIP Conf. Proc., No. 165, AVS Ser. 3, 114, 1988; **Barr, T. L., Krauss, A., and Auciello, O.,** to be published.

39. **Wagner, C. D., Six, H. A., Jensen, W. T., and Taylor, J. A.,** *Appl. Surface Sci.,* 9, 263, 1981; see the numerous Tables in Reference 1.

40. **Barr, T. L., Chen, L. M., Mohsenian, M., and Lishka, M. A.,** *J. Am. Chem. Soc.,* 110, 7962, 1988.

41. **Goodenough, J. B.,** in *Proc. Solid State Chemistry,* Vol. 5, Reiss, H., Ed., Pergamon Press, Elmsford, NY, 1971, 145.

42. **Levine, B. F.,** *J. Chem. Phys.,* 59, 1453, 1973.

43. **Barr, T. L., Brundle, C. R., Klumb, A., Liu, Y. L., Chen, L. M., and Yin, M. P.,** in *High T_c Superconducting Thin Films, Devices, and Characterization*, Vol. 6, Margaritando, G., Ed., (AVS/AIP Proc. Ser.), 1989, 216.

44. As indicated, the practice is commonly employed in texts, both elementary: **Smith, W. F.**, *Principles of Materials Science and Engineering*, 2nd ed., McGraw-Hill, New York, 1990, chap. 10, and more advanced: **Wyatt, O. H. and Dew-Hughes, D.**, *Metals, Ceramics and Polymers*, Cambridge University Press, Cambridge, 1974. It is even a part of the classic literature, i.e., L. Pauling does not use it in *Nature of the Chemical Bond*, 3rd ed., Cornell University Press, Ithaca, NY, 1960, whereas A. F. Wells does on occasion in *Structural Inorganic Chemistry*, Clarendon Press, Oxford, 1945, chap. 14. It seems to be a relatively common practice in some aspects of the theoretical literature, see, for example, **Tossell, J. A.**, *J. Electron, Spectrosc. Rel. Phenomena*, 8, 1, 1976.

45. **Shackelford, J. F.**, *Introduction to Materials Science for Engineers*, 2nd ed., MacMillan, New York, 1988, chap. 5.

46. **Barr, T. L.**, *Zeolites*, 10, 760, 1990.

47. **Wells, A. F.**, *Structural Inorganic Chemistry*, Clarendon Press, Oxford, 1945, 102.

48. **Mason, M. G. and Baetzold, R. C.**, *J. Chem. Phys.*, 64, 271, 1976.

49. **Siegal, R. W. and Eastman, J. A.**, *Mat. Res. Soc. Proc. Issue*, 132, 3, 1989.

50. **Onodera, Y. and Toyozawa, Y.**, *J. Phys. Soc. Jpn.*, 24, 341, 1968.

51. **Jones, W. and March, N. H.**, *Theoretical Solid State Physics*, Vol. 2, Dover, New York, 1985, 1084.

52. **Barr, T. L.**, to be published.

53. **Wagner, C. D.**, *Practical Surface Analysis*, 2nd ed., Briggs, D. and Seah, M. P., Eds., John Wiley & Sons, Chichester, England, 1990, app. 5.

54. **Kittel, C.**, *Introduction to Solid State Physics*, 4th ed., John Wiley & Sons, New York, 1971.

55. **Barr, T. L. and Lishka, M. A.**, *J. Am. Chem. Soc.*, 108, 3178, 1986.

56. **Barth, G., Linden, R., and Bryson, C.**, *Surface Interface Anal.*, 11, 307, 1988.

57. **Fadley, C. S.**, *Electron Spectroscopy: Theory, Techniques and Applications*, Vol. 2, Brundle, C. R. and Baker, A. D., Eds., Academic Press, New York, 1978, 1.

58. **Hedin, L. and Johansson, A.**, *J. Phys. B*, 2, 1336, 1969.

59. **Castle, J. E.**, *Applied Surface Analysis*, Barr, T. L. and Davis, L. E., Eds., (ASTM STP 699), American Society for Testing and Materials, Philadelphia, 1980, 182.

60. **McIntyre, N. S.**, *Practical Surface Analysis*, Briggs, D. and Seah, M. P., Eds., John Wiley & Sons, Chichester, England, 1983, chap. 10.

61. **McIntyre, N. S. and Chan, T. C.**, *Practical Surface Analysis*, 2nd ed., Briggs, D. and Seah, M. P., Eds., John Wiley & Sons, Chichester, England, 1990, chap. 10.

62. **Barr, T. L.**, *Chem. Phys. Lett.*, 43, 89, 1976.

63. **Barr, T. L. and Hackenberg, J. J.**, *J. Am. Chem. Soc.*, 104, 5390, 1982.

64. **Gilbert, P. T.**, *Corrosion*, Vol. 1, Shreir, L. L., Ed., Butterworths, London, 1976.

65. The now classic solutions to the problem of dezincification of brass resulted in the production of brasses with small addendums of key dopants, such as Sn or Sb. The extreme importance of these solutions to military matters resulted in the name "Naval Brass" for the most common of these altered materials.

66. **Baer, D. R.**, *Appl. Surface Sci.*, 7, 69, 1981.

67. **Seah, M. P.**, *Practical Surface Analysis*, 2nd ed., Briggs, D. and Seah, M. P., Eds., John Wiley & Sons, Chichester, England, 1990, chap. 7.

68. **Medima, A.**, *Z. Metallkd.*, 69, 455, 1978.

69. **Williams, F. L. and Nason, D.**, *Surface Sci.*, 45, 377, 1974; **Wynblatt, P. and Ku, R. C.**, *Surface Sci.*, 65, 511, 1977.

70. **Vasquez, R. P., Lewis, B. F., and Grunthaner, F. J.**, *Appl. Phys. Lett.*, 42, 293, 1983; **Grunthaner, F. J., Grunthaner, P. J., Vasquez, R. P., Lewis, B. F., and Maserjan, J.**, *J. Vac. Sci. Technol.*, 16, 1443, 1979; **Lopez de Ceballos, I., Munoz, M. C., Goni, J. M., and Sacedon, J. O.**, *J. Vac. Sci. Technol. A*, 4, 1621, 1986.

71. **Wilmsen, C. W.,** *J. Vac. Sci. Technol.,* 19, 179, 1979.

72. **Many, A., Goldstein, T., and Grover, N. B.,** *Semiconductor Surfaces,* North-Holland, New York, 1965.

73. **Barr, T. L., Greene, J. E., and Eltoukhy, A. H.,** *J. Vac. Sci. Technol.,* 16, 517, 1979; **Natarajan, B. R., Eltoukhy, A. H., Greene, J. E., and Barr, T. L.,** *Thin Solid Films,* 69, 217, 1980.

74. **Natarajan, B. R., Eltoukhy, A. H., Greene, J. E., and Barr, T. L.,** *Thin Solid Films,* 69, 201, 1980; **Barr, T. L. and Liu, Y. L.,** *J. Phys. Chem. Solids,* 50, 657, 1989.

75. **Barr, T. L.,** presented in part at the Department of Chemistry, University of Connecticut, April 1984.

76. **Levine, R. P.,** *Chemistry in the Environment,* W. H. Freeman, San Francisco, 1983, chap. 7.

77. Prepared by Prof. Richard Wool, University of Illinois, Champaign.

78. **Barr, T. L. and Yin, M. P.,** *J. Vac. Sci. Tech. A,* 10, 2788, 1992.

79. **Barr, T. L.,** Presentation FACSS Conference, Detroit, November 1977.

80. **Miller, D. C., Fowler, D. E., Brundle, C. R., and Lee, W. Y.,** AVS Conference on High T_c Superconductivity, AIP Conference Proceedings, New York, Vol. 3, February 1989, 336.

81. **Barr, T. L., Brundle, C. R., Klumb, A., Liu, Y. L., Chen, L. M., and Yin, M. P.,** in *High T_c Superconducting Thin Films, Devices and Applications,* Margaritondo, G., Joynt, R., and Orellion, M., Eds., (AVS Series, Vol. 6), AIP Conference Proceedings, 162, 1989, 216.

82. **Engel, T. and Ertl, G.,** *Adv. Catal.,* 28, 1, 1979.

83. **Bryson, C. E., III,** private communication.

84. **Clark, D. T. and Dilkes, A.,** *J. Polymer Sci. Polymer Chem. Ed.,* 14, 533, 1976.

85. **Siegbahn, K. and Gelius, U.,** private communication.

86. **McFeeley, F. R., Kowalczyk, S. P., Ley, L., Cavell, R. C., Pollack, R. A., and Shirley, D. A.,** *Phys. Rev. B,* 9, 5268, 1974.

87. **Galuska, A. A., Madden, H. H., and Allred, R. E.,** *Appl. Surface Sci.,* 32, 253, 1988.

88. **Xie, Y. and Sherwood, P. M. A.,** *Chem. Materials,* 1, 427, 1989; **Xie, Y. and Sherwood, P. M. A.,** *Chem. Materials,* 2, 293, 1990; **Xie, Y. and Sherwood, P. M. A.,** *Chem. Materials,* 3, 164, 1991; **Xie, Y. and Sherwood, P. M. A.,** *Appl. Spectrosc.,* 44, 1621, 1990.

89. **Barr, T. L., Mohsenian, M., and Chen, L. M.,** *Appl. Surface Sci.,* 51, 71, 1991.

90. **Dilks, A.,** in *Electron Spectroscopy — Theory, Techniques and Applications,* Vol. 4, Brundle, C. R. and Baker, A. D., Eds., Academic Press, London, 1981.

91. **Pireaux, J. J., Riga, J., Caudano, R., and Verbist, J.,** in *Photon, Electron and Ion Probes of Polymer Structure and Properties,* Dwight, D. W., Fabish, T. J., and Thomas, H. R., Eds., (ACS Symposium Series 162), American Chemical Society, Washington, DC, 1981, 169.

92. **Clark, D. T. and Harrison, A.,** *J. Polymer Sci. Polymer Chem. Ed.,* 19, 1945, 1981.

93. **Yin, M. P., Koutsky, J. A., Barr, T. L., Rodriguez, N. M., Baker, R. T. K., and Klebanov, L.,** *Chemistry of Materials,* 5, 1024, 1993.

94. **Baker, R. T. K. and Harris, P. S.,** in *Chemistry and Physics of Carbon,* Vol. 14, Walker, P. L., Jr. and Thrower, P. A., Eds., Dekker, Marcel, New York, 1978, 83.

95. **Baker, R. T. K., Barber, M. A., Harris, P. S., Feates, F. S., and Waite, R. J.,** *J. Catal.,* 26, 51, 1972.

96. **Baker, R. T. K., Harris, P. S., and Terry, S.,** *Nature,* 253, 37, 1975; **Baker, R. T. K. and Chludzinski, J. J.,** *J. Catal.,* 64, 464, 1980; **Kim, M. S., Rodriguez, N. M., and Baker, R. T. K.,** *J. Catal.,* 134, 253, 1992.

97. (a) **Barr, T. L.,** *J. Catal.,* submitted. (b) **He, H., Alberti, K., Barr, T. L., and Klinowski, J.,** *J. Phys. Chem.,* to be published, 1993. (c) **He, H., Alberti, K., Barr, T. L., and Klinowski, B.,** *Nature,* 365, 429, 1993.

98. **Breck, D. W.,** *Zeolite Molecular Sieves,* Wiley-Interscience, New York, 1974.

99. (a) **Flanigen, E. M.,** *Zeolite Chemistry and Catalysis,* Rabo, J. A., Ed., (ACS Monograph 171), American Chemical Society, Washington, DC, 1976, chap. 2. (b) **Ward, J. W.,** in *Zeolite Chemistry and Catalysis,* Rabo, J. A., Ed., (ACS Monograph 171), American Chemical Society, Washington, DC, 1976, chap. 3.

100. **Angell, C. L.,** *J. Phys. Chem.,* 77, 222, 1973.

101. **Smith, J. V.,** *Zeolite Chemistry and Catalysis,* Rabo, J. A., Ed., (ACS Monograph 171), American Chemical Society, Washington, DC, 1976, chap. 1.

102. **Barrer, R. M.,** *Zeolite and Clay Minerals as Sorbents and Molecular Sieves,* Academic Press, London, 1978.

103. **Kasai, P. H. and Bishop, R. J., Jr.,** *Zeolite Chemistry and Catalysis,* Rabo, J. A., Ed., (ACS Monograph 171), American Chemical Society, Washington, DC, 1976, chap. 6.

104. **Maroni, V. A., Martin, K. A., and Johnson, S. A.,** *Perspectives in Molecular Sieve Science,* Frank, W. H. and Whyte, T. E., Jr., Eds., (ACS Symposium Series 368), American Chemical Society, Washington, DC, 1988, chap. 5.

105. (a) **Thomas, J. M. and Klinowski, J.,** *Adv. Catal.,* 33, 199, 1985; (b) **Klinowski, J.,** *Spectroscopic and Computational Studies of Supermolecular Systems,* Davis, J. E. D., Ed., Kluwer Academic, Amsterdam, 1992, 115.

106. **Barr, T. L.,** *Div. Petroleum Chem.,* ACS, 23, 82, 1978.

107. **Barr, T. L.,** *Practical Surface Analysis,* Briggs, D. and Seah, M. P., Eds., John Wiley & Sons, Chichester, England, 1983, chap. 8.

108. **Madey, T. E., Wagner, C. D., and Joshi, A.,** *J. Electron. Spectrosc. Rel. Phenomena,* 10, 359, 1977.

109. Private communications from ESCA scientists at UOP, Union Carbide, and Englehard.

110. **Barr, T. L.,** presented 12th Int. Vac. Congress, The Hague, 1992, to be published.

111. **Dwyer, J., Fitch, F. R., Quin, G., and Vickerman, J. C.,** *J. Phys. Chem.,* 86, 4574, 1982.

112. **Barr, T. L., Chen, L. M., and Mohsenian, M.,** to be published.

113. **Argauer, R. J. and Landolt, G. R.,** U.S. Patent, 3702886, assigned to Mobil Oil Co., 1972.

114. **Flanigen, E. M. and Pattan, R. L.,** U.S. Patent, 4073685, assigned to Union Carbide Co., 1978.

115. **Mortier, W. J.,** *J. Catal.,* 55, 138, 1977.

116. **Jacobs, P. A.,** *Catal. Rev. Sci. Eng.,* 24, 415, 1982.

117. **Barthoneuf, D. J.,** *J. Phys. Chem.,* 83, 249, 1974.

118. **Barr, T. L.,** to be published.

119. **Barr, T. L.,** submitted to Microporous Materials.

120. **McDaniel, C. V. and Maher, P. K.,** *Zeolite Chemistry and Catalysis,* Rabo, J. A., Ed., (ACS Monograph 171), American Chemical Society, Washington, DC, 1976, chap. 4.

121. **Corma, A., Formes, V., Palleta, D., Cruz, J. M., and Ayerba, A.,** *J. Chem. Soc. Commun.,* 333, 1986.

122. **Burch, R., Ed.,** Pillared clays, *Catal.Today,* 2, 185, 1988.

123. **Pines, H.,** *The Chemistry of Catalytic Hydrocarbon Conversions,* Academic Press, New York, 1981, 221.

124. **Barr, T. L., Fries, C. G., Cariati, F., Bart, J. C. J., and Giordano, N.,** *J. Chem. Soc. Dalton Trans.,* 1825, 1983.

270

125. **Barr, T. L.,** in *Preparation and Characterization of Catalysts,* (ACS Book Ser., Vol. 411), Bradley, S. A., Bertolacini, R. J., and Gattuso, M. J., Eds., 1989, 203.
126. **Heinemann, H.,** *Catalysis: Science and Technology,* Anderson, J. R. and Boudart, M., Eds., Springer-Verlag, Berlin, 1981, 1.
127. **Burch, R.,** *J. Catal.,* 71, 348, 360, 1980.
128. **Clarke, J. K. A. and Creamer, A. C. M.,** *Ind. Eng. Chem. Prod. Dev.,* 20, 575, 1981.
129. **Bouwman, R. and Biloen, P.,** *J. Catal.,* 48, 209, 1977.
130. **Biloen, P., Helle, T. N., Verbeek, H., Dautzenberg, F. M., and Sachtler, W. M. H.,** *J. Catal.,* 63, 112, 1980.
131. **Sexton, B. A., Hughes, A. E., and Foger, R.,** *J. Catal.,* 88, 466, 1984.
132. **Cocke, D. L., Johnson, E. Q., and Merrill, R. P.,** *Catal. Rev.,* 26, 1984.
133. **Czaran, E., Finster, J., and Schnabel, K. H.,** *Z. Anorg. Ablg. Chem.,* 443, 175, 1978.
134. **Dautzenberg, F. M. and Walters, H. B. M.,** *J. Catal.,* 51, 267, 1984.
135. **Andrade, J. D., Hlady, V., Herron, J. and Lin, J.-N.,** Symposium on Surfaces in Biomaterials, Minneapolis, MN, October 1991.
136. **Gardella, J. A., Grobe, G. L., III, and Salvati, L.,** Symposium on Surfaces in Biomaterials, Minneapolis, MN, October 1991.
137. **Cooper, S. L.,** Symposium on Surfaces in Biomaterials, Minneapolis, MN, October 1991.
138. **Gassman, P. G.,** Symposium on Surfaces in Biomaterials, Minneapolis, MN, October 1991.
139. **Castner, D. G. and Ratner, B. D.,** Symposium on Surfaces in Biomaterials, Minneapolis, MN, October 1991.
140. **Larsson, N., Steiner, P., Eriksson, J. C., Maripuu, R., and Lindberg, B.,** *J. Colloid Interface Sci.,* 90, 127, 1982; **Maripuu, R.,** Ph.D. thesis, University of Uppsala, 1983.
141. First Symposium on Surfaces in Biomaterials, Minneapolis, MN, 1991.
142. **Ranieri, J. P., Bellamkonda, R., and Hebischer, P., Jr.,** Symposium on Surfaces in Biomaterials, Minneapolis, MN, October 1991.
143. **Healy, K. E. and Ducheyne, P.,** Symposium on Surfaces in Biomaterials, Minneapolis, MN, October 1991.
144. **Clemmer, C. R., Leavitt, A. J., and Beebe, T. P., Jr.,** Symposium on Surfaces in Biomaterials, Minneapolis, MN, October 1991.
145. **Linton, R. W., Guarisco, V. F., and Lee, J. J.,** Symposium on Surfaces in Biomaterials, Minneapoilis, MN, October 1991.
146. **Gassman, P. G. and Winter, C. H.,** *J. Am. Chem. Soc.,* 110, 6130, 1988.
147. **Giroux, T. G. and Cooper, S. L.,** *J. Appl. Polymer Sci.,* 43, 145, 1991.
148. **Ratner, B. D.,** *Concise Encyclopedia of Medical and Dental Materials,* Williams, D. F., Ed., Pergamon Press, Oxford, 1989.
149. **Ratner, B. D.,** *Ann. Biomed. Eng.,* 11, 313, 1983.
150. **Ratner, B. D. and McElroy, B. J.,** *Spectroscopy in the Biomedical Sciences,* Gendreau, R. M., Ed., CRC Press, Boca Raton, FL, 1986.
151. **Goddard, E. D. and Harris, W. C.,** *J. Soc. Cosmet. Chem.,* 38, 233, 1987.
152. **Twu, D. and Barr, T. L.,** unpublished results, 1988.
153. **Barr, T. L.,** unpublished.
154. **Fowler, D. E., Brundle, C. R., Lerczak, J., and Holtzberg, F.,** *J. Electron Spectrosc. Rel. Phenomena,* 52, 323, 1990.
155. **Meyer, H. M., Hill, D. M., Wagener, T. J., Gao, Y., Weaver, J. H., Capen, D. W., II, and Gozetta, K. C.,** *Phys. Rev. B,* 38, 6500, 1988.
156. **Larson, P. E.,** *J. Electron Spectrosc. Rel. Phenomena,* 4, 213, 1974.

157. **Brundle, C. R.,** *Faraday Discuss. Chem. Soc.,* 60, 159, 1975.
158. **Wertheim, G. K. and Hufner, S.,** *Phys. Rev. Lett.,* 28, 1028, 1972.
159. **Nelson, D. L., Whittingham, M. S., and George, T. F., Eds.,** *Chemistry of High-Temperature Superconductors I,* (ACS Symp. Ser. 351), American Chemical Society, Washington, DC, 1987.
160. **Nelson, D. L. and George, T. F., Eds.,** *Chemistry of High-Temperature Superconductors II,* (ACS Symp. Ser. 377), American Chemical Society, Washington, DC, 1980.
161. **Harper, J. M., Colton, J. H., and Feldman, L. C., Eds.,** *Thin Film Processing and Characterization of High-Temperature Superconductors,* (AVS Ser. 3), AIP, New York, 1988.
162. **Pauling, L.,** *Phys. Rev. Lett. A,* 59, 225, 1987.
163. **Anderson, P. W., Basksian, G., and Hou, T. C.,** *Phys. Rev. Lett.,* 58, 2790, 1987.
164. **Hubbard, J.,** *Proc. Roy. Soc. London A,* 276, 238, 1963; 227, 237, 1964; 281, 401, 1964.
165. **Van der Laan, G., Westra, C., Haas, C., and Sawatsky, G. A.,** *Phys. Rev. B,* 23, 4369, 1981; **Sawatsky, G. S. and Allen, J. W.,** *Phys. Rev. Lett.,* 55, 418, 1985.
166. **Shen, Z.-X., Lindberg, P. A. P., Spicer, W. E., Lindau, I., and Allen, J. W.,** in *High T_c Superconducting Thin Films, Devices and Characterization,* Margaritondo, E., Joynt, R., and Onellion, M., Eds., (AVS/AIP Pro. Ser. 182), 1989, 330.
167. **Veal, B. W. et al.,** *Physica C,* 158, 276, 1989.
168. **Chen, L. M., Barr, T. L., Krauss, A. R., Gruen, D. M., and Gady, B.,** *J. Vac. Sci. Technol. A,* 9(3), 394, 1991.
169. **Gelius, U., Wannberg, B., Baltzer, P., Fellner-Feldegg, H., Carlsson, G., Johansson, C.-G., Larsson, J., Münger, P., and Vegerfors, G.,** *J. Electron Spectrosc. Rel. Phenomena,* 52, 747, 1990.

Chapter 9

SURVEY OF ADDITIONAL PROMISING NEW METHODS

I. MODERN ESCA AND THE CONTINUING UPPSALA CONNECTION

Interest in the phenomena of photoelectron spectroscopy and its use has now spread to nearly every corner of the globe. A variety of laboratories throughout the world have made and are continuing to make substantial contributions to the development of this technology. The most varied, innovative, and productive center remains, however, in the Department of Physics and related facilities of the University of Uppsala, Sweden (UUIP), where ESCA first began under the guidance of Prof. Kai Siegbahn.

Because of particular points of emphasis, some of the recent Uppsala developments are not described in this book, including substantial innovations in high-resolution, versatile ultraviolet photoelectron spectroscopy (UPS) by Siegbahn and his colleagues,[1] and the implementation of a variety of novel methods for handling and examining organic and biological systems via (XPS).[2] In addition, a strong tradition of using high-resolution XPS in conjunction with detailed quantum calculations to study the characteristics of the photoemission of inert gases began in the early 1970s at UUIP (largely under Gelius[3]), and continues today under the combined direction of Gelius and Martensson.[4] In these areas, implementation of a special high-resolution ESCA system (see below) has permitted the detailed analysis of loss spectra, including shake-up and shake-off lines, as well as correlation effects, for inert gases and related systems.[5] Extensions of these methods to other gases such as Hg are now being published.[6]

Despite these omissions, Uppsala still has played a major role in many of the areas described in this book. For example, the development of the new, expanded, high-resolution XPS system described in Chapter 3 has been primarily led by Ulrik Gelius, Professor of Physics at the University of Uppsala.[6] In addition, some of the results described in Chapters 6 and 7 were developed by the author during a visiting professorship at the UUIP. Several sections in this chapter describe developments that have substantially (in some cases, primarily) evolved from the research efforts of UUIP scientists (i.e., surface-to-bulk photoemission, theory, and experimentation, under N. Martensson, and liquid and solution phase ESCA, under H. Siegbahn).

II. INVERSE PHOTOEMISSION AND TWO-PHOTON PHOTOEMISSION

A. INTRODUCTION

A problem of substantial importance (particularly to scientists involved with semiconductors, adsorption phenomena, and sensor technology) is the character of the electronic energy level region between the Fermi level and the top of the conduction band. Certain surface states, some color centers, and a variety of other types of trap states are often found in this region, as are the critical onset features of the conduction band. Most of these features are as important to the microelectronic behavior of these materials as are the valence band and assorted discrete states found below the Fermi edge, E_F. It is also important to note that, while the states above E_F sometimes resemble mirror images of the corresponding (bonding) states found below E_F, generally they do not — and often it is this dissimilarity that is critical to the electronic behavior of the material. For these and related reasons it would be very useful if the type of spectral generation produced by photoelectron spectroscopy for states below E_F were also possible for those states above

E_F. Unfortunately, the nature of the process obviously makes that type of spectroscopy impossible. It turns out, however, that it is possible to change the nature of the process in a way that permits the monitoring of some features of the states located in energy between the vacuum level, E_v, and the Fermi level, E_F. There are several methods for doing this, but we will describe the two methods that are already producing useful results. These are Bremsstrahlung isochromat or inverse photoemission spectroscopy (IPES)[7] and two-photon photoemission spectroscopy.[8]

B. INVERSE PHOTOEMISSION

As pointed out by several researchers, IPES actually originated in the early part of this century[9] during attempts to thoroughly test and exploit the full nature of the Einstein relation:

$$hv = \frac{1}{2}mv^2 + E_{BE} + q\phi \qquad (9\text{-}1)$$

During these studies, it was discovered that the process could be run in reverse, i.e., the sample could be bombarded with electrons, some of which are captured by the sample atoms, with their energy lowered into certain unoccupied states. When this occurs, photons are generally emitted. If this process is generated by the procedure described in Figure 9-1, the bombarding electrons could be swept in energy, E, while the photon energy is maintained at a constant value. After taking proper account of the work function, the measurement will produce a spectrum that reveals the density pattern of the (before experiment) empty energy states for the material system. In this process, of course, capture of the electron by the system requires that the measured electron states are located in the energy region between the vacuum and Fermi levels. With work functions generally between 3 and 8 eV, measurements of this type will usually provide a "picture" of the densities of state in the critical region of interest.

It is common to run IPES with ultraviolet, rather than X-ray excitation energy.[10] In addition, one often resolves the momentum or k vectors of the photon emitting solid.[7,11] This form of spectroscopy (labeled as KRIPES) is the inverse equivalent of angular-resolved photoemission spectroscopy (ARPES). Fairly extensive explorations of the theoretical aspects of IPES have already been made.[12]

The IPES photon output is often detected using iodine-filled Geiger-Mueller counters.[7] On the other hand, some researchers have employed monochromator-based (generally Seya-Namioka) grating detectors.[13] A low-energy gun, often of the Pierce type, is usually employed as the electron source.[7,13] It is possible to organize the system with a source producing spin-polarized electrons, using, for example, the photoemission from a negative affinity GaAs surface.[7] This produces the inverse form of spin-polarized photoemission (SPIPES). Just as spin-polarized photoelectron diffraction (SPPED) is revealing significant features of the valence regions of magnetic materials, SPIPES is doing the same for states above E_F for these materials (see below).

The first uses of IPES (and particularly, KRIPES) were to map the band features above E_F for good conductors [such as Cu(001)[14] or Co(0001)[13]] in a manner analogous to that accomplished in the valence region by ARPES. Extensions to include band distortions induced by adsorbates [e.g., CO on Ni(111)[13] and oriented O on Ni(100) and on Ni(111)[15]] were also rapidly achieved. More detailed adsorption studies recently have been performed, including oxygen[16] and hydrogen[17] on W(100). In these cases, evidence of reconstruction and extensive band structure changes were noted. The mapping of the unoccupied states formed by metals such as Ag and Au, when adsorbed on a semiconductor, e.g., Ge(111), has also been accomplished.[18]

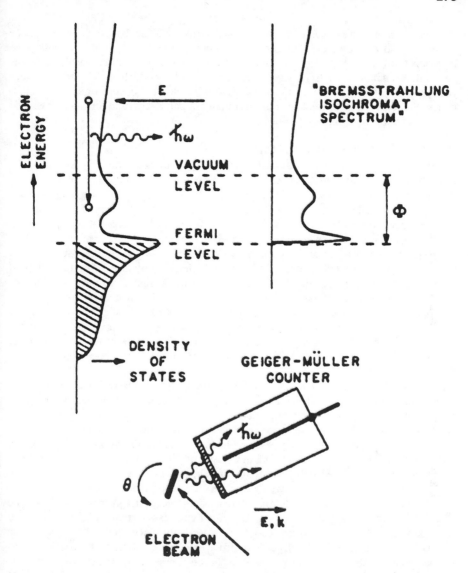

Figure 9-1 Energetics of inverse photoemission, indicating how the Bremsstrahlung isochromat spectrum ($\hbar\omega$ constant, E swept) should replicate features in the unoccupied density of states (the lower portion of the figure is a schematic diagram of one of the possible experimental arrangements).

Obviously, one of the challenges of a method such as IPES is to detect and characterize defects and surface states for metals and semiconductors. An early example of this prospect was described by Johnson and Smith,[19] who mapped the disappearance of a surface state of Pd following Cl_2 adsorption. This type of analysis is now fairly routine.[20]

There are several obvious problems inherent in any attempt to extend the use of IPES and its companion techniques to analyses of more practical materials without, for example, the use of oriented single crystal surfaces. No one has, of course, solved or perhaps even considered the problem of charging and lack of Fermi edge coupling during IPES-type measurements.

A problem of interest would be to try to achieve above E_F analyses related to those of (valence) band structures and band gaps of well-formed group A oxides and of these same materials following alterations to induce various discrete and/or continuous defect states and/or nonstoichiometries (see Chapter 7, Section II). Many of these cases should exhibit corresponding states above E_F, and an examination of the trailing edge of the conduction bands should permit determinations of the important band gaps, and thus ionicities, and defect states for these oxide materials.

C. TWO-PHOTON SPECTROSCOPY

Some of the previously mentioned problems with inverse photoemission may be circumvented in two-photon photoemission spectroscopy.[8] The advantage of this procedure is that it makes (better) use of high-energy sources and, due to the use of lasers, may be employed to follow certain temporal features. It may also be employed in an angular-resolved mode, particularly since the lasers employed may be polarized. A typical two-photon photoemission arrangement is presented in Figure 9-2.[21] Note that intense pulses of lasers are employed to excite some of the electrons below E_F to select unoccupied states. This method may be employed to examine relaxation dynamics,[8] and a variety of features of the image potential.[21] One may also use this method to measure resonant excitations and relaxations into lower unoccupied states. In this manner, angular-resolved two-photon photoemission was employed to examine both the bulk and surface states of Pd(111).[21] The technique also has been used to examine the empty states of various symmetries of carbon (diamond).[22] Thus, for the first time, some of the features of the important reconstruction of the surfaces of the latter material have been mapped.

D. EMPTY STATE ANALYSIS — GENERAL CONSIDERATIONS

Due to a number of problems, none of these empty state methods has, as yet, been applied to the analysis of semiconductors or other applied systems. Several of these problems seem to relate to the resulting complexity of the multifaceted band structure and also to the fact that most of these systems will exhibit some form of the charging problem and Fermi edge decoupling. In addition, in the case of certain insulators a finite-sized gap opens up between the (pseudo) Fermi edge and the onset of the conduction band.

There is no reason why the charging and decoupling problems cannot be handled in exactly the same manner as with conventional photoelectron spectroscopy. Thus, the placement of an appropriate electron or UV flood gun on a system with IPES, and/or a laser-generated two photon system, should be reasonably successful. A band gap on the conduction band side of E_F actually should prove beneficial in these types of analyses, as surface states, defects, and reconstructions become essentially easier to observe as discrete states (in the gap). This should be particularly useful in studies of select cases of the relaxation dynamics of excited surface states.

III. PHOTOELECTRON DIFFRACTION (PED)

Photoelectron diffraction (PED)[23] (or X-ray photoelectron forward scattering[24]) is an adjunctive procedure to photoelectron spectroscopy that has recently become very popular with several major research groups. Its popularity is, in part, due to the realization that a reasonable version of PED can be accomplished with a conventional XPS system endowed with a capability to perform good angular-resolved (AR) XPS[25,26] and, in part, to the ability of PED to not only complement "low energy electron diffraction" (LEED), but to exceed LEED in the information provided about the short-range order (SRO) effects that may exist in a solid system.[24,27] In addition to these features, it turns out that by adding an appropriate spin-polarization capability one can extend the technique to a spin-polarized form of PED (SP-PED).[28] As with so many procedures in this field, PED

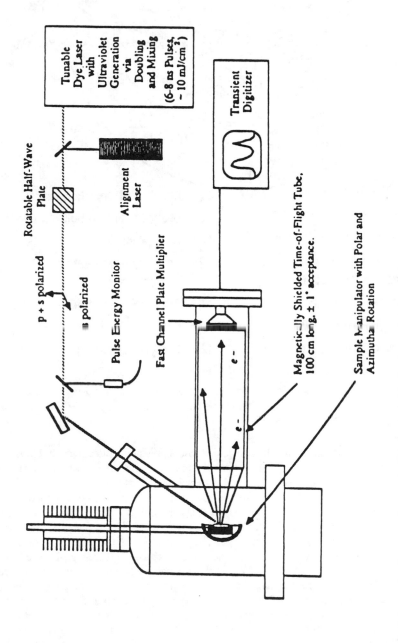

Figure 9-2 Schematic diagram of experimental apparatus for two-photon spectroscopy.

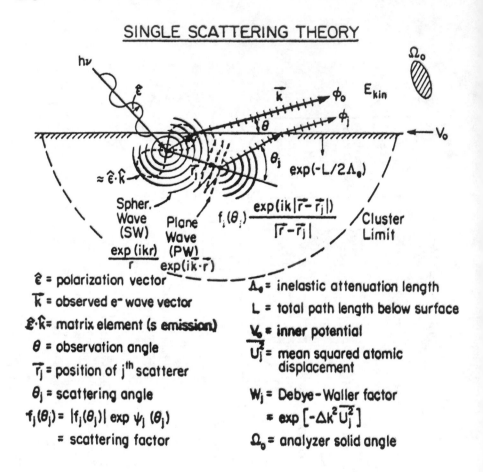

Figure 9-3 Schematic illustration of the features of a typical forward-scattering (photoelectron diffraction) experiment. The terms are arranged to indicate the possible single scattering cluster (SSC) models.

had its origins in the research laboratories of Siegbahn and colleagues in Uppsala, Sweden.[29] This group was the first to note that, if one employs angular resolution to create core level photoelectrons in a single crystal material, one finds that the spectral peak of those photoelectrons that are channeled to be directed at neighboring atoms is enhanced in intensity by forward scattering (Figure 9-3), whereas photoelectrons directed off-axis are, in essence, retarded. Thus, a characteristic undulating interference-type pattern is produced for a particular key core electron peak as a function of the particular angle of orientation of the single crystal and/or of the angle of emission (see, e.g., Figure 9-4). Since the directions involved depend upon the crystal geometry, rather complicated mixtures and interferences may ensue. However, it seems apparent that the great majority of the enhanced scattering occurs from atoms immediately adjacent to the emitter.[23] The latter may be visualized semi-classically using Wentzel-Kramers-Brillouin type multiple-scattering theory (Figure 9-5), in which case the zeroth order constructive interference is enhanced by an attractive deflection of the emitted wave in the direction parallel to the internuclear axis.[30] Considering contributions from higher orders, constructive interfer-

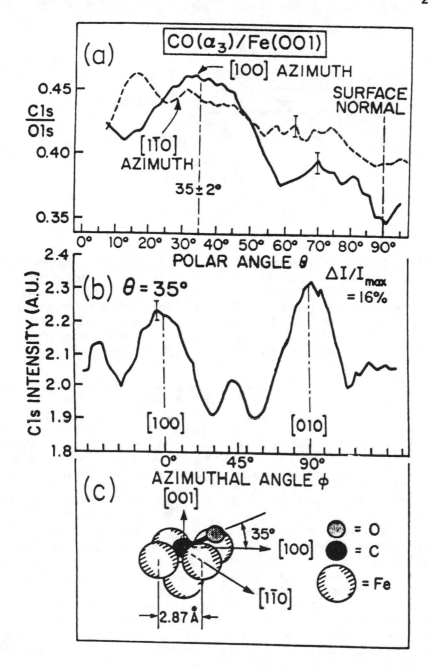

Figure 9-4 (a) Experimental polar scans of the C(1s)/O(1s) intensity ratio for the α state of CO on Fe (100); the C(1s) kinetic energy is 1202 eV; curves are shown for two azimuths: [100] (solid curve) and [1, −1, 0] (dashed curve). (b) Experimental azimuthal scan of C(1s) intensity for the α_3 state of CO at a polar angle of 35°, chosen to coincide with the peak in the [100] data of (a). (c) Bonding geometry as deduced from these data.

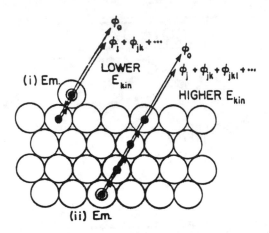

Figure 9-5 Illustration of assumptions used in the single scattering cluster (SSC) model, with various important quantities indicated. Two types of multiple scattering corrections to the SSC model that may be significant for certain energies and geometries: (i) at lower energies of <200 eV, back scattering from a nearest-neighbor atom behind the emitter, and then forward scattering by the emitter; (ii) multiple forward scattering along lines of closely spaced atoms that leads to a reduction in the expected intensity enhancement, particularly at higher energies of >500 eV.

ences fall off rapidly after the first order. Originally, there was a significant controversy surrounding the need to utilize either single[23,31] or multiparticle scattering theory to describe this phenomena,[30,32,33] but it now seems established that single particle theory is generally adequate[23] (at least for qualitative purposes). Important details from some experiments appear to be enhanced by multiple scattering considerations,[24,34] and may, in some instances, be improperly accounted for without due consideration of the latter.

PED experiments may be carried out in a variety of modes, including (1) variable polar angle, (2) variable azmuthal angle, or (3) fixed angle, with variable photon energy (see Figure 5-2 in Chapter 5).[23,26,28,31,35] The first two approaches are particularly useful in determining the bonding site location and geometry for adsorbates,[23,26,31] whereas the latter is useful in determining certain adsorbate-substrate features, such as bond length.[31,35] It appears that the variable energy approach also depends upon the recognition of surface-to-bulk photoemissions (see Section VII).[31,36] Because of its surface sensitivity, the variable energy technique is generally employed to examine monolayer adsorptions or epitaxial growths, and not the character of the substrate.[35,36]

Because of the need to either vary the photoelectron energy or the angular orientation (polarization) of the sample vis-a-vis the photon source, PED is often accomplished at a synchrotron radiation center.[23,26,31,35,37,38] For certain applications this is not necessary, however, and excellent examples of the angular-resolved form have been achieved with commercial ESCA systems.[23,24,26–29,34]

As mentioned above, many of the PED applications achieved so far have emphasized the superior nature of this approach compared to LEED in the determination of SRO.[23,27] Thus, in the determination of epitaxial overlayers and sandwich structures of Au, and Ag on Ni, and Cu on Ni(100),[39] Egelhoff found it useful to employ LEED to determine the long-range order, and PED to determine the SRO.[24,27,39] The latter technique is particularly useful in demonstrating the presence of surface segregations. Egelhoff and Steigerwald[40] have recently extended this approach to include the effects of certain molecular and atomic gases on these metal-on-metal epitaxial systems. At or around room temperature, some of these gases are found to "float out" over the growing metal surface — thus, these gases do not affect the evolving epitaxy. At lower temperature, the more strongly bound gases tend to suppress the agglomeration and interdiffusion of the transition metals controlling epitaxial growth.

Extensions of this procedure are obviously limited by the need to start with single crystal orientations. This means that the method is very useful in examinations of monolayer adsorptions and epitaxial film growths on metals and semiconductors, but would appear to have little or no direct applicability for applied problems in corrosion, polymers, or catalysis. It may, of course, be the crucial method in certain chemisorption and single crystal catalysis studies. The techniques also should be important in applications to device growth areas, such as those related to complicated junctions and quantum well problems. It is important to reiterate that the method provides an important adjunct to LEED and ion-scattering methods by providing short-range chemical, as well as structural, information.

It should be noted that one might also exploit Auger electrons for many of the diffraction processes being described herein.[25]

In addition, an interesting, and potentially very important, extension of photoelectron diffraction has recently been made that involves the use of holographic transformations. The process was theoretically evolved by Szoeke[41] and Barton,[42] and has recently been experimentally realized in a study of various Cu crystals by Harp et al.[43] With this procedure, one may determine (among other things) the positions of near-neighbor atoms to within 0.5 Å or less.[44] Thus, the accentuation of short-range information proposed by Egelhoff may be much more dramatic than first realized. These observations suggest an obvious interconnection between photoelectron diffraction and photoelectron microscopy (see Section V).

IV. SPIN-POLARIZED PHENOMENA

As pointed out by Pierce and Celotta,[45] a number of methods now exist that are capable of exploiting the presence of polarized spin situations at or near the surface of certain systems. Systems with a non-zero spin may exhibit finite spin orbit effects (particularly high Z elements) or finite exchange effects. Several of these approaches involve some form of photoemission, including spin-polarized photoemission, spin-polarized inverse photoemission, and spin-polarized photoelectron diffraction.

A. SPIN-POLARIZED PHOTOEMISSION SPECTROSCOPY (SPPES)[45,46]

This technique has been utilized in a variety of modes to determine such effects as magnetic ordering and changes in the saturation magnetization. A typical example includes the change from the antiferromagnetic to the ferromagnetic state experienced by the Cr(100) surface following the incorporation of a few percent of oxygen.[47]

It is also possible to exploit resonance photoemission (Section VIII) in the spin-polarized mode. Thus, for example, the unique satellites often achieved by this process[48] have been explained in terms of the generation of a virtual bound state of coupled holes. According to Feldkamp and Davis,[49] one should be able to spin-polarize these resonant satellites,[45] and subsequent studies have verified these expectations.

B. SPIN-POLARIZED INVERSE PHOTOEMISSION SPECTROSCOPY (SPIPES)[50]

This method permits the detection of the spin polarization of states in the region between the Fermi edge and the vacuum level. The spin polarity may be registered through the use of a polarized electron source (e.g., polarized GaAs). Angular-resolved versions of this process have demonstrated majority/minority unfilled state populations. The method seems relatively easy to apply.[45]

C. SPIN-POLARIZED PHOTOELECTRON DIFFRACTION (SPPED)[51]

This is a method that has been championed by Fadley and his co-workers,[51] who have exploited the multiplet splittings of the S states that exist for many transition metal systems. Thus, for example, the spin orbit split, Mn 3s emission from single crystal $KMnF_3$, has been forward scattered with a substantial enhancement of the short-range magnetic order.[52] The latter was the first case of a successful SPPED experimental result, and it has subsequently been exploited in many other examples.

Much like the now-classic electron spin techniques, such as ESR, the spin-polarized surface methods are obviously quite selective in their use areas. However, just like ESR, these various selective methods are providing unique information about elemental systems in extremely important recently developed areas involving magnetic materials.[53] The unique nature of the spin-oriented effects realized from these materials greatly enhances the potential importance of these spin-polarized methods.[54] Just as we have seen in many other rather esoteric areas, however, most applications to analyses of practical materials await further developments.

V. PHOTOELECTRON MICROSCOPY

The use of photoelectrons in microscopy was simultaneously explored with their employment in spectroscopy. In fact, during the 1930s and 1940s, a number of groups, particularly in Germany,[55] tried to develop a photoelectron emission microscopy (PEEM) in parallel with the development of the (now) conventional, transmission electron microscopy (TEM). The advantages of the use of relatively high-energy electron beams in TEM (particularly their ability to be focused and rastered), led to an eventual acceptance in the 1950s and 1960s of commercial transmission and scanning electron microscopy (SEM) systems — systems that are now used by the tens of thousands in laboratories around the world. During this period, on the other hand, photoelectron microscopy was seeing only modest development and use.

As displayed in Figure 9-6, a photoelectron microscope often provides an electron collection lens assembly that is fairly similar to that employed in conventional TEM.[56] In the example depicted, a typical three-stage lens arrangement (objective, intermediate, and projection) is specified — although some of the earlier models employed just one lens.[55] The feature that makes the PEEM unique, of course, is the source of the photoelectrons. In the case displayed, a UV lamp is suggested.[57] The relatively "soft" nature of the latter (particularly compared to a focused beam of high-energy electrons) provides one of the major advantages of the technique — a substantial reduction in possible damage of the sample by the beam. Thus, PEEM has proven to be much more amenable to examinations of biological species and other sensitive surfaces, particularly when compared to TEM and SEM.[58] The relatively soft excitation source in this form of PEEM produces photoelectrons only from the valence and near valence states of atoms (generally <40 eV) that are located very near the surface of the samples involved. SEM and TEM, on the other hand, produce scattered primary and secondary electrons that often exhibit energies of several thousand electronvolts. As shown in Figure 9-7, the latter may initiate several thousand angstroms inside the material being examined, whereas UV PEEM is a relatively *surface-sensitive microscopy.*[57,58]

Because of the similarities in electron optics, several PEEM systems have been developed as alterations of, or attachments to, conventional TEM systems.[56] One of the major disadvantages of this arrangement is that it often does not lend itself well to the introduction of UHV technology. As implied above, PEEM is generally quite surface sensitive. If this feature is to be exploited for studies of metals, alloys, and semiconductors (most of these systems are readily subject to oxidation in air), it is important to preserve the integrity of the surface of the samples being examined — hence, UHV is required.[60]

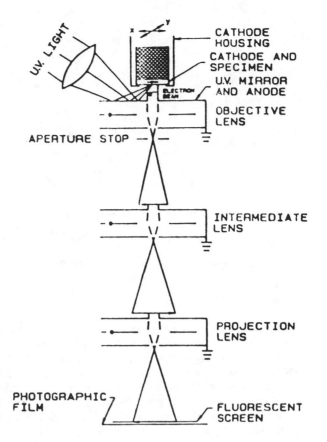

Figure 9-6 Schematic diagram of a photoelectron microscope. The lenses can either be electrostatic (shown here) or magnetic.

Due to the relatively low energy of the photoelectron emission, a better vacuum is also needed in the lens column of the PEEM, compared to the TEM, in order to facilitate the path of the low-energy electrons to the detector.

Recently, UV photon source PEEM has seen substantial increase in its interest and development. Some groups have attempted employing UV lasers (better potential focusing),[61] since it is obvious that, for many uses, enhanced brightness and focusability would be beneficial. For systems with more conventional sources, several research groups have spearheaded this development, noting particularly the efforts of Griffith and colleagues.[57–60] This group has noted that UV source PEEM (with UHV) provides not only less beam damage, but often superior contrast.[58,60] Thus, the topological valleys and hills characteristic of most biological samples may be readily resolved.[59] Recent improvements in sample treatment and mounting have been combined with this enhancement of contrast and lack of sample damage to provide a superior method for biological study.[58,62] For many of these cases, the coherent lack of spatial resolution in UV PEEM, compared to the best TEM or SEM, may not be a major flaw (as the high-resolution modes of these latter techniques may produce too much beam damage for their employment). The enhanced contrast of UV PEEM and its surface sensitivity also may be very useful in analyses of corrosion problems. A very interesting example produced by Brode et al.[63] is shown in Figure 9-8, illustrating the enhanced surface sensitivity of PEEM compared to SEM in an investigation of the oxidation of aluminum bronze.

The major unexploited feature of the UV PEEM system described so far is the omission of sufficient spectroscopic energy resolution to simultaneously try to discern

284

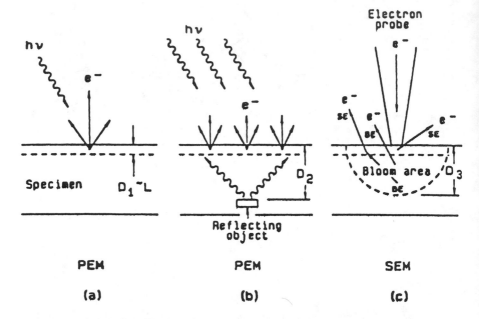

Figure 9-7 Schematic comparison of depth of information in PEEM and SEM: (a) in PEEM, light is absorbed over a relatively large range of depths in the specimen, but only those electrons that are very close to the surface that are emitted within a depth D_1 (approximately equal to the escape depth, L) can contribute to the image; (b) an example of increased depth of information (D_2) in PEEM caused by reflected light; information from within or beneath the specimen (depth D_2) is usually not present but, if present, such information is readily identified because it appears as a diffuse background on which the surface information from depth D_1 is imaged at high resolution; (c) in SEM, the backscattered electrons (BE) carry information from below the escape depth, and some of this signal is converted into secondary electrons (SE) near the surface so that in either mode, the depth of information, D_3, is larger than D_1.

quantification and chemistry of the materials under study. This combination of microscopy and spectroscopy has been the goal of a number of groups exploring photoelectron microscopy with X-ray or synchrotron radiation sources.

One of the first successful efforts with X-ray sources, however, was more concerned with the development of a scanning capability than spectral resolution. Considering the difficulties involved in direct attempts to focus X-rays, Cazaux et al.[64] in the early 1980s ingeniously realized that, if a high energy source of electrons were bombarded on an anode, oriented as a backing onto thin samples, X-rays may be generated that tend to scatter forward through the attached sample from the front of the anode. When produced with the right energies, the latter X-rays will produce photoelectrons in the sample — some of which escape from the top of sample and may, in turn, be energy analyzed in a cylindrical mirror analyzer. If the original electron source is a focused beam from a (converted) SEM, it is possible to employ this technique to generate some focusing and scanning of the outgoing photoelectrons. The spatial resolution achieved by this procedure has been reduced to ~200 to 300 Å. The method does not yet achieve high spectral resolution, but its principal drawback would appear to be the substantial restriction placed upon the type and thickness of the materials that may be examined as samples.

Several approaches to ESCA have attacked the microscopy problem from the opposite extreme, i.e., beginning with good to very good methods for spectral analyses, they have

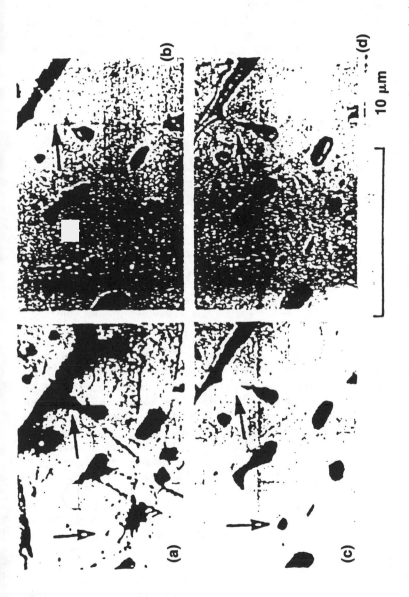

Figure 9-8 Experimental comparison of the relative depths of information in SEM and PEEM using an aluminum bronze sample: (a) SEM (primary electron energy, 30 keV); (b) to (d) PEEM; (b) surface cleaned by ion etching; (c) after prolonged ion etching; (d) after more prolonged etching. The SEM micrograph shows more information from below the surface which does not appear in the PEEM micrograph (b), but is revealed during successive stages of etching in (c) and (d).

attempted to reduce the spatial resolution into the microscopic regime. Procedures that focus the input X-ray beam[65] as well as methods that focus the subsequent photoelectron output[66] have been commercially exploited. As described in Chapter 3, both methods have permitted ready reduction of the analyzed spot to ~150 μm, but particular small spot analysis in the former case was generally achieved with slow, rather tedious, optical microscope identification. Further reduction in size seems most easily achieved by a method most closely related to UV PEEM, i.e., photoelectron focusing.[66] This process (often labeled as a "defined photoelectron collection system" or DCS) had to overcome the natural tendency to simply segment the photoelectron output, often with the corresponding compromise of both signal intensity and spectral resolution. Fine focusing (only) of the X-rays has proved to be limited spatially, but quite superior in retaining good intensity and spectral resolution.[65] As mentioned above, the latter so-called "defined source system", DSS,[65] approach had certain advantages, in that superior spectral resolution could be achieved down to ~100 μm with excellent retained intensity — but enhanced microscopy by this approach appears unlikely. On the other hand, this may not be true for the previously mentioned DCS technique.

Further design has demonstrated the ability of DCS to segment and scan across the typical 150-μm signal and transfer the results as a segmented line, which can be given the semblance of a two-dimensional spatial rendition by simultaneously scanning the sample stage. When this has been coupled with a superior, fine focus monochromator lens using the DSS technique with a relatively high intensity input dramatic point-by-point two-dimensional mappings in both the spectral and spatial modes may be achieved. Such results are being produced by the present-generation Scienta 300 system with a spectral resolution of <0.30 eV (Ag $3d_{5/2}$) and spatial resolution of <20 μm.[67]

As was mentioned earlier, superior spatial resolution is expected from the focused collection of the photoelectrons.[66] Recently, the VG company has demonstrated the ability to employ a series of interrelated lenses, located both before and after the entry of the photoelectrons into a hemispherical analyzer.[68] In the case of the VG system, three interconnected lenses are employed to initially focus and image enhance the photoelectron output. The final input into the hemispherical analyzer is actually the diffracted, Fourier transform of the lens input. Energy dispersion of this photoelectron signal is achieved in the analyzer and subsequently collected on an array of channeltrons located around the entrance of the last (output) lens. The latter lens acts then to invert the Fourier transformation of the third (input) lens for the transmitted central part of the signal, recreating the real two-dimension, energy-filtered image. This spatially oriented output is detected on a two-dimensional position-sensitive detector (p.s.d.). In this manner, the spatial and spectral parts of the analysis are rendered separately. These two features are then convoluted through the adroit use of computer software. This system, designated by VG as the ESCASCOPE,[68] has obtained 2-d spatial maps with resolution of <10 μm (see Figure 9-9).

The spectral analysis of insulators with the ESCASCOPE is aided by the use of a standard X-ray source, since this generally permits successful use of a conventional electron flood gun for charge shift removal. Spectral analyses, in general, however, are compromised by this use of a "routine" X-ray source. The enhanced resolution achieved by monochromator-based systems demonstrates some obvious advantages here that have been extended by recent improvements in charge neutralization techniques demonstrated by researchers at SSI (the select wire grid).[69] It would appear that additional advantages are possible by integrating both DSS and DCS, as is partially done in the Scienta 300.[67] An obvious direction to consider would be to try to combine the unique multilens, ESCASCOPE, DCS approach with the multicrystal, monochromator, DSS method of Scienta.

Recently, several research groups have initiated serious attempts to employ relatively high-energy photon inputs, with at least moderate spectral resolution, while achieving

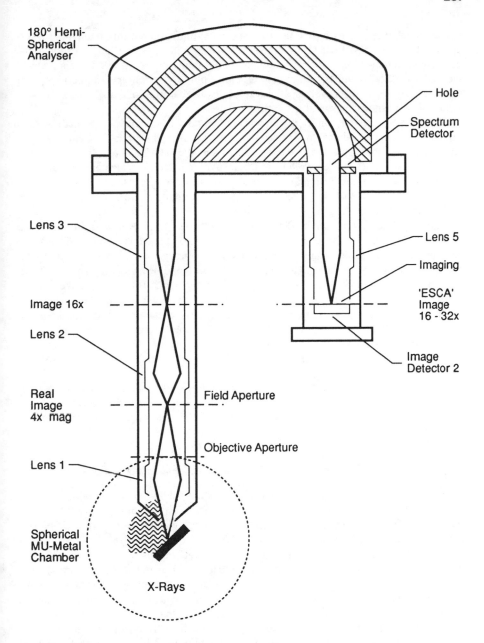

180° Hemi-Spherical Analyser

Hole

Spectrum Detector

Lens 3

Lens 5

Imaging

Image 16x

'ESCA' Image 16 - 32x

Lens 2

Image Detector 2

Real Image 4x mag

Field Aperture

Objective Aperture

Lens 1

Spherical MU-Metal Chamber

X-Rays

Figure 9-9 Schematic representation of ESCASCOPE electron optics.

maximum spatial resolution (in the manner of an electron microscope). Many of these approaches have been constructed around synchrotron radiation centers in order to take advantage of their optimized intensity and monochromatic input, with the possibility of also exploiting the variable energy feature.

An excellent example of this approach is the MAXIMUM system, constructed and tested by G. Margaritando and colleagues at the University of Wisconsin (Stoughton), in association with scientists from LBL, Xerox, and the University of Minnesota.[70] MAXIMUM is an acronym standing for "multiple-application X-ray imaging undulator microscope". It makes extensive use of a special symmetrotonic ring undulator, developed at LBL and SSRL. It also employs Schwarzschild objective lenses to focus the generated photoelectrons. Margaritando points out that, depending upon one's point of view, the method might be referred to as either photoemission spectromicroscopy or microspectroscopy. At present, spatial resolutions in the order of 2 to 3 μm have been achieved, with modest spectral resolution.[71]

The research program under the direction of Tonner has also recently made several important advances related to PEEM.[72] In this regard, this group has employed an electrostatic cathode lens objective to spatially focus the electron output from a properly irradiated sample. The objective lens utilizes a Wehnelt focusing element. Initially, the system has been separated from the photoelectron output to detect the "X-ray absorption near edge structures" (XANES).[72] Relatively tight machining requirements are necessary (see Figure 9-10), and the group has improved their specification to the point of expecting eventual x-y microscopy in the 1000-Å range. Eventually, this is to be achieved in the XPS mode with the simultaneous registration of fairly good energy resolution. This group has not yet coupled their system through a superior hemispherical or other type of β-spectrometer with the retarded fields needed for careful binding energy registration. Thus, a combined microscope-spectrometer has yet to be fully tested. Tonner et al. have recently provided a theoretical workup for the electron optics necessary to achieve these goals with their present experimental setup.[73] Until the latter is experimentally realized, the group has made excellent use of the tightly controlled, variable energy capability of both the Aladdin and National Synchrotron light sources to achieve excellent (~1 μm) microscopy for select photoelectrons, thus permitting the distinction of relatively exact microscopic adjacency and interfacial situations.[74]

The advantages provided by synchrotron radiation sites, with their tunable energies and superconducting magnets, are now being combined by others with commercial-type electron optics to achieve photoelectron microscopy in the 5-μm range, with spectral resolution.

VI. PHOTOEMISSION OF ADSORBED XENON (PAX)

Many of the methods described in this chapter have positive attributes with respect to the description of adsorption phenomena. Most of these procedures, however, are not locally specific. It is true that both the differential charging and photoelectron microscopy methods detect certain localized features of the adsorbate, but this only works for relatively macroscopic clusters, and most of these techniques are essentially blind to many local variations in the substrate. It has been pointed out that, to describe local features about either a specific substrate or adsorbate, it is advantageous to combine that material not with a chemically active species, but rather with an inert substance that will register the attribute in question (and not the multifaceted chemistry that rapidly and irretrievably regresses beyond the desired species locality). Thus, a procedure is sought that will register one feature and be inert to everything else. It turns out that either the adsorption of (or on) xenon often may offer this specific character. The first of these two areas (often labeled as "photoemission of adsorbed xenon", or PAX[75]) is the more generally studied — although a number of surface scientists have also provided interesting photoemission studies in which xenon is employed as an inert substrate onto which specific clusters of metals are adsorbed. The latter method, of course, requires very low temperatures (~15°K) to maintain the "lattice" of condensed Xe for adsorbing clusters of metals, such as Sm.[76]

In the PAX method,[77] adsorption of Xe is rendered onto the surface of a metal at low temperature (usually ~65°K). This is generally followed by the recording of UPS spectra for the substrate-adsorbate system,[75,77,78] although XPS also has been tried. A represen-

(a)

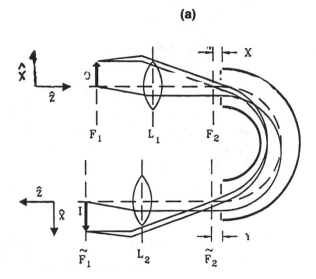

(b)

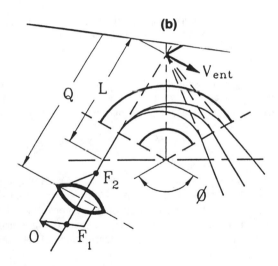

Figure 9-10 (a) Optical elements for energy-filtered imaging with a hemispherical sector capacitor (the focal planes of the lenses are measured from the lens reference plane and are considered to be positive); (b) Illustration of the location of the achromatic image planes of a spherical sector of arbitrary deflection angle. The image formed by the entrance lens must lie at the entrance virtual image plane, V_{ent}; this will produce a diverging virtual object at the plane, V_{exit}, which is achromatic.

tative example of the characteristic nature of the electronic energetics of the Xe plus metal systems is described below from some plots developed by Wandelt[75] (see Figure 9-11). Some researchers have suggested that these features represent dramatic chemical shifts in the binding energy region. Indeed, there seems to be a substantial chemical-type shift registration on the part of the xenon. Close scrutiny, however, reveals that, as expected, little or no chemical change is occurring. In fact, it is this lack of sensitivity of Xe to the chemistry of the substrate that allows it to be so sensitive to changes in the "physical environment", on, and around, the substrate.[75] In particular, it has been shown by Wandelt that Xe dramatically senses any change in the work function of the substrate, Φ_{sub}, even if the latter change is very local.[75] This occurs because the xenon has no Fermi edge, and is thus coupled at its E_o to the *vacuum* level of the substrate. If all substrates are good conductors, they will all have their Fermi levels coupled together, and to that of the spectrometer. Most variations in the *type* of conductor employed as the substrate from, for example, materials 1 to 2 will cause a shift in the vacuum level, i.e., a change in the

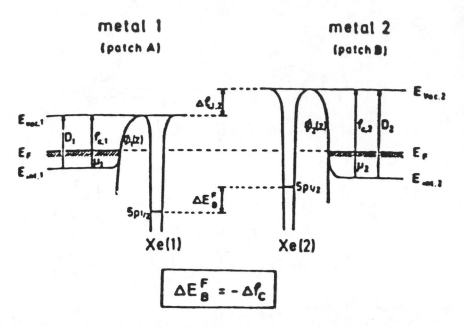

Figure 9-11 Potential energy diagram for xenon atoms Xe(1) and Xe(2) being sorbed on two semi-infinite metals of different work functions, Φ_{c1} and Φ_{c2}, or two patches, A and B, on one Φ-heterogeneous surface (Wandelt model).

resulting work functions, $\Delta\Phi_{1\to2}$. Since all of the vacuum levels involved are coupled to the E_o for xenon, any energy level of the latter (e.g., E_2) must shift by an amount, ΔE_2, that closely replicates this change in work function, i.e.,

$$\Delta E_2 \approx -\Delta\Phi_{1\to2} \tag{9-2}$$

It is not necessary that 1 and 2 represent two different metals. Two localized, unique states of the same metal (e.g., Pd steps as opposed to Pd kinks) may do. As Wandelt has pointed out,[75] this method has the advantage of permitting the examination of relatively small clusters because of the unique nature that generally arises for the "local" work function. As we shall describe below, this feature may also be related to the chief limitation of the technique.

In successful studies, PAX has been applied to identify such features as steps on Pt,[79] Pd,[80] and Ru.[81] It also has been used to describe binary Pt/Pd surfaces as a function of coverage,[82] and also employed to describe the texture (purportedly to atomic scale) of Pt adsorbed on TiO_2 surfaces.[83]

Although the utility of this procedure for select cases of analysis seems assured (particularly for single crystal situations), its liabilities, however, may prevent its widespread use in cases involving semiconductor and/or insulating systems. Wandelt describes the advantages of a non-chemisorption (Xe) adsorbate, but also notes that charging (and correspondingly Fermi edge decoupling) must be avoided.[75] In this regard, one must be cognizant of the fact that a lattice work of xenon results in an insulator and, therefore, one must not exceed the space charge capabilities of the conductive substrate. Semiconductor or insulating substrates will produce the types of differential charging effects described above,[84] and also introduce the difficulties inherent in the Mason,[85] Wertheim,[86] and Bagus[87] models when trying to examine the effects of a small metal cluster on insulating supports. These features probably complicated the above-mentioned interpretations regarding Pt on TiO_2. There is also a problem inherent to a lack of consideration of the influence of photon effects on these results — although the delocalization necessary for the latter may be removed in the localized cases under study. One should also consider the attributes of the use of other inert gases in select cases.

In general, given a liquid refrigerator capability attached to one's UHV setup, the PAX method is a potentially unique method, useful for the analysis of many systems, particularly their supports. This author feels that the method has yet to be pushed to its limit, and may find significant application with numerous applied problems. It is worth noting that a related magnetic resonance technique, xenon-doped NMR, has also recently appeared. Interestingly, this method also introduces a unique "chemical" shift with utility in describing the micromorphology of mixed solid systems.[88]

An explanation of these PAX-type shifts based upon $\Delta\Phi_{1\rightarrow2}$ is not, however, without controversy. Earlier, Kaindl et al.[77b] discovered that a significant shift in E_g resulted for physisorbed xenon between that observed following monolayer coverage and that following adsorption of the next two layers. Based upon the details of their results, they employed a final state metal substrate screening (with distance) argument to explain these shifts. On the other hand, as indicated above, Wandelt employed an extension of the $\Delta\Phi_{1\rightarrow2}$ argument to explain these same layer effects.[75]

Originally, the research group led by Jacobi[78,89] investigated a number of related rare gas coverage problems, and concluded that the results seemed to support the $\Delta\Phi_{1\rightarrow2}$-Wandelt argument.[75] Recently, however, Jacobi has used angular-resolved UPS[90] and synchrotron (BESSY source) radiation[91] to study the binding energy shifts for a variety of environments of rare gases on various single crystal metal substrates. In many of the cases involving physisorbed layers, Jacobi has not found evidence suggesting $\Delta\Phi_{1\rightarrow2}$. Instead, they proposed ΔE_B effects based upon changes in final state relaxation, ΔE_R. It is implicit (but unstated) in one of the most recent publications of Jacobi[91] that they are proposing that not only those shifts observed for multilayered systems (Figure 9-12), but other gas coverage shifts (perhaps including localized adsorption effects), are all situations that result from final state relaxations (Figure 9-13). Recent observations by Wandelt, however, suggest that his group still supports the $\Delta\Phi_{1\rightarrow2}$ model;[92] hence, interested observers and/or prospective users of PAX and PAX-related methods should probably retain an open mind until this lively (and useful) controversy is settled. It should be noted, however, that although a detailed understanding of the *causes* of PAX-type shifts is, as yet, unattained, the *reproducible presence* of these shifts and their analytical potential is unchallenged.

VII. PHOTOELECTRON SHIFTS DUE TO COMPONENT ORIENTATION

A. INTRODUCTION

Although it is suggested herein and in most other general descriptions that ESCA is a surface analysis tool, several features about the spectroscopy strongly suggest that, to be exact, this statement needs to be somewhat amended. The first point to note is that, if a conventional commercial ESCA is employed *at its most common incidence and reception angles*, then, although not completely blind to the outer monolayer or two, most of these systems actually seem to exhibit maximum detection of features in the range of a 5- to 25-Å depth, falling off on either side of this range in almost exponential fashion. Second, since the early days of ESCA it has been argued that, if photoelectron spectroscopy is sensitive to details of chemical bonding, then it should often register some special surface and/or interfacial bonding properties. The simple argument of this effect suggests that an atom, A, should experience different bonding forces if surrounded entirely by a solid matrix of (for example) B atoms (as in the bulk of a solid) than would that same A

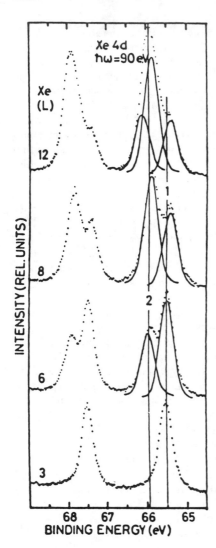

Figure 9-12 Angle-resolved photoelectron spectra of the Xe 4d levels taken in *normal emission*. The binding energy is in reference to the vacuum level. The Ni(III) 7 × 7 Pb substrate was exposed to four different Xe doses in units of L (10^{-6} Torr-s) as indicated. The $4d_{5/2}$ levels are decomposed into Gaussians, and the components of the first (1) and the second (2) layer are given by solid curves (the third layer also is indicated for the 12-L curve).

atom sitting on or near a surface of this B matrix, exposed above to a vacuum (or in the case of an interface, exposed to a different solid matrix).

Both of these features have been explored to some extent using variable and/or low energy sources and also angular resolution (AR) photoemission. In these cases, it has been observed that, with some exceptions, the first assumption is valid. Thus, as pointed out previously in our discussion of the use of AR (particularly for quantitative effects), one must be aware that, at conventional incidence, the outermost layers may not be the most intense features detected.

Initial investigations of surface-to-bulk photoemission,[93-99] on the other hand, discovered that, while select surface shifts do occur, they are often quite subtle and particular to the species and structures involved. For these and related reasons, it became apparent that these types of investigations were generally most feasible when using good angular resolution and/or variable source energy along with superior resolution. This has led most experimentalists in this area to employ synchrotron radiation-based photoemission or high-resolution UPS[97,98] (although some of the initial observations of this effect were made using monochromator-based XPS and angular resolution).[93,94]

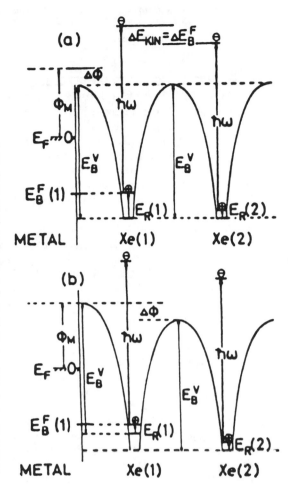

Figure 9-13 Electrostatic-potential diagram comparing $\Delta\phi$ and ΔE_R models for the first two Xe layers physisorbed at a metal surface. The diagram is sketched for photoemission from a core level, leaving behind the photohole ($\oplus$) relaxed in energy by the amount of E_R. ϕ_M is the work function of the clean metal. E_B is the Xe gas phase binding energy in reference to the vacuum level. ΔE_B^F is the work function change induced by the Xe atoms in the first layer. Diagram (a) is for the E_R model, which explains the change in binding energy by a distance-dependent change of the relaxation energy E_R in the final state. Diagram (b) is for the $\Delta\phi$ model, which explains the change in binding energy by the Xe-induced $\Delta\phi$.

Considering all of the specialized methods and advanced procedures described in this book, perhaps none is as potentially important (or as difficult to explain in detail) as the resulting shifts in the binding energy positions of select species that sometimes occur when such species are restricted to certain environments in the surface-near surface region of a composite system. Thus, for example, photoelectron spectroscopists have found that the spectrum emitted from the outermost monolayer of such materials, e.g., pure rare earth metals may be shifted significantly (at least 0.4 eV) from the binding energy position exhibited by those same rare earth species when they are located (by XPS examination) in the bulk (near-surface) region of the material under investigation.[93,94,98] Although this type of shift has been primarily monitored using photoelectron peaks in the near valence region, the general label for this "new branch" of photoemission spectroscopy is either surface-to-bulk shifts or surface core-level spectroscopy.[93-99] In addition to these effects, innovative studies have also found cases of discernible binding energy shifts for a select species, A, when "immersed" as clusters of different sizes in a certain matrix, B.[100] Special cases of binding energy shifts also have been reported for peaks from a select species A, when localized at certain interfaces,[101] and/or alloyed in a matrix B. As we shall describe below, the rationale for most of these individual effects is still in dispute, so there is no certain demonstration of their interrelationship; however, as described by Citrin and

Wertheim,[102] *it seems natural to expect that many, if not all, of these effects are somehow related to one another*. We have, therefore, borrowed in a loose manner this concept of an interrelationship, and collectively refer to these potentially very important effects as *"orientation shifts"*.

B. SURFACE CORE-LEVEL SHIFTS: DISCOVERY AND THEORETICAL CONTROVERSY, 1978 TO 1983

The existence of finite surface shifts was predicted from the beginning of ESCA.[103] However, as pointed out in 1983 by Citrin et al.,[104] "numerous early attempts to detect such differences between surface-and-bulk atom photoemission have led to conflicting or negative results." The year 1983 proved to be pivotal for the detection of surface shifts in photoemission, since it featured not only the paper by Citrin et al.,[104] but also a companion article by Citrin and Wertheim,[102] and a related key contribution from Johansson and Martensson.[105] These three papers appeared to have provided unequivocal proof that detectable surface shifts exist for numerous systems. These papers not only detailed these new studies, but also provided well-constructed compilations of the previous existing data. Numerous interesting progressions were documented by both groups of investigators for a wide range of metals. In addition, Citrin and Wertheim[102] also suggested that these metallic surface ESCA shifts were probably directly related to shifts that appeared to be produced by the surfaces of select semiconductors,[103b,106] and even those realized by small clusters in insulating matrices.[100,107] In addition to their compilations, both groups also provided explanations for these surface shifts; unfortunately, these two hypotheses are not in agreement.

1. Surface Core-Level Shifts — Metals

As reported above, three key 1983 papers documented a variety of surface shifts for metals.[102,104,105] These observations have since been confirmed and extended, particularly by the research group of Martenson.[101] A typical set of results is presented in Table 9-1.[105] Additional data are provided in the plots in Figure 9-14. As implied by these data, the presence of finite surface-core level shifts for metals generally grows with their atomic number. The shifts generally are much larger for group B metals (particularly the 5d transition metals and the lanthanides). Both the size and sign of the surface shifts seem to be influenced by the number of d electrons in the valence region. Thus, for the transition metals, the surface shifts are finite and positive when the d electron part of the valence band has only one electron. The magnitude of such a shift seems to fall toward zero as that metal's d shell adds electrons; it essentially reaches zero near the half-filled state, only to rise in magnitude (but with negative sign) with further filling, peaking again at d (the single hole case!).[102,105] The lanthanide surface shifts, on the other hand, are generally uniformly large across the periodic table row, and are always positive — a fact that seems to be related to the presence in most of these metals (from Ce to Lu) of a valence band that contains, in part, one d electron.[102,105]

Viewed from an even more general stance, one should note that the magnitude of these surface-core level shifts seems to depend not only upon the number of unpaired valence electrons, but also very much on the type of electron involved. Perhaps this fact can best be illustrated by noting the surface shift behavior detected for the noble metals.[104] All the values are negative, suggesting an influence from holes rather than electrons. Further, the magnitude of these shifts is reasonably large for Cu, is reduced to almost zero for Ag, and is elevated again for Au. How can one explain this strange progression for a series of metals that are all from the same (IB) group? We suggest that the explanation lies in a close examination of the respective valence bands for these metals (see Figures 9-15A through C).[6,108,109] Close scrutiny reveals that all three bands may be qualitatively partitioned into sections due primarily to large densities of their d sub-bands, and into other

TABLE 9-1 **Experimental surface core-level binding energy shifts (in eV) for metallic elements**

Lanthanides

Ce	≤0.4	Dy	0.55(5)
Pr	0.5(1)	Ho	0.63(5)
Nd	0.5(1)	Er	0.65(5)
Eu	0.60, 0.63(3)	Tm	0.70(5)
Gd	0.50(5)	Yb	0.62(3), 0.60(3)
Tb	0.55(5)	Lu	0.70(5)

5d Transition Series

Yb	0.62(3), 0.60(3)	Ir(111)	−0.50
Lu	0.70(5)	Ir(110) − (5 × 1)	−0.49
Hf	0.44(5)	Ir(100) − (1 × 1)	−0.68
Ta	0.3	Pt(110)	−0.35(2)
Ta(111)	0.40	Pt(111)	−0.4
W(111)	−0.43	Au	−0.40(2)
W(100)	−0.35	Au(111)	−0.35
W(110)	−0.30	Au(110) − (2 × 1)	−0.35
		Au(100) − (1 × 1)	−0.38
		Au(100) − (5 × 20)	−0.28

Noble Metals

Cu	−0.24(2)	Au	−0.40(2)
Ag	−0.08(3)		

Simple Metals

Na	0.22	Al(100)	−0.12, −0.06
Mg	0.14	Al(111)	~0.0

Refer to Johansson, B. and Martensson, N., *Helv. Phys. Acta,* 56, 405, 1983.

regions due to much weaker densities of primarily s origin. The latter are, of course, the parts of the total valence band that make contact with the Fermi edge, and thus provide the *modus operandi* for conductivity. It should also be noted that *each s-dominated sub-band is composed of a highly delocalized electron density, whereas the d sub-band is substantially localized*. With these general features in mind, one should next consider the subtle (but characteristic) differences between the band structures for the three metals in

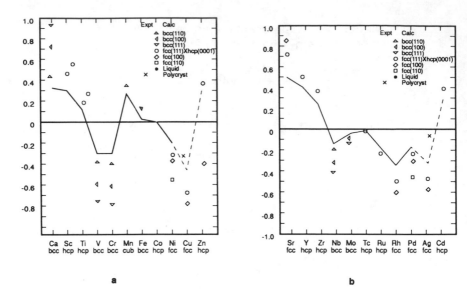

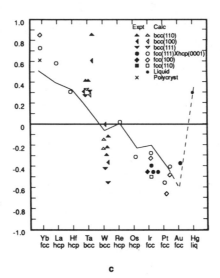

Figure 9-14 Calculated (—, using Equation 9-6, p. 299) and experimental values of surface-atom core-level shifts for: (a) 3d transition metal series; (b) 4d transition metal series; (c) 5d transition metal series.

question (Figures 9-15A through C).[6,108,109] Compared to Ag, the d sub-bands for both Cu and Au are substantially shifted toward their Fermi edges (i.e., the d onset is at ~3.4 eV for Ag, whereas it is at ~1.5 eV for Cu and ~1.7 eV for Au). Further, the density of states in the s-like part of the Ag band seems to be (relatively) less populated than the s sub-band for the other two elements. This suggests that the band structures for Cu° and Au° contain substantial d^9s^2 character, with extensive d mixing near the Fermi edge, whereas Ag° seems to be primarily $d^{10}s^1$, with very little d involvement at E_F.

It is our conclusion, therefore, that *surface core-level shifts result from the extensive presence of unpaired, localized electrons (or their equivalent holes) near the Fermi edge*, and d electrons provide excellent examples of these localized situations. Thus, (d < 5)

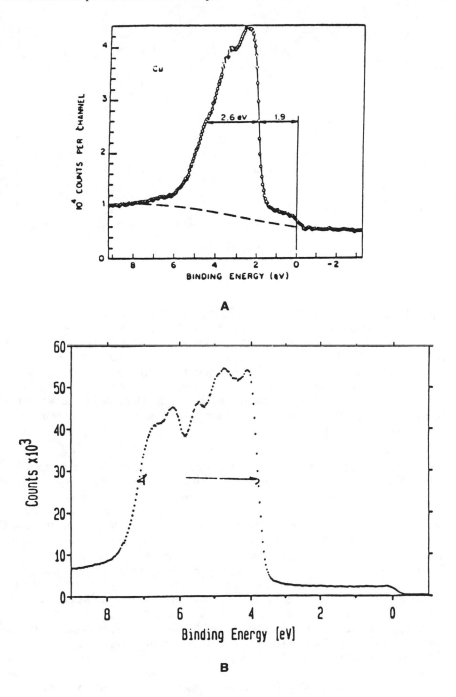

Figure 9-15A & B

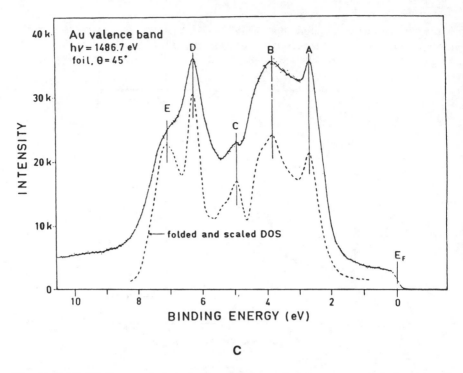

Figure 9-15 (A) X-ray photoemission spectrum of vacuum-evaporated copper film taken with monochromatized AlKα radiation on the spectrometer. Raw data are shown. The approximate background due to inelastically scattered electrons is indicated by the dashed line; (B) valence band spectrum of silver recorded at high-energy resolution with a step size of 0.03 eV. The individual energy channels are plotted here in order to more clearly show the details in the spectrum. The sharpness of the Fermi edge indicates that the energy resolution has been 0.27 eV or slightly better. At this resolution and statistics, fine structures appear which have not been observed earlier; (C) cleaned valence band spectrum of gold.[93-195]

electrons produce positive shifts from the corresponding bulk binding energy positions, and (d > 5) holes produce negative shifts. *The size of these shifts seems to depend, not only on the presence of nearly empty or nearly full d sub-bands, but also on their proximity to the Fermi edge, and how extensively they are* (or are not) *compromised by* (i.e., *mixed with*) delocalized (s-like) electrons. Hence, if the (unpaired) d electrons (or holes) are adjacent to the Fermi edge and retain their localized identity, then the surface shifts are relatively large, (e.g., Cu and Au); if not, these shifts are muted (e.g., Ag).

With these facts in mind, *we can even make predictions for band structure features we cannot detect.* Thus, in the case of those transition metals in the d^1-d^3 part of the periodic table (e.g., Sc, Y, and Hf), for the d^1 cases we find an increase in the magnitude of the surface-core level shift as n increases for these elements. If the suggested model is necessary and sufficient, then we would predict that the (as yet, unconfirmed) sub-band structures for some of these metals will be characterized by d^1 density near the Fermi edge that "grows" as n increases. Further, the s and d bands should be less separated for those elements with larger n in a particular column (e.g., Zr compared to Ti). Thus, we should find for these transition metals that the ready presence of multiple valencies runs counter to the existence of relatively large surface shifts. Unfortunately, this two-dimensional argument is too simple. Hence, many exceptions arise, as other factors apparently come into play. In any case, any explanation requires a determination of the features of these localized electrons (and/or holes) that create the surface shifts. This leads us to the conflicting arguments of Citrin et al.[102,104] and the Martensson group.[98,101,105]

Citrin and Wertheim[102] realized the importance of the near Fermi edge s to d ratio in the presence of these shifts. They also noted that, for those metals experiencing shifts, the delocalized parts of their s and the d bands are apparently muted at (or near) the surface, and the localized parts of their d bands are accentuated. They argue this effect based upon differences between the resulting bulk and surface cohesive energies exhibited by a particular metal in its initial (premeasurement) state. They then show that, since the total density in the valence band must remain the same whether atom A is at the surface or in the bulk, then surface to bulk movement by A should just produce a shift in the distribution of density. Their adjustment for this difference is presented in Figure 9-16, where we see the positive (d < 5) and negative (d > 5) surface shifts arising from the establishment of a consistent Fermi edge. These arguments appear to be persuasive, and seem to extend quite nicely through many of the species that exhibit these surface shifts.

The Martensson group, on the other hand, has taken an entirely different approach.[98,105] They argue that these surface shifts should be examined in the context of the "detailed picture" for the photoelectron experiment (as presented previously in Chapter 4). In that discussion, we noted that the photoelectron measurement process affects the resulting binding energies due to the creation of relaxation shifts experienced by the disturbed system as the photoelectron is extracted. This led to Equation 4-3 in Chapter 4. Johansson and Martensson[105] further partitioned this equation into its (apparently reasonable) components, and considered how these components differ for an atom A in the bulk of a matrix, as opposed to that same atom at the surface. This led to Figure 9-17, and the following equation:

$$\Delta E_A = E_S^{z*} - E_S^z - \left[E_{z*}^{imp}(z) - E_{z*}^{imp\,surf}(z) \right] \tag{9-3}$$

where z refers to the charge, z* to the final state charge, s the surface, "imp" to the fact that the atom in question acts like an impurity in the matrix in question. Due consideration also is given by Johansson and Martensson to the concept developed originally by Jolly and Hendrickson,[110] that the relaxed final state measurement of atom A, with charge z*, resembles the now-disturbed detection of that same energy for an atom of charge z + 1 (i.e., the z + 1 or equivalent core approximation). As a result, according to Martensson, the surface-core level shifts may be estimated from the heats of surface segregation in a Miedema type argument.[111] This leads to

$$\Delta E_A = E_S^{z+1} - E_S^z - \left[E_{z+1}^{imp}(z) - E_{z+1}^{imp\,surf}(z) \right] \tag{9-4}$$

Empirically, Johansson and Martensson[105] noted that

$$E_S \simeq 0.2\, E_{coh} \tag{9-5}$$

where E_{coh} is the corresponding cohesive energy. This assumption and further truncations led Johansson and Martensson to

$$\Delta E_A \cong 0.2 \left[E_{coh}^{z+1} - E_{coh}^z \right] \tag{9-6}$$

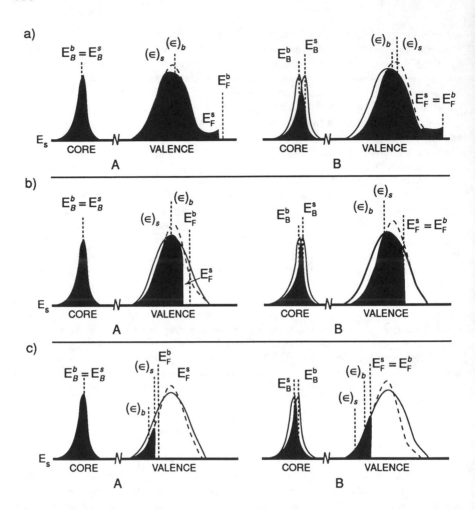

Figure 9-16 Model for explaining the surface-atom core-level shifts (SCS) for (a) noble metals and transition metals with more than (b) and less than (c) half-filled d bands. The surface and bulk are denoted by s and b, respectively. In hypothetical state A (left), the Fermi level of narrowed surface DOS (dashed line, lightly shaded) falls below or above bulk Fermi level E_F due to layerwise neutrality. Core-level binding energies E_B for surface and bulk atoms in state A are the same. In true state (right), Fermi levels of surface and bulk DOS are the same, and core-level binding energy for surface atoms falls above or below bulk atom binding energy. Centers of gravity (ϵ) for surface and bulk DOS have shifts of similar magnitude and sign as corresponding core-level binding energies.

The plots in Figures 9-14, 9-17, and 9-18 were in part realized employing Equation 9-6.

Thus, Johansson and Martensson also attributed the surface shift to differences in binding energy of the core electrons (bulk to surface), but these differences were assumed to be related to a final state relaxation effect.[105]

C. INTERFACIAL SHIFTS

Recently, the Martensson group completed a variety of new studies on these orientation shifts that greatly expanded the concept. This is particularly well represented in two

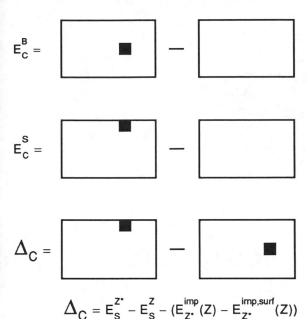

$$\Delta_C = E_s^{Z^*} - E_s^Z - (E_{Z^*}^{imp}(Z) - E_{Z^*}^{imp,surf}(Z))$$

Figure 9-17 Illustration of the correspondence between the surface core-level shift ΔE_A and the surface heat of segregation of the Z^* "impurity".

Martensson-directed 1989 Ph.D. theses: "Core Level Electron Spectroscopy Studies of Surfaces and Adsorbates" by Anders Nilsson[101] and "Surface Studies of Thin Epitaxial Lanthanide Films" by Anders Stenborg.[112] This combination has produced 23 publications that not only further confirmed the presence of select surface shifts, but also unequivocally demonstrated the presence of a closely related feature, labeled as "*interfacial shifts*".

Perhaps the best illustration of the presence of select surface, bulk, and interfacial shifts is exemplified by the epitaxial growth of Yb on Mo(110), as demonstrated in Figure 9-19.[101,112,113] These results were achieved on a synchrotron radiation source (Max-lab) using a photon energy of 100 to 150 eV. As the growth evolves from its onset, several features of these results should be noted, including the initial presence of only surface and interfacial peaks. As the monolayers evolve, the bulk peak begins to grow and the interfacial peak to disappear. The surface peak is seen to remain roughly the same throughout the evolution from two to four monolayers. As the authors have pointed out, these studies amplify the utility of Yb as an excellent "medium" for these types of shifts;[101,112,113] however, these authors also have simultaneously demonstrated (for the first time) that La may often be substituted for Yb in these studies.[101,114]

Perhaps of even greater importance has been the diverse shift results achieved when the same overlayer (Yb) was examined under the same physical conditions (670°K and room temperature) with high-resolution UPS (see Figure 9-20A) and $Al_{k\alpha}$ X-rays (see Figure 9-20B).[101,115] Both methods detect the presence of surface-type shifts following the initial deposition of Yb, but as the layers of the latter species grow the XPS result soon loses "touch" with the surface effect, which becomes dominated by the corresponding bulk peak formation. The UPS spectrum, on the other hand, still demonstrates a substantial surface effect. It is apparent that these results are, in fact, expressions of the universal curve for emitted electrons and the previously stated preferential detection by (conventional incidence) XPS of features immediately below the outermost surface. Thus, in Figure 9-20A, the He(II)-generated photoelectrons strongly emphasize the surface Yb, whereas the $Al_{k\alpha}$-induced results almost entirely miss this same feature. Despite this "deficiency", the XPS results do have several demonstrated attributes, as they tend, for example (see Figure 9-

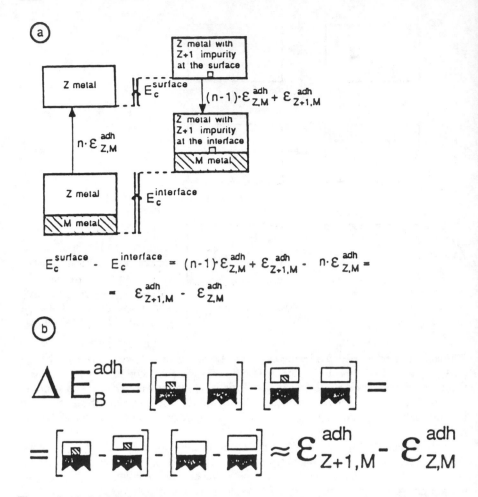

Figure 9-18 (a) A Born-Haber cycle, describing the difference between a surface and interface photoionization process; (b) the same process descibed as a total energy difference equation.

20B), to produce a better record of the evolving chemistry [in this case, the initial adsorption of Yb(II), followed by the growth of mixed Yb(II) and Yb(III) sites]. In addition, the XPS signature tends to provide a larger "panorama" of the data which helps confirm any preliminary assumptions. There is no doubt, however, that even with adroit angular resolution XPS may overlook some surface features.[115]

Another area of substantial importance that was discovered and studied in some detail in these investigations by the Martensson group concerns the effect of those shifts on the substrate employed, and also the influence of various substrates. Thus, a Yb overlayer analysis has been reported for Au,[114] Ag,[114] Ni(100),[115] and Mo(110)[113,116,117] substrates. In addition, examples of overlayers of Al,[117] Si,[117] and Sm[118] on Mo(110), La on Au,[114] and Cu on Ni(100)[118] have also been reported.

As expected, the surface and interfacial shifts realized by the different kinds of overlayer systems were shown to vary substantially for each material, with an obvious "proficiency" in these shifts for the rare earth systems.[105]

Substantial importance should also be attached to the fact that Nilsson et al.[114] have unequivocally demonstrated that, during the selective segregations of Yb and La (inter-

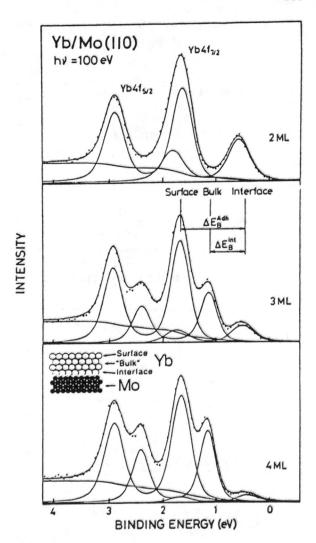

Figure 9-19 Yb 4f photoelectron spectra for two, three, and four monolayers of Yb deposited on Mo (110).

face ↔ bulk ↔ surface) on the substrates Ag and Au, there is obvious evidence for the transfer of electrons between the substrate and overlayer species. In fact (see Figures 9-21 and 9-22), as the overlayer of Yb grows on both Au and Ag, electron density is apparently extracted from the d sub-band of the valence density of the coinage metals, and "added" to the interfacial and/or bulk states of the rare earth overlayer. This feature, in part, may explain the negative binding energy shift direction experienced by the interfacial peaks, relative to the corresponding bulk and surface states (see, e.g., Figure 9-19).[113] It should be noted that, due to the interference of the near valent (overlayer) rare earth peaks, one cannot comment on the degree of simultaneous contribution of the delocalized s-type valence density (of, for example, Au and Ag), but it is readily apparent that the localized d states of the noble metals are intimately involved. We will comment below about the possible implications of these observations, adding, at present, only that the shift induced by Au seems to be somewhat larger than that induced by Ag, whereas the density depletion effects seem to be reversed (see Figures 9-21 and 9-22).[114]

As with the previous studies reported by both the Martensson[98,105] and Wertheim and Citrin groups,[104] the rare earth metals seem to occupy a special position for these

304

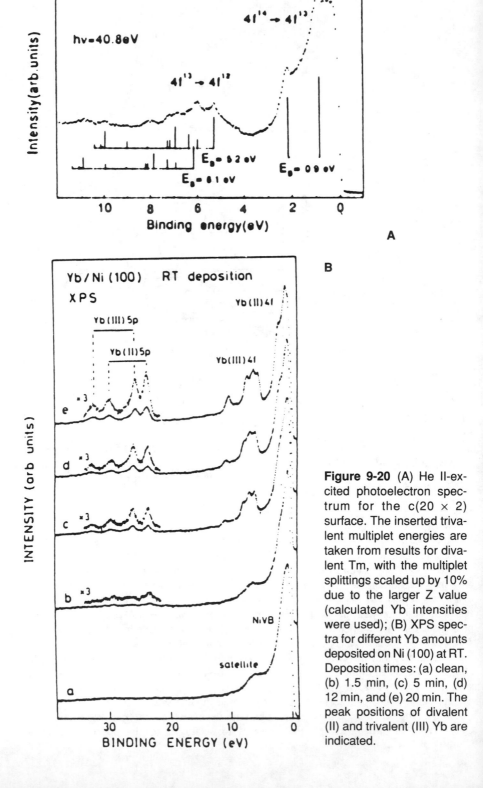

Figure 9-20 (A) He II-excited photoelectron spectrum for the c(20 × 2) surface. The inserted trivalent multiplet energies are taken from results for divalent Tm, with the multiplet splittings scaled up by 10% due to the larger Z value (calculated Yb intensities were used); (B) XPS spectra for different Yb amounts deposited on Ni (100) at RT. Deposition times: (a) clean, (b) 1.5 min, (c) 5 min, (d) 12 min, and (e) 20 min. The peak positions of divalent (II) and trivalent (III) Yb are indicated.

orientation shifts; however, that position is more a matter of degree than type. Thus, Nilsson and co-workers[117] have noted that both Si and Al also seem to exhibit these same types of interfacial and surface core-level shifts. Once again, these effects have been "optimized" at low photon energies (in this case by employing the Max-lab source). The spectra reported in that paper are exclusively those for Al(2p), perhaps suggesting that those for the higher energy Si(2p) are less well resolved.[117] Although these shifts are quite small (~0.24 eV for both Al and Si), they do appear to continue the previously established patterns.

Despite all of the statements made in these paragraphs that may be construed to denote effects of initial state chemistry, one should be aware that the Martensson group has retained its conviction that the principal cause of these shifts (both surface and interfacial) is due to final state relaxation effects. Thus, the cited papers[113–118] and the two previously mentioned theses[101,112] present further refinements of the long-held final state segregation arguments. In these cases, extensive use is made of both the Born-Haber cycle (see, e.g., Figure 9-23) and surface and interfacial segregation arguments provided earlier by Miedema[111] and others.[119] In the latter, one notes that the surface may be visualized (with the vacuum) as a

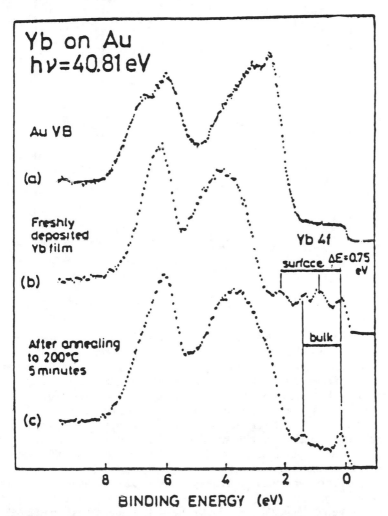

Figure 9-21 Spectra of ytterbium evaporated onto a gold substrate.

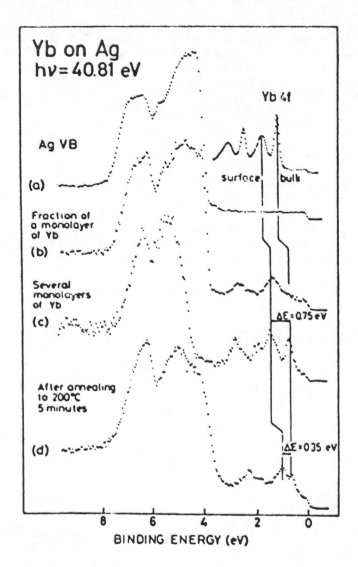

Figure 9-22 Spectra of ytterbium evaporated onto a silver substrate.

special type of interface on which two different metals with different crystal structures (and/or sizes) create a special interaction due to "geometric mismatch". In addition, there is a (unique) chemical interaction between any two metals, that may be realized from their different heats of solution in each other's matrices. The latter factors are, hence, related to differences in surface tension, work functions, and/or heats of segregation.

These Miedema-type arguments are employed by the Martensson group[101,112] to describe the detected surface and/or interfacial shifts; however, the segregants purported to be producing the photoelectron shifts are not the dopant metals (e.g., Yb), but rather their relaxed final state forms. Further, Martensson et al.[105,112] demonstrate that these relaxed states may be reasonably approximated by substituting into a Miedema-type (impurity segregating in matrix) argument, the Jolly-type[110] $z + 1$, *equivalent atom* approximation (e.g., Lu for Yb).

There can be no doubt that this final state shift argument has substantial merit. Recently, Martensson et al.[120] have expanded upon this concept and recomputed many of the realized surface and interface shifts. In an interesting addendum, they also employed the same concept to calculate the possible two-hole shifts that would be realized in corresponding surface-oriented Auger studies.[120]

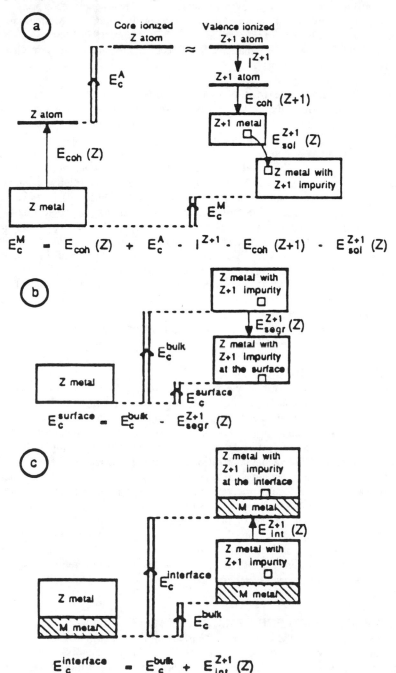

Figure 9-23 Born-Haber cycles of (a) a bulk core photoionization process, (b) a surface ionization process, and (c) an interface ionization process.

D. ALLOY FORMATION AND SURFACE RECONSTRUCTION

In previous statements we mentioned (and Citrin and Wertheim have implied[102]) that, in addition to surface and interfacial effects, there should be other cases of specific orientation shifts. In this regard, several research groups have explored aspects of alloy formation and the important question of surface reconstruction. The Martensson group has been equally productive in these areas, notably in a third 1989 Martensson-directed Ph.D. thesis by Ericksson,[121] "Photoelectron Spectroscopy Studies of Rare Gases and Lanthanide Overlayer Systems".

1. Alloy Formation: Intermediate Compounds

One of the most intriguing questions asked of ESCA results (and indeed, of any spectroscopy employed to analyze solids) is its ability to distinguish specific situations in alloy formation. It sometimes seems that little concrete progress has been achieved in the fundamental areas of heterogeneous equilibria in the more than 100 years since the magnificent studies of Gibbs, and, although this assertion is not true, it is still often difficult to distinguish exactly between alloy cases forming (1) partially insoluble mixtures, (2) solid solutions, and (3) intermediate compounds. During its investigations of metallic overlayer systems, the Martensson group discerned that it was detecting substantial evidence for intermediate, mixed-valent compound formation. For example, the two-dimensional, intermediate compounds, Yb_2Ni and $YbNi_2$, were identified during studies of the deposition and interaction of Yb overlayers on Ni. As a result, a phase diagram was proposed (Figure 9-24). This observation was preceded by the detection of mixed valencies for these systems using XPS (Figure 9-20b), and the use of AES, ISS, and LEED to confirm these features and document the nature of the resulting structures.[122,123] In view of the growing importance of general alloy descriptions and, specifically, film deposition-driven two-dimensional characteristics, one should immediately see the potential value of these (as-yet, limited) observations.

2. Surface Reconstruction

Many surfaces seem to form in one manner, and then, following additional deposition or agitation, seem to dramatically alter their structure, or reconstruct. This may occur under a variety of conditions that often precede conventional intermediate stages of the deposition processes. This appears to be particularly true for compound semiconductors and metals, and may occur more often for pristine systems than for impure forms.[124,125] It is vital to achieve detailed characterizations of these processes. The Martensson group reinforced the earlier observations of others[126] in noting that, in addition to obvious shifts in LEED patterns, this selective orientation effect often produces a detectable shift in the resulting photoemission spectra.[127] For example, in the investigation of the epitaxial growth of single crystal Sm(001) onto Mo(100), a 25% expansion of the interatomic distances of the hexagonal surface layer was detected (see Figure 9-25).[128] It seems obvious that many more examples of this phenomenon should be detected, since reconstruction is, in fact, the detection of selective surface modifications (relaxations) achieved to accommodate the *differences between surface and bulk effects* under different physical conditions.

E. SURFACE SHIFTS FOR SEMICONDUCTORS

As discussed above, the discovery of surface core-level shifts for metals was soon accompanied by the detection of similar effects for semiconductors. First observed in detail by the Eastman group for GaAs(110) and GaSb(100),[106] this phenomenon was found to occur in elemental Si, as well as in compound forms of this type of (sp^3-tetrahedral) material.[129,130] The obvious importance of surface reconstruction and such companion effects as preferential top site occupancy for AB systems drives these

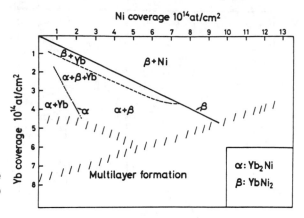

Figure 9-24 Proposed phase diagram of Yb-Ni on Mo (100).

observations. In the silicon study various surface structures of Si(111) and Si(100) were selectively studied, and then selective shifts for Si(111)-(1 × 1)H (see, e.g., Figure 9-26) were also reported. The results seemed obvious and the potential for gaining new information seemed great, and several groups have employed this approach to attack the viability of the controversial "buckled surface model".[131] Unfortunately, interpretations of these data have proven to be questionable, and even empirical curve fitting has been found to be complex.[132]

The major difficulty with these results has not been their existence or impact, but rather, their interpretation. Most groups investigating semiconductor surface shifts have invoked some form of the initial state, charge-transfer argument employed by Eastman that is essentially identical to the initial state concepts of Citrin et al.[94] Watson et al.,[133] however, have argued that, based upon reasonable initial state concepts, the resulting shifts must be primarily due to differences between the bulk and surface Madelung potentials for these materials, *and (unfortunately) the Madelung potentials for the first few atomic layers must be very similar.* Thus, strict interpretation of Watson's arguments suggests that the initial state shifts for these semiconductors should not exist! *Even more disturbing is the fact that the Watson concepts should have a negative impact on parts of all of the previous explanations, for all of the orientation shifts.* This suggests that a total explanation of these shifts is, as yet, not formulated. We further explore this aspect below, using, in part, a feature that may seem to be only a tiny ripple in this rather expansive river — the intriguing result obtained with hydrogen on silicon.

F. SMALL CLUSTER SHIFTS

The special importance of small clusters of metals and other materials, free-standing,[134] on substrates,[135] and in insulating matrices[100] have been amply demonstrated in catalysis and device technology. This has led to a variety of unique methods for production of these systems,[100,134–136] and to separate areas for both theory and materials characterization.[100,102,134,137,138] As a result, numerous novel approaches to surface characterization have been spawned, including several unique aspects of photoelectron spectroscopy.[100,102,139]

Viewed from one perspective, the possibility of unique core and valence band ESCA binding energy positions for small clusters does not seem to be related to the other shifts described in this section. The resulting cluster shifts that apparently occur in some systems do suggest effects related to the "orientation" of a particular "dopant" in a variety of matrices or substrates; however, it takes a careful perspective to see (as did Citrin and Wertheim[102]) that small cluster shifts may be just a special version of the same general effect that creates surface-core level shifts.

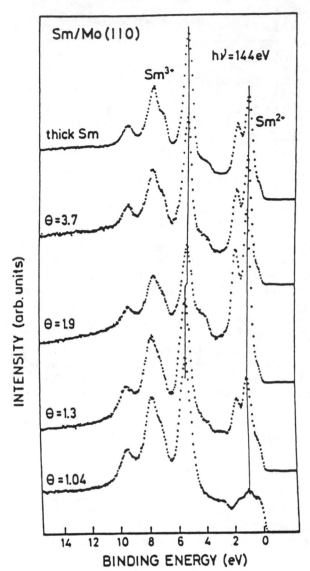

Figure 9-25 Photoelectron spectra of Sm/Mo (100) at a photon energy of 144 eV. The coverage range is >1 ML. The trivalent structures appear at binding energies of 5 to 12 eV and the divalent at 0 to 5 eV. The straight lines indicate peak positions for the spectra at the top and at the bottom of the figure.

The observation of small cluster shifts appears to have originated in studies of the research group of Mason, where the first cases considered involved different dispersions of Ag in C.[103] These studies were quickly followed by related investigations from the Egelhoff group[140] of Pd, Cu, and Ni in C, and examinations by Abbrati et al.[141] of dispersions of select metal dopants in different metallic substrates. The Mason group[142] and others[143] continued to expand on these studies, discovering that the most obvious examples seemed to consist of cases involving *noble metals (e.g., Ag or Au) dispersed in common insulating substrates (e.g., SiO_2 or Al_2O_3).* A representative example of this effect is presented in Figure 9-27.

As pointed out by Citrin and Wertheim,[102] nearly all of the realized cluster shifts are cases in which the core level binding energies of the agglomerate *increase with decreasing cluster size,* compared to the apparent positions of the same core level peaks for the "infinite cluster" (or bulk system). Citrin and Wertheim[102] also noted that the

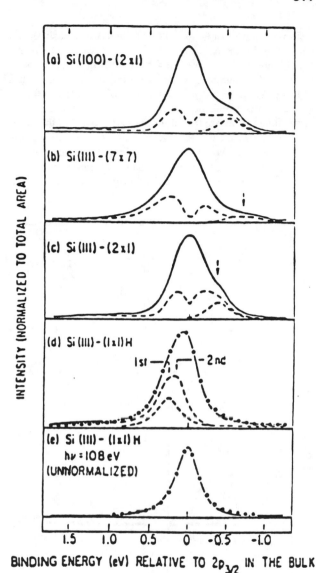

Figure 9-26 Si($2p_{3/2}$) core-level spectra for various surfaces (full lines) (hv = 120 eV). The dashed lines (without crosshatching) show the contribution of the outer two surface layers, after subtraction of the Lorentzian bulk contribution [dotted curve in (e)] which was obtained via a three-Lorentzian fit [dotted curve in (d)] to Si(III)-(1 × 1)H data. Certain surface core levels are crosshatched.

core level effects are usually accompanied by corresponding shifts and *contractions* in the valence band region, generally in the d electron part of that band. Viewed from an initial state-only perspective of ESCA, the latter changes seem to be appropriate, since, for example, removal of valence electrons generally suggests a positive shift in binding energy for core peaks (however, as we shall describe below, this feature can be deceptive).

Furthermore, one can visualize that the ESCA-generated valence region for a metal cluster should evolve from spectral features that are primarily composed of relatively discrete states (the "single atom cluster") to broadened regions containing the full "character" of the (valence) *band* structure for that metal (as the number of atoms in the cluster →∞). Thus, as Mason et al.[100,137,142] have described, as the cluster size shrinks, it is natural to expect the total valence density of states to shrink as well. (It should be noted that the valence atomic "density" [number of d electrons/atom of the metal (e.g., Ag)] is the same for the single atom case and the bulk cluster.)

A very important feature for all of this is obviously the point of zero energy reference. Citrin and Wertheim[102] have provided a very carefully reasoned discussion of this aspect (see Figure 9-28), noting, among other things, that *the principal point of zero reference is obviously dominated by the major ingredient.* This means that, for the infinite cluster of metal A, the zero is the Fermi edge of A; whereas for small clusters of A in (matrix) B, the zero is established by the Fermi (or pseudo Fermi) edge of B (see Figure 9-28). (One should be aware that all of the participants in this research rather indifferently note that arguments such as this are made complicated by the presence of a charging shift.) Such statements would seem to minimize substantially a significant potential problem, since most of the substrates of interest not only produce substantial charging shifts, but, further, do not have easily realized Fermi edges. This means that the coupling arrange-

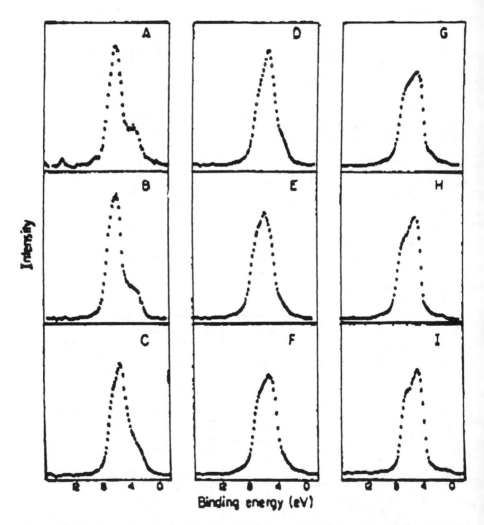

Figure 9-27 An early representation of small cluster shifts as generated by the Mason group.[133] Shown here are ESCA spectra of silver (after subtraction of the carbon background and inelastic scattering tail) at different silver coverages. The coverages (atoms/cm²) are A, 5×10^{13}; B, 2.18×10^{14}; C, 3.75×10^{14}; D, 7.5×10^{14}; E, 1.5×10^{15}; F, 2.5×10^{15}; G, 5×10^{15}; H, 1×10^{16}; I, 4×10^{16}.

ment depicted in Figure 9-28 can only be approximated. In addition (and of substantial significance in this area of analysis), all of the systems under consideration are prime candidates for *differential charging* (see Section I), and nothing in the above description will "cope" with this feature. The implications of these "referencing problems" are considered in more detail below in Section VII.G.

Unfortunately, (even if one ignores charging) it is at this point that the concept of small cluster shifts begins to break apart into several conflicting views. The basis for the difficulty seems to have its genesis in the uncomfortable observation by Citrin and Wertheim,[102] that is, although on the one hand there seems to be a direct relationship between surface-core level shifts and small cluster shifts, there is, in fact, a very significant disparity. Thus, they note that while the surface-core level shifts for the d^8, d^9, and d^{10} metals are all negative, the corresponding, measured (!), small cluster shifts for these same metals seem always to be positive, with values between ~0.5 and 1.5 eV.

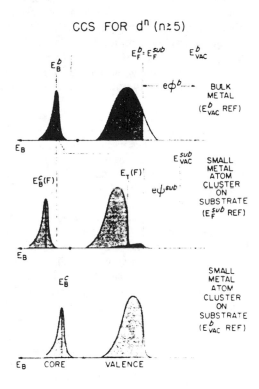

Figure 9-28 Model for explaining the clustered-atom core-level shift (CCS) for transition metals with more than half-filled d bands, where c denotes small-clustered metal atoms evaporated on a substrate (in the limit, single atoms) and b denotes evaporated bulk metal on the same substrate. The work function $e\phi$ is the difference between Fermi and vacuum levels, E_F and E_{vac}. The substrate is assumed to be a conductor, and the total atom cluster coverage (and, thus, the change in $e\phi^{sub}$) is assumed to be negligible. Substrate conduction electrons, shown in the center panel shaded above and below "threshold" energy E_T (F) of metal atom d states define E_F^{sub}. With this reference level, the core-level binding energy E_B^c (F) of the cluster is measured at higher binding energy relative to that in the bulk metal. Using the physically meaningful and common reference level E^b_{vac}, the core-level binding energy in the cluster E_b^c is actually lower than that in the bulk.

Bagus and his group have considered this problem in detail with both experiments and elaborate quantum chemical calculations.[144] Experimentally, they feel they have confirmed that the previously mentioned positive shift in binding energy (relative to the bulk position) appears to increase regularly as the size of the cluster decreases (e.g., for Pt in C or SiO_2 [Figure 9-29]). Theoretically, they carried out *ab initio* SCF calculations of the core ionization potentials for variously sized, small clusters of some of the metals in question (e.g., Li and Pt), and noted that the latter strongly indicate that the *initiate state* binding energy positions for these species should shift to increasingly larger values as the size of the cluster decreases. In addition to the relative direction, the Bagus group[144] also has noted that the magnitudes of their calculated shifts are comparable to those realized in the experimental observations (e.g., 0.5 to 2.0 eV).

Beginning with the studies of Citrin and Wertheim[102] and continuing in a well-conceived series of added experimental observations,[139,145] the Wertheim group also observed that the cluster shifts for noble (and other) metals on insulating substrates seem to always be positive, and that they grow as the size of the cluster dwindles. In addition, almost as an afterthought in one of their studies, Wertheim et al.[102,139] examined the effect of cluster size for a noble metal (Au) on both insulating (amorphous C) and conductive (metallic glass) substrates. *They concluded that the positive cluster shift that increased with decreasing cluster size in cases involving insulating supports seemed almost nonexistent for the conductive substrate.* (We will return to these important observations in the next section.)

Citrin and Wertheim[102] further noted that the referencing (zero point) for these various systems changes during the experiments (recall Figure 9-28). Thus, they pointed out that, when the cluster size of metal A approaches its bulk dimension, the binding energy zero position is the Fermi edge of A. As the size of the A cluster shrinks, however, the zero point has to switch to that of the dominant (B) matrix. This, they argued, tends to push the d band of the A metal away from the (new) Fermi edge (see Figure 9-28), as electron density is naturally forced out of the valence region of the less and less metallic A. In the view of the findings of Citrin and Wertheim,[102] this creates *the semblance* of a (0.5- to 2.0-eV) positive binding energy shift. If, however, one realigns the zero to the leading edge of the (now somewhat depleted) d band of the A species, the shifts exhibited by the correspondingly realigned core level peaks would actually be somewhat negative (see Figure 9-28). Thus, Citrin and Wertheim[102] argue that this brings the direction of the cluster shift into "alignment" with the negative shift generally exhibited by the surface A (d > 5)-type species (see Section VII.B of this chapter).

In order to rationalize the positive direction of the detected cluster shift, Citrin and Wertheim originally suggested that this effect arose due to two effects: the zero energy reference level alignment problem and *a final-state relaxation energy*.[102] In later studies, the Wertheim group has concentrated on an explanation for the positive shift that emphasizes the contribution of the "unit positive charge that remains on the cluster in the photoemission final state".[139,145]

As a result of the different conclusions reached by their respective groups, one finds the Wertheim[139,145] and Bagus[144] explanations completely polarized.[146] Both groups conceded that initial and final-state effects both play a role in these shifts, but *the Bagus group has employed calculations to argue that the principal cause is initial state,[146a] whereas the Wertheim group has employed comparative analysis (and some calculations) to argue in favor of, primarily, final-state effects.[146b]*

Examination suggests that both arguments have merit, but closer scrutiny finds that there may still be key missing features. Perhaps the major problem is one concerned with the influence of charging. This is described in the next section.

It should be noted that attempts to marry in their entirety all of the various orientation shifts may be headed for the rocky cliffs of divorce. Thus, we find the Wertheim group[102] arguing

(a)

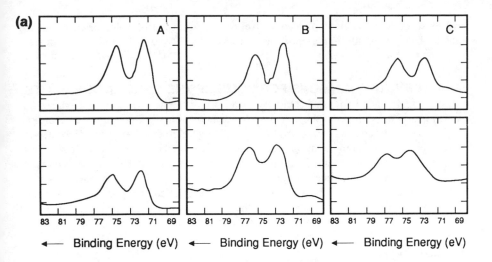

(b)

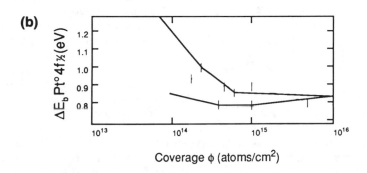

Figure 9-29 (a) XPS spectra in the Pt 4f region: (A) for 1.0×10^{15} atoms/cm² Pt coverage on SiO_2 and 5.9×10^{14} atoms/cm² coverage on Teflon™; (B) for 3.8×10^{14} atoms/cm² Pt coverage on SiO_2 and 2.3×10^{14} atoms/cm² Pt coverage on Teflon™; (C) for 9.0×10^{13} atoms/cm² Pt coverage on SiO_2 and 7.9×10^{13} atoms/cm² coverage on Teflon™. (b) Changes in the binding energies of the Pt $4f_{7/2}$ peak on the Pt-Teflon™ and Pt-SiO_2 systems as a function of coverage.

that the critical component of the surface core-level shifts are initial state effects (in contrast to the final state arguments of the Martensson group,[98,101,105,112]) only to have the Wertheim group reverse itself[145,146b] and suggest a primacy for final state effects in small cluster shifts (against the initial state arguments of the Bagus group[144,146a]). Further, the Wertheim group notes that the d density for noble metals at the surface is enhanced compared to the equivalent density in the bulk of a noble metal lattice, whereas they note that in the case of small cluster shifts the d density is reduced. They find that the latter should be true despite their well-conceived argument that both surface and small cluster shifts should be negative if all situations are aligned to identical reference zeroes (see Figure 9-28).

All of this suggests that, although extensively examined and theoretically argued by some of the most capable photoemission groups (e.g., those of Wertheim, Martensson, Bagus, Eastman, Mason, Egelhoff, etc.), a number of unexplored aspects and problems still exist, both individually and collectively, concerning all of these orientation shift effects. Two potentially important areas that have largely avoided previous scrutiny are examined in the next section.

G. ALTERNATIVE REASONS FOR SOME OF THE ORIENTATION SHIFTS

As we elaborated above, substantial well-argued differences of opinion have accompanied nearly every form of the so-called "orientation shifts". Thus, in the case of both *surface core-level* and *small cluster shifts*, competing arguments have been presented for the predominance of both initial state and final state origins for these effects. Both of these approaches have been presented with such persuasiveness that it currently seems inappropriate to favor either side. Interestingly enough, while we have not taken sides, in several of these areas of analysis we have come up with entirely different explanations for at least modest parts of these effects. Because of the novel and seemingly consistent nature of these alternative explanations, they are also discussed herein.

1. Charging

In one of the previous sections of this review (see Section I in Chapter 7), the author described many of the facets related to the charging problem. In particular, it was noted that the problem may actually have a bright side, i.e., that under certain circumstances it may be useful as an auxiliary tool in situations involving the same chemical components in different morphologies. It also was noted that, in these cases, one of the primary aspects of interest is differential charging.[84] Thus, it was noted that differential charging usually ensues from some combination of the following: (1) two or more detectable components (or two or more morphological arrangements of the same component) must be present; (2) these various "ingredients" must either exhibit significant differences from one another in their lack of conductivity; and/or (3) the components must differ substantially from one another in the degree to which either is Fermi edge coupled to the spectrometer (and correspondingly coupled to each other).

Thus, for example, the two-component systems containing Si° and SiO_2 provided a variety of different charging behaviors depending upon their morphological arrangement.[84,147] SiO_2 is a broad band-gapped insulator. A thick wafer of it will produce a relatively large charging shift, with all peaks shifted to about the same extent. (This shift can then often be removed with proper use of a flood gun.) Si°, on the other hand, is a relatively narrow band-gapped semiconductor that produces, at room temperature, little or no charging shift. In the case of a thin layer of SiO_2 on Si° [both detectable in the same $Si(2p)$ scan], the former charge-shifts, and the latter does not. On the other hand, when small islands of Si° are deposited *in* an SiO_2 matrix, the Fermi edge of the former (if formed at all) is decoupled from that for the spectrometer by the surrounding insulating matrix. As a result, $Si^\circ(2p)$ and $Si(2p)O_2$ peaks both initially produce "incorrect" binding energies, and both respond to the electron flow from a flood gun (albeit, to different extents). Thus, the case represented by Figure 7-4 in Chapter 7 results; since Fermi edge coupling to the spectrometer was never achieved, none of the (spectrometer referenced) binding energies produced are true readings (also shown in Figure 7-4 in Chapter 7). Interestingly, although the resulting binding energies were apparently not the conventionally accepted values, they did change magnitude when the morphology of the components was altered, and these changes seemed to follow certain patterns.[84,147]

As described in sections 7.I.D.2 and 7.I.D.3, the most common practical occurrence of these morphological variations arises during attempts to disperse metals (e.g., Pt°) in classic support materials (e.g., Al_2O_3) to achieve common heterogeneous supported metal catalysts.[84,148] The extent of this dispersion seems to be one of the pivotal features in the useful functionality (activity, selectivity, and lifetime) of these catalysts that are of such paramount importance to the petroleum industry. Catalytic performance appears to substantially depend on having the metal uniformly dispersed in the support in tiny (<20-Å^2) units. These units are so small and the field of display so unreceptive that generally a proper functioning (0.5 wt %)Pt/Al$_2$O$_3$ catalyst exhibits no Pt$^\circ$ at all in the best TEM. Malfunction, on the other hand, is often characterized by the clustering of the tiny Pt units into relatively large crystallites, ranging 50 to 100 Å^2 or larger. These clusters are generally detectable in a TEM.

The latter is, of course, the classic practical example of the "small cluster" situation that is considered in Section 9.VII.F. The dispersion of Pt in Al$_2$O$_3$ has been extensively examined by the author using high-resolution ESCA (as described in detail in Section 7.I.D.3.[84,148,149] The insulating matrix (Al$_2$O$_3$) always produces a charging shift. When the Pt$^\circ$ is in the form of fine particles, uniformly dispersed throughout the Al$_2$O$_3$ (the optimal catalyst), the platinum also tends to experience the "full thrust" of the charging shift of the matrix. With charging removed (e.g., through use of a flood gun), and the Al$_2$O$_3$ peaks coupled to the Fermi edge of the spectrometer (perhaps through the use of adventitious carbon, see Section 7.I), the binding energies realized by the Pt ESCA peaks range from ~ +1.0 eV to ~0.0 eV, compared to Pt$^\circ$ for a bulk platinum foil (e.g., Pt 4d$_{5/2}$ ~314.0 eV — see, e.g., Figure 7-8 in Chapter 7). If, on the other hand, the metal dopants are "abused", and clustering is produced, then the differential charging-floating Fermi edge situation results (see, e.g., Figure 7-8 in Chapter 7).

None of these statements necessarily change the arguments by the other small cluster investigators (as outlined in Section 9.VII.F), but one should be aware that there are extensive charging "problems" looming on the horizon of almost every investigation of this type. It also should be noted that we discovered that layers of SiO$_2$ or Al$_2$O$_3$ of only ~20 Å are sufficient to reduce to zero the space charge layer from a conductive probe, and thus bring charging into play. It is our guess that some degree of charging was present in nearly every measurement reported in Section 9.VII.F. One must be very careful here because, as discussed earlier in Section 2.III, some spectrometers merely "hide" this situation while others provide flood guns that are far from adequate. For example, we anticipate that if a system is subject to certain forms of, final state extra-atomic charge adjustment, it is also subject to macroscopic charging.

It is our view that XPS cluster analysis is a potentially very useful form of the XPS orientation shifts, and that all cases of interest will prove to exhibit interesting balances of three closely related effects: (1) initial-state cluster size,[144,146] (2) final state charge relaxation,[102,139,145] and (3) differential charging shifts.[84] The composition, cluster size, and evolving morphology will all affect the resulting balance, and premature lack of consideration of any one of these three effects may negate correct analysis.

2. Hydrogen Contamination

The previously described surface shifts (Section 9.VII.B) have been suggested to be intimately connected to both interfacial (Section 9.VII.C) and small cluster (Section 9.VII.F) shifts. Close scrutiny, however, reveals that there are some obvious differences. For both interfacial and small cluster shifts we can see evidence of electronic involvement between the key species and the "adsorbing" matrix.[101] In other words, the latter two shifts are obviously of chemical (electron transfer) origin, and we see evidence of how that chemistry evolves. The problem with the surface core-level shifts is that we also see tantalizing snippets of its chemistry,[102] but it is difficult to see how it fits together. To assist in trying to document the latter, it may be helpful to consider a general review:

1. Surface core-level shifts connote the transfer of electrons (either initial[102] or final state[101]) *to or from* a surface component (to *another species*).
2. Although the electron transfer may be achieved to some degree by any valence band region (or available orbital), it is greatly facilitated by localized parts of bands (or localized available orbitals).[102]

3. Thus, the most "useful" electrons (or available orbitals) for this purpose are either d or f electrons (orbitals),[102] or hybrid-localized (directed) s-p states.[106]

4. The prospect for surface core-level shifts is further optimized if these localized states (d electrons or orbitals) are "readily available" — i.e., are relatively close to the Fermi edge. (Thus, the Cu and Au metals, which are mostly d^9, exhibit larger surface shifts, than the d^{10} Ag, and the latter has its d band origin almost twice as far from the Fermi edge.)[101,102,105]

5. The more d electrons or empty shell states available, the poorer the surface shift — i.e., "too much of a good thing creates confusion."[101,102,105]

If localized chemical bonding does play the principal role in these surface processes, then we must ask: "localized bonding to what?" Certainly, one of the answers to this is implied in both the arguments of Citrin and Wertheim[102] and Johansson and Martensson,[105] in that the fields induced by the "retarded" coordination at the surface are going to impose different component coupling criteria in both initial and final state situations, compared to related effects in the bulk. Further, it should be apparent that the lack of continuation of the lattice should make the surface bonding situations much more susceptible to localized and selective effects. (Thus, we have the interconnection to surface reconstruction.[125,150]) However, despite these possible reasons for surface shifts, another feature may be involved that has, so far, only been occasionally hinted at — which we present here as a suggestion. *It may be that part of the XPS effects realized as surface core-level shifts are due to the chemical coupling of the detected surface species and various chemical forms of hydrogen.*[151]

Before we consider this somewhat startling proposal, we should quickly review the status of hydrogen in metals and other matrices, particularly as analyzed by XPS and other forms of surface analysis. The reader should not need to be reminded that XPS (as well as all other forms of photoelectron spectroscopy) and Auger are blind to hydrogen. However, this is a handicap that also comes equipped with a form of Braille in that one can often use XPS to detect the "effect" of hydrogen on other elements. Thus, many hydroxides can be XPS differentiated from oxides;[152,153] complex Bronsted acidic systems can be differentiated from their corresponding oxide salts;[154] and certain hydrocarbon and related organics also can be distinguished.[155,156] Substantial evidence has also been provided that the ambient air-induced oxidation process — dubbed by us as "natural passivation" — is generally capped by an adsorption reaction of oxidized species of hydroxide and/or aquated oxide layers.[152,153] The latter features seem to be part of a somewhat disturbing pattern that establishes that all atmospheres (analyzed using a residual gas analyzer) are mixtures of numerous species, dominated by N_2, O_2, CO_2, $H_2O_{(g)}$ and H_2, but that *evacuation down to high (and even ultra-high) vacuum generally tends to produce some imbalance of these species, often in favor of an atmosphere dominated by H_2O and H_2*. Given that these gases are continually bombarding the sample surface, one may well ask: "Where does all the hydrogen go?"[151]

Ion scattering spectroscopy (ISS) provides a somewhat disquieting answer. All users of that methodology know that the initial registration of most ion bombardment data do not detect the known outer surfaces, i.e., top monolayer, constituents *because of the apparent detection of sputtered hydrogen.*[157] This is true even for quite pristinely handled (single crystal in vacuum) surfaces.

This suggests that the outer surface of most materials (M) systems are at least partially "involved" with hydrogen, either in the form of hydroxide or M-H, quasihydride, species. In fact, to varying degrees, these features may be *chemically* involved with surfaces in the region where researchers are trying to detect such features as surface core-level shifts. *Thus, it seems reasonable to ask if, at least, some finite part of many of the detected surface shifts are not due to chemisorbed hydrogen (or hydroxide).* There is little direct evidence to support this possibility, but the circumstantial evidence is intriguing.[151]

If chemisorbed hydrogen is involved, then there should be some indication of this effect in the known chemistry of hydrides and related systems. Thus, the characteristic features of hydrides and other hydrogen-involved chemistry should provide a signpost.

First of all, we need to ask if there is any indication in the existing literature of a periodic tendency to form hydrides and other hydrogen-containing systems that seems to replicate those progressive features for surface-core level shifts that have been found in many parts of the periodic table.

In order to address this question we turn to the surprisingly sparse studies of hydride formation. Several of these have been summarized, based on the periodic table. Among the most interesting are those produced by Gibb (see Table 9-2).[158] These tables, which originated with the work of Luder and Zuffanti,[159] classify the various "hydrides" in terms of their "forming ability" and type of hydride formed [saline, metallic, molecular (covalent), etc.]. In addition, these various hydrides are listed along with their first ionization potential, with the supposition that low values of this term signify ease of saline-type

Table 9-2 Relation of periodic table to hydride-forming ability

Saline hydrides (s-metals)		Molecular and polymetric hydrides (p-metals, metalloids) (nonmetals omitted)				Metallic hydrides (d-metals)
LiH	BeH$_2$	(BH$_8$)$_3$	—	—	—	
121	215	191				
NaH	MgH$_2$	AlH$_3$	SiH$_4$	—	—	ScH$_3$
118	176	138	188			151
KH	CaH$_2$	GaH$_3$	GeH$_4$	AsH$_5$	—	YH$_3$
100	141	138	182	226	—	147
RbH	SrH$_2$	InH$_3$	SnH$_4$	SbH$_5$	TeH$_2$	LaH$_3$
96.3	131	133	169	199	208	129
CsH	BaH$_2$	TlH$_3$	PbH$_4$	BiH$_5$	PoH$_2$	
89.7	134	141	171	167	189	

Lanthanides (metallic)	LaH$_3$	CeH$_3$	PrH$_3$	NdH$_{3.8}$	Pm	SmH$_{2.3}$	EuH$_2$	GdH$_{3.3}$
	129	151	134	145	—	129	130	142
Actinides (metallic)	AcH$_3$	TbH$_3$	PaH$_3$	UH$_3$	NpH$_3$	PuH$_3$	AmH$_3$	CmH$_3$
	—	—	—	—	—	—	—	—

Metallic hydrides (continued)							Unspecified types	
TiH$_2$	VH$_3$	(CrH$_1$)	Mn	Fe	Co	NiH$_2$	(CuH)	(ZnH$_2$)
157	155	156	171	182	181	176	178	216
ZrH$_2$	NbH$_{1.2}$	Mo	T$_c$	Ru	Rb	PdH$_{<1}$	Ag	(CdH$_2$)
157	156	165	166	173	177	192	176	207
HfH$_2$	TaH$_{<3}$	W	Re	Os	Ir	Pt	Au	(HgH$_2$)
160	177	183	181	200	212	206	212	240
TbH$_3$	Dy	Ho			Er	Tm	YbH$_3$	LuH$_3$
155?	157	—			—	—	143	142

Note: Known hydrides are formulated. Those known to be thermally unstable under ordinary conditions are in parentheses. The numbers given are first ionization potentials (rounded, in kcal).

formation, and are thus a partial indication of the stability of this form (and correspondingly, the other hydride types). Thus, we find some indication of enhanced hydride formation for the alkali metals, and also enhancement as the type of metal considered moves down a particular column of the periodic table. We also see indications that those transition metals with only one or two d electrons are generally good hydride formers, as are the lanthanides, and (according to Cotton and Wilkinson[160]) several of the actinides. Blanket continuation of this type of argument beyond these general statements, or to other elemental groups, is not easily accomplished. However Gibb does point out[158] that the relative size of the ionization potentials (or, for that matter, their electronegativities) depends very much on the type of hydride formed, and there is substantial evidence that *many metals* do not form conventional hydrides (with H^- units), but rather achieve some form of modest covalent bonding in which the hydrogen may be *positive* in charge. Interestingly, the elements that seem most inclined to $H^{\delta+}M^{\delta-}$ formation come primarily from the right (d > 5 electrons) side of the transition metals, and from some of the III A and IV A species. In this model, the lack of sufficient electrophilicity on the part of some metals may be more than compensated by the resulting large dielectric constant. In any case, one cannot be sure that classic M^+H^- units form in any but the saline and most ionic metallic hydrides.[158]

We are thus left with far too little information to reach any firm conclusion, but one must be intrigued by the observation that the same periodic propensity exists for surface-core level shifts and hydride formation, and there is also some evidence that the type of bond formed by the M_x-H_y unit may change (both in type and sign) at about the same place in the periodic table where the surface core-level shifts go from positive to negative.[151]

All of this obviously may still be argued as happenstance, since the Gibb study[158] and most related observations[159] refer to bulk, primary hydride formation and retention, in other words, how does this affect selective *surface* situations? We have already noted that most surface techniques are blind to hydrogen, and those that see it find much more of it on surfaces than expected. However, we must also note that many of the surface core-level studies have involved the use of purportedly pristine ultra-high vacuum techniques, in which alternating sputter-cleaning, annealing, and cooling steps are employed to provide supposedly very clean, controlled surfaces. The problem may be that the very processes designed to clean these surfaces may, under certain circumstances, be making them at least somewhat hydrogen contaminated. This may occur, because, as described by Roberts and McKee,[161] many metals are excellent absorbers of hydrogen into their bulk lattices. Thus, when these systems are treated by techniques designed to remove adsorbed species and reconstruct the surface to some prescribed status, *the process may also promote segregation of hydrogen to the surface, and its selective attachment in the form of hydrides (or quasi-hydrides)*. Thus, in this form, hydrogen, which may be the most difficult adsorbate to keep off, also may be insidiously arising from the depths of selective metals. Interestingly, though there are still only a few systematic studies of this segregation process, Roberts and McKee[161] do suggest which metals may be most/least proficient at this, and, once again, these patterns seem to closely resemble those for proficiency/deficiency of surface core-level shifts.

Finally, it should be noted that, due to its size and single orbital type, hydrogen occupies a special status in chemical bonding that seems to favor the present argument. Thus, hydrogen generally tends to form localized-directed bonds with metals (and other elements) that are often able to circumvent more delocalized orbital situations (e.g., multiple bonds in olefins) as the hydrogen reaction develops.[151] Yet, despite this apparent lack of bond delocalization, surface hydrogen on metals seems to exhibit a special form of mobility ("spill over").[162]

In conclusion, we must admit that evidence for the possible involvement of hydrides in surface core-level shifts is, at the least, very incomplete. We know, however, that

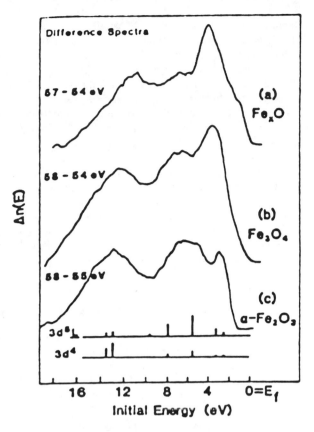

Figure 9-30 Distributions of Fe 3d-derived final states in (a) Fe$_x$O(100), (b) Fe$_3$O$_4$ (110), and (c) α-Fe$_2$O$_3$ (1012)4 determined by the difference between valence spectra measured at photon energies just above and just below the p to 3d resonance; an inelastic background has been removed from the spectrum. The vertical lines in (c) represent the relative intensities of calculated ground states.

surface shifts are most pronounced in situations that strongly favor localized bonding with d electrons (or holes) and s-p-hybrids.[102] These same situations seem to promote strong hydride-type bonding, with an interesting balance in the direction of the resulting shifts.[151,158] *Therefore, we suggest that surface core-level shifts probably have substantial initial[102] and final state[105] components simply due to the presence of a solid vacuum surface — but we feel that current evidence supports the possibility that some forms of hydride (on hydroxide) formation may at least contribute in some cases to the total shifts.[151]*

VIII. RESONANT PHOTOEMISSION

Another supplementary technique that appears to be receiving increasing attention in a variety of areas is resonant photoemission spectroscopy (RES-PES).[163] This method requires a well-controlled, variable energy source in the low-to-medium energy range. For these purposes, a synchrotron radiation system is ideal. In the resonant procedure, the photon energy is swept through the range of a transition from a near core state to an empty state in the valence band (see Figure 9-30). In a study of Fe, Lad and Henrich[164] swept the energy range of the 3p to 3d transition. They then employed an electron detector to collect the low-energy photoelectrons emitted from the valence (3d) band, including the original valence electrons and those excited from the core state. At the photon energies employed, the latter are resonant-enhanced by such transitions as the $3p^6 3d^n \rightarrow 3p^5 3d^{n+1}$, etc. As the photon energy is swept into the latter transition region, the 3d spectrum is noticeably altered from one containing only the "normal" 3d photoemissions to one featuring the addition of 3p to 3d satellite enhancements. Employing these procedures, one is able to separate out many of the subfeatures of complex valence bands that are often found hopelessly jumbled together in conventional photoemissions. In this manner, for example, Lad and Henrich[164,165] were able to designate those parts of the valence bands of iron and various iron oxides that are due to O(2p) from those that are due to Fe(3d).

Resonant photoemission has also been employed in studies of the new high T_c oxide superconductors. This type of study has been conducted for the Y-Ba-Cu (1-2-3) system by Shen et al.[166] who employed RES-PES to try to determine the degree of Cu-O correlation. The latter is a reflection of the degree of covalency in these compounds.

In all of these cases, the key is to have access to a low, variable-energy photon source that may be accurately tuned to the energy range around that of a near-core transition. The way the latter influences the valence band (as a series of satellite peaks) is a strong expression of the particular chemical and structural aspects of the system. The disadvantage of this method (from the point of view of conventional ESCA) is the need for the additional low-energy source. Otherwise, the method has significant merit since it produces unique, informative transitions that are essentially independent of the previously described Fermi-edge problems.

IX. LIQUID PHASE ESCA

Liquid phase ESCA is one of the most promising, but presently also one of the most frustrating areas of photoelectron spectroscopy. As with so many techniques in this spectroscopy, initiation and development in liquid phase (XPS) ESCA have been primarily achieved at the University of Uppsala. In this case, work has been centered in the laboratories of Hans Siegbahn, one of the productive sons of the ESCA founder, Kai Siegbahn.[167]

The initial Uppsala studies quickly realized that a key feature of this new type of ESCA was its ability to detect the interface between the liquid state and any vapor material that is evaporating from the surface of the latter.[168] Since surface evaporation is continuous, a constant regeneration of the liquid-vapor interface is needed. These requirements have influenced the designs of liquid phase XPS systems, resulting in a substantial dependence on both unique "sample presentation" and differential pumping.[168] Both continuous wire rotation[169] and a rotating drum[170] have been successfully used as the sample stage. In both cases, the speed of rotation was often sufficient so that the resulting liquid film was thin enough to reveal photoelectrons from the subsurface metal drum or wire. With this technology, a single spectrum can be produced that simultaneously contains information on all three phases. The sample presentation techniques were eventually perfected to the point where monochromatic XPS could be employed.[170] Some critics have complained that a resulting liquid film that is less than 100 Å thick is not representative of the liquid itself, but it should be obvious that the Uppsala methodology reveals information from a region that is at least as representative of a liquid as the surface (subsurface) of a solid is of the bulk of that phase. The beauty of the Uppsala liquid phase ESCA techniques is its obvious interconnection with many (as-yet unanswered) questions in electrochemistry and catalysis (e.g., the key processes occurring at the fluid/gas and fluid/solid interfaces[171]). A general representation of one of the systems employed by Siegbahn and his group is shown in Figure 9-31.[171]

Siegbahn and his associates[171,172] have used an acceleration potential to bias mixed gas and liquid results. These studies have shown that the gas phase peak may be completely removed by broadening. On the other hand, the liquid phase peak simply shifts (biases) with the applied voltage (see Figure 9-32). This demonstrates that the (now separate)

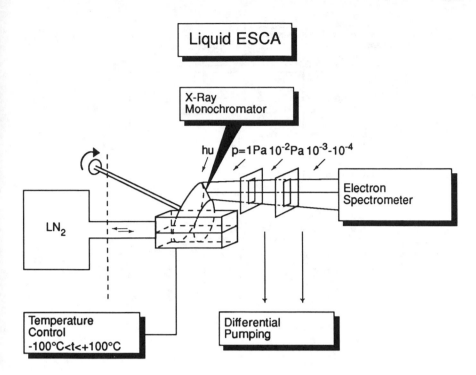

Figure 9-31 Liquid ESCA arrangement based on a wetted metal surface in the form of a rotating trundle. The arrangement includes regulated cooling of the sample and excitation with monochromatized X-rays.

liquid phase result is coupled to the Fermi edge of the spectrometer. As a result, it is possible to evolve an energy level scheme for the liquid system as shown in Figure 9-32.

Liquid phase photoelectron emission *spectroscopy* was actually first reported in the ultraviolet range in 1972 by Paul Delahay and his group at New York University.[173,174] These studies have continued to evolve to include "energy distribution curves" (EDC's) under a variety of analysis conditions (see, e.g., Figure 9-33, where a spectrum is displayed for varying amounts of potassium in liquid ammonia). Actually, just as with the gas and solid phase forms, there are nonspectroscopy liquid phase examples of photoemission that date from the late 1880s.[175] An excellent discussion of all these features was presented in a review by Prof. Delahay in 1984.[176]

Delahay et al.[176,177] have recently proposed a major improvement in cell design (Figure 9-34) that employs a rotating disk that brings a film of the liquid in question into the photon beam in much the same manner as the drum and wires utilized by the Siegbahn group.

Successful studies also have been achieved by the Delahay group in examinations of solutions and polar liquids.[178] In particular, they have evolved select gas-liquid shifts, Δ_{gl}, for certain polar liquids that seem to correlate with electronic polarizations, P_{el}.[176,179] Apparently, some of the other effects that should contribute to Δ_{gl} seem to cancel out (see, e.g., Table 9-3). The limitations of the ultraviolet form of liquid phase ESCA are the same as those observed in the XPS method. Hence, the type of liquids successfully examined does not include those with significant vapor pressures (i.e., P must be less than ~10^{-3} torr),[176] and *the method has yet to be successfully applied to water.*

As might be expected, most of the detailed studies in liquid phase ESCA have been primarily concerned with the chemical and physical behavior of solute species (generally

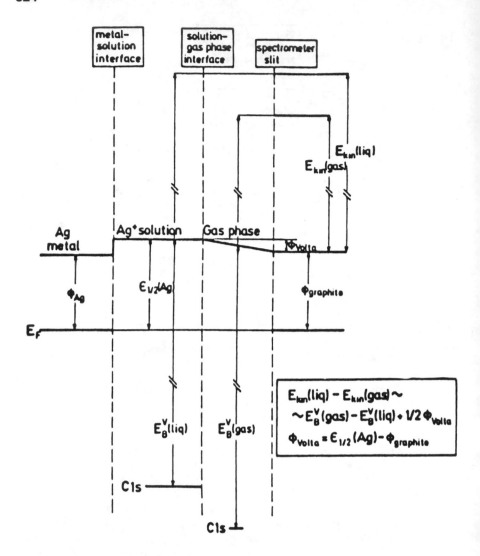

Figure 9-32 Relationship between liquid and gas ESCA based on $E_F(Ag)$, $E^V(liq\ Ag)$, E^V (gas Ag), $E^{1/2}$ (Ag), and ϕ (graphite).

solute ions), as opposed to investigations of the liquidus (solvent) species itself. This places liquid phase ESCA on a par with bulk phase liquid studies, where there is a similar preponderance of interest in what the phase does to dissolved species. In these types of studies, Siegbahn has been able to utilize solution thermodynamics to follow the behavior of both positive and negative ions.[171] For example, in a study of negative ions in various alcohols, Siegbahn and his colleagues have detected several different solution effects exhibited by a set of simple ions (halides) compared to more complex species (e.g., NO_3^-).[171,180]

In many cases, the effect of the solid sample holding material is reflected in shifts in the resulting gas and liquid phase spectra. In Figure 9-35, we see examples presented by the Siegbahn group of C(1s) spectra realized with three different half-cell (trundle) materials.[171,181] Note that the peak positions do not shift uniformly, but the resulting shifts can be argued from simple electrochemical principles.

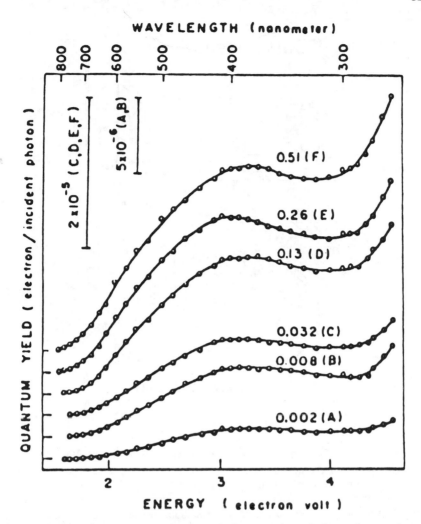

Figure 9-33 Emission spectrum of solvated electrons for solutions of potassium of different molar concentrations in liquid ammonia at −60°C (zero yield indicated at left for each curve). Yields measured at 6.85 V/cm-torr.

In some cases, the intermolecular forces involved in the liquid phase have been investigated with ESCA.[171] In Figure 9-36, we see that both the C(1s) and O(1s) peak positions for methanol are reduced in binding energy in the liquid phase, compared to their positions in the gas phase. The Siegbahn group has analyzed these shifts and considers them to be of multiple origin, with measurement-dependent relaxation effects suspected to provide the largest contribution.

A few simulated applications to solutions in water have been described by the Siegbahn group[171,182] (e.g., the comparison of the O(1s) spectra achieved for liquid and vapor water in the presence of a strong electrolyte, as in Figure 9-37). These studies, however, are still limited by experimental difficulties. These features must be overcome in order to propel liquid phase ESCA into its justifiable position alongside the solid and gas phase methodologies.

Among the recent advances that have proven to be very promising are liquid phase extensions to meaningful angular resolution (AR) photoemission.[183]

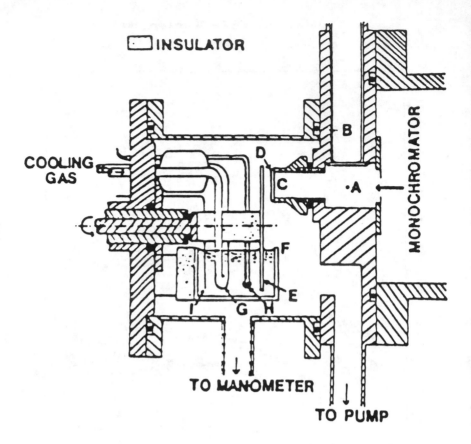

Figure 9-34 Schematic diagram of rotating disk target assembly, designed by Delahay and colleagues for the determination of emission spectra. Codes: A, wire covered with sodium salicylate; B, photomultiplier; C, lithium fluoride window; D, gold grid mesh (80% transparency); E, rotating (30 to 240 rpm) quartz disk (52 mm diameter, 2 mm thick); F, reservoir containing the liquid or solution being studied; G, glass tube for cooled gas; H, thermistor; and I, platinum wire electrode.

Table 9-3 Gas-liquid shifts Δ_{gl} and electronic polarization P

Substance	I for gas	E_1 for liquid at 11.7 eV	E_{max} for liquid at 21.2 eV	Δ_{gl} at 11.7 eV	Δ_{gl} at 21.2 eV	P_{cl}
	(eV)	(eV)	(eV)	(eV)	(eV)	(eV)
N-Methylaniline	7.73	6.3	7.1	1.4	0.6	1.2
	9.04	7.9	8.6	1.1	0.4	
	10.24	9.3	9.8	0.9	0.4	
N,N-Dimethyl-*p*-toluidine	7.48	6.1	6.8	1.4	0.7	1.0
Formamide	10.33	8.9	9.8	1.4	0.5	1.4

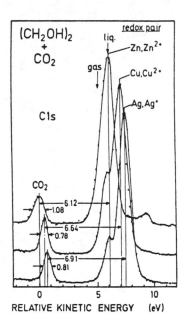

Figure 9-35 C(1s) core lines in ethylene glycol with CO_2 as reference gas (three different half cells).

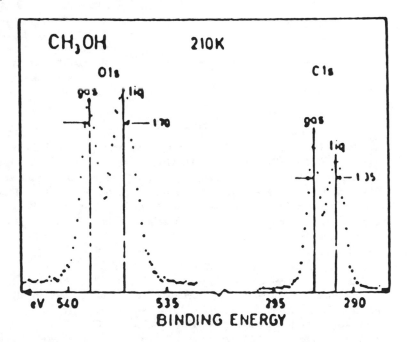

Figure 9-36 O(1s) and C(1s) signals from liquid methanol obtained at a temperature (210 K) where the vapor and liquid lines are of similar intensities. The indicated shifts have not been corrected for the difference between liquid and vapor reference levels (the binding energy scale refers to the liquid). This may affect the shifts by a few tenths of an electron volt (C(1s) and O(1s) shift equally).

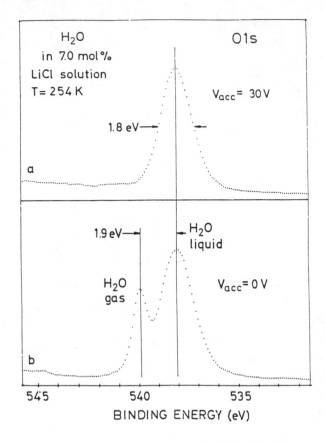

Figure 9-37 Spectrum of liquid water. The O(1s) water gas line is removed by applying a potential on the backing of the liquid samples. The remaining O1s is from liquid water alone (upper part of figure).

REFERENCES

1. **Werme, L. O., Grennberg, B., Nordgren, J., Nordling, C., and Siegbahn, K.,** *Phys. Rev. Lett.,* 30, 523, 1973; **Werme, L. O., Grennberg, B., Nordgren, J., Nordling, C., and Siegbahn, K.,** *Nature,* 242, 453, 1973; **Wannberg, B., Veenhuizen, H., Mattsson, L., Norell, K.-E., Karlsson, L., and Siegbahn, K.,** *J. Phys. B,* 17, L259, 1984a; **Siegbahn, K.,** *J. Electron Spectrosc. Rel. Phenomena,* 51, 11, 1990.
2. **Larsson, N., Steiner, P., Eriksson, J. C., Maripuu, R., and Lindberg, B.,** *J. Colloid Interface Sci.,* 90, 127, 1982.
3. **Gelius, U.,** *J. Electron Spectrosc. Rel. Phenomena,* 5, 985, 1974.
4. **Svensson, S., Martensson, N., and Gelius, U.,** *Phys. Rev. Lett.,* 58, 2639, 1987; **Ericksson, B., Svensson, S., Martensson, N., and Gelius, U.,** *J. Physics B,* 21, 1371, 1988; **Svensson, S., Ericksson, B., Martensson, N., Gelius, U., and Wendin, G.,** *J. Electron Spectrosc. Rel. Phenomena,* 47, 327, 1988.
5. **Eriksson, B. and Martensson, N.,** U. Uppsaala Inst. Phys. Report 1215, to be published.
6. **Gelius, U., Wannberg, B., Baltzer, P., Fellner-Feldegg, H., Carlsson, G., Johansson, C.-G., Larsson, J., Münger, P., and Vegerfors, G.,** *J. Electron Spectrosc. Rel. Phenomena,* 52, 747, 1990.
7. **Woodruff, D. P., Johnson, P. D., and Smith, N. V.,** *J. Vac. Sci. Technol. A,* 1, 1104, 1982.
8. **Geisen, K., Hage, F., Himpsel, F. J., Riess, H. J., and Steinmann, W.,** *Phys. Rev. Lett.,* 55, 300, 1985; **Bokor, J., Haight, R., Storz, R. H., Stark, J., Freeman, R. R., and Bucksbaum, P. H.,** *Phys. Rev. B,* 32, 3669, 1985.

9. Duane, W. and Hunt, F. L., *Phys. Rev.*, 6, 166, 1915.
10. Dose, V., *Prog. Surface Sci.*, 13, 225, 1983; Johnson, P. D., Hulbert, S. L., Garrett, R. F., and Howells, M. R., *Rev. Sci. Instrum.*, 57, 1324, 1986.
11. Denninger, G., Dose, V., and Bonzel, H., *Phys. Rev. Lett.*, 48, 299, 1982.
12. Pendry, J. B., *Phys. Rev. Lett.*, 45, 1356, 1980; Pendry, J. B., *J. Phys. C.* 14,1381, 1981.
13. Fauster, T. and Himpsel, F. J., *J. Vac. Sci. Technol. A*, 1, 1111, 1983.
14. Woodruff, D. P. and Smith, N. V., *Phys. Rev. Lett.*, 48, 283, 1982.
15. Dose, V., Glöbi, M., and Scheidt, H., *J. Vac. Sci. Technol. A*, 1, 1115, 1983.
16. Krainsky, I. L., *J. Vac. Sci. Technol. A*, 6, 780, 1988.
17. Krainsky, I. L., *J. Vac. Sci. Technol. A*, 5, 735, 1987.
18. Knapp, B. J. and Tobin, J. G., *J. Vac. Sci. Technol. A*, 6, 772, 1988.
19. Johnson, P. D. and Smith, N. V., *Phys. Rev. Lett.*, 49, 290, 1982.
20. Gao, Y., Grioni, M., Smandek, B., Weaver, J. H., and Tyrie, T., *J. Phys. E. Sci. Instrum.*, 21, 489, 1988.
21. Kubiak, G. D., *J. Vac. Sci. Technol. A*, 5, 731, 1987.
22. Kubiak, G. D. and Kolasinski, K. W., *J. Vac. Sci. Technol. A*, 6, 814, 1988.
23. Fadley, C. S., Physica Scripta, T17, 39, 1987.
24. Egelhoff, W. F., *Crit. Rev. Solid State Mat. Sci.*, 16, 213, 1990.
25. Fadley, C. S., *J. Electron Specrosc. Rel. Phenomena*, 5, 725, 1974.
26. Fadley, C. S., in *Synchrotron Radiation Research: Advances in Surface Science*, Bachrach, R. Z., Ed., Plenum Press, New York, 1993, in press.
27. Egelhoff, W. F., Jr., *J. Vac. Sci. Technol. A*, 6, 730, 1988.
28. Sinkovic, B. and Fadley, C. S., *Phys. Rev. B*, 31, 4665, 1985.
29. Siegbahn, K., Gelius, U., Siegbahn, H., and Olsen, E., *Phys. Lett. A*, 32, 221, 1970.
30. Tong, S. Y., Poon, H. C., and Snider, D. R., *Phys. Rev. B*, 32, 2096, 1985.
31. Fadley, C. S., *Prog. in Surface Sci.*, 16, 275, 1984; Sinkovic, B., Orders, P. J., Fadley, C. S., Trehan, R., Hussain, Z., and Lecante, J., *Phys. Rev. B*, 30, 1833, 1984; Sagurton, M., Bullock, E. L., and Fadley, C. S., *Phys. Rev. B*, 30, 7333, 1984; Sagurton, M., Bullock, E. L., Saiki, R., Kaduwela, A., Brundle, C. R., Fadley, C. S., and Rehr, J. J., *Phys. Rev. B*, 33, 2207, 1986.
32. Poon, H. C. and Tong, S. Y., *Phys. Rev. B*, 30, 6211, 1984; Poon, H. C., Snider, D. R., and Tong, S. Y., *Phys. Rev. B*, 33, 2198, 1986.
33. Xu, M. L., Barton, J. J., and Van Hove, M. A., *J. Vac. Sci. Technol. A*, 6, 2093, 1988; Barton, J. J., Xu, M. L., and Van Hove, M. A., *Phys. Rev. B*, 37, 10475, 1988; Xu, M. L. and Van Hove, M. A., *Surface Sci.*, 207, 215, 1989.
34. Egelhoff, W. F., Jr., *J. Vac. Sci. Technol. A*, 5, 1684, 1987; Egelhoff, W. F., Jr., *Phys. Rev. Lett.*, 59, 559, 1987; Egelhoff, W. F., Jr., *Mater. Res. Soc. Symp. Proc.*, 83, 189, 1987.
35. Barton, J. J., Bahr, C. C., Robey, S. W., Tobin, J. G., Klebanoff, L. E., and Shirley, D. A., *Phys. Rev. Lett.*, 51, 272, 1983; Barton, J. J., Bahr, C. C., Hussain, Z., Robey, S. W., Klebanoff, L. E., and Shirley, D. A., *J. Vac. Sci. Technol. A*, 2, 847, 1984; Barton, J. J. and Shirley, D. A., *Phys. Rev. B*, 32, 1892, 1985.
36. Leckey, R. C. G., *J. Electron Spectrosc. Rel. Phenomena*, 43, 183, 1987.
37. Woodruff, D. P., Norman, D., Holland, B. W., Smith, N. V., Farrell, H. H., and Traum, M. M., *Phys. Rev. Lett.*, 41, 1130, 1978.
38. Kevan, S. D., Rosenblatt, D. H., Denley, D., Lu, B.-C., and Shirley, D. A., *Phys. Rev. Lett.*, 41, 1565, 1978.
39. Egelhoff, W. F., Jr., *Phys. Rev. B*, 30, 1052, 1984.
40. Egelhoff, W. F., Jr. and Steigerwald, D. A., *J. Vac. Sci. Technol. A*, 7, 2167, 1989.

41. Szoeke, A., *Short Wavelength Coherent Radiation: Generation and Applications,* Attwood, D. T. and Bokor, J., Eds., (AIP Conference Proc. No. 147), American Institute of Physics, New York, 1986, 361.

42. Barton, J. J., *Phys. Rev. Lett.,* 61, 1356, 1988; Barton, J. J., *J. Electron Spectrosc. Rel. Phenomena,* 51, 37, 1990.

43. Harp, G. K., Saldin, D. K., and Tonner, B. P., *Phys. Rev. Lett.,* 65, 1012, 1990.

44. Fadley, C. S., *15th Int. Conf. on X-Ray and Inner Shell Processes,* (AIP Conference Proc. No. 215), American Institute of Physics, New York, 1990, 296.

45. Pierce, D. T. and Celotta, R. J., *J. Vac. Sci. Technol. A,* 1, 1119, 1983.

46. Alvarado, S. F., Eib, W., Meier, F., Siegmann, H. C., and Zurcher, P., *Photoemission and the Electronic Properties of Surfaces,* Feuerbacher, B., Fitton, B., and Willis, R. F., Eds., John Wiley & Sons, London, 1978, chap. 18.

47. Meier, F., Pescia, D., and Schriber, T., *Phys. Rev. Lett.,* 48, 645, 1982.

48. Guillot, C., Ballu, Y., Paigné, J., Lacante, J., Jain, K. P., Thiry, P., Pinchaux, R., Pétroff, Y., and Falicov, L. M., *Phys. Rev. Lett.,* 39, 1632, 1977.

49. Feldkamp, L. A. and Davis, L. C., *Phys. Rev. Lett.,* 44, 673, 1980.

50. Unguris, J., Seiber, A., Celotta, R. J., Pierce, D. T., Johnson, P. D., and Smith, N. V., *Phys. Rev. Lett.,* 48, 1047, 1982.

51. Sinkovic, B. and Fadley, C. S., *Phys. Rev. B,* 31, 4665, 1985; Fadley, C. S., *Synchrotron Radiation Research: Advances in Surface Science,* Bachrach, R. Z., Ed., Plenum Press, New York, 1993, in press.

52. Sinkovic, B., Hermsmeier, B., and Fadley, C. S., *Phys. Rev. Lett.,* 55, 1227, 1985; Sinkovic, B., Hermsmeier, B., and Fadley, C. S., *J. Vac. Sci. Technol. A,* 4, 1477, 1986.

53. Celotta, R. J., Unguris, J., and Pierce, D. T., *J. Vac. Sci. Technol. A,* 6, 574, 1988; Idzerda, Y. U., Lind, D. M., Prinz, G. A., Jonker, B. T., and Krebs, J. J., *J. Vac. Sci. Technol. A,* 6, 586, 1988.

54. Johnson, P. D., *J. Electron Spectrosc. Rel. Phenomena,* 51, 249, 1990.

55. Bruche, E., *Z. Phys.,* 86, 448, 1933; Mahl, H. and Pohl, J., *Z. Technol. Phys.,* 16, 219, 1935.

56. Wegmann, L., *Praktische Metallogr.,* 5, 241, 1968.

57. Dam, R. J. and Griffith, O. H., *Soc. Photo-Opt. Instr. Eng.,* 78, 143, 1976.

58. Griffith, O. H., *Adv. Opt. Electron Microsc.,* 10, 269, 1987.

59. Rempfer, G. F., Nadakavukaren, K. K., and Griffith, O. H., *Ultramicroscopy,* 5, 449, 1980.

60. Griffith, O. H., Lesch, G. H., Rempfer, G. F., Birrell, G. B., Burke, C. A., Schlosser, D. W., Mallon, M. H., Lee, G. B., Stafford, R. G., Jost, P. C., and Marriott, T. B., *Proc. Natl. Acad. Sci. U.S.A.,* 69, 561, 1972.

61. Massey, G. A., Plummer, B. P., and Johnson, J. C., *IEEE J. Quant. Electron.,* 14, 673, 1978.

62. See, for example, Dam, R. J., Kongsile, K. F., and Griffith, O. H., *Biophys. J.,* 14, 933, 1974; Houle, W. A., Brown, H. M., and Griffith, O. H., *Proc. Natl. Acad. Sci. U.S.A.,* 76, 4180, 1979; Griffith, O. H., Houle, W. A., Kongslie, K. F., and Sukow, W. W., *Ultramicroscopy,* 12, 299, 1984.

63. Brode, M., Pfefferkorn, G., Schur, K., and Wegmann, L., *J. Microscopy,* 95, 323, 1972.

64. Cazaux, J., *Ultramicroscopy,* 12, 321, 1984; Cazaux, J., *Scanning Electron Microscopy III,* Joshi, A., Ed., SEM Inc., AMF, Chicago, 1984, 1193; Cazaux, J., Gramari, D., Mouze, D., Nassiopoulos, A. G., and Perrin, J., *J. Phys.,* C2, 45, 271, 1984.

65. (a) Chaney, R. L., *Surface Interface Anal.,* 10, 36, 1987. (b) Brochure, SSL X-Probe Series 5000, Surface Science Laboratory, Inc., 1984.

66. (a) Yates, K. and West, R. H., *Surface Interface Anal.,* 5, 217, 1983. (b) VG ESCALAB II of V.G. Scientific Inst., East Grinstead, U.K., 1984.

67. Gelius, U., Wannberg, B., Baltzer, P., Fellner-Feldegg, H., Carlsson, G., Johansson, C.-G., Larsson, J., Münger, P., and Vegerfors, G., *J. Electron Spectrosc. Rel. Phenomena*, 52, 747, 1990.

68. Coxon, P., Krizek, J., Humperson, M., and Wardell, I. R. M., V.G. ESCA-SCOPE — A New Imaging Photoelectron Spectrometer, to be published.

69. SSI-M-Probe brochure, SSI Division of V.G., 1988.

70. Cerrina, F., Margaritando, G., Underwood, J. H., Hettrick, M., Green, M. A., Brillson, L. S., Franciosi, A., Hochist, H., Deluca, P. M., Jr., and Gould, M. N., *Nucl. Instrum. Methods A*, 266, 303, 1988.

71. De Stasio, G., Ng, W., Ray-Chaudhuri, A. K., Cole, R. K., Gao, Z. Y., Wallace, J., Margaritando, G., Cerrina, F., and Underwood, J., *Nuc. Instrum. Methods Phys. Res. Sect. A*, A294, 351, 1990.

72. Tonner, B. P. and Harp, G. R., *J. Vac. Sci. Technol. A*, 7, 1, 1989; Tonner, B. P. and Harp, G. R., *Rev. Sci. Instrum.*, 59, 853, 1988.

73. Tonner, B. P., *Nucl. Instrum. Methods A*, 291, 60, 1990.

74. Harp, G. R., Han, Z.-L., and Tonner, B. P., *J. Vac. Sci. Technol. A*, 8, 2566, 1990; Harp, G. R., Han, Z.-L., and Tonner, B. P., *Phys. Scripta*, T31, 23, 1990; see also, Pianetta, P., King, P. L., Borg, A., Kim, C., Lindau, I., Knapp, G., Keenlyside, M., and Browning, R., *J. Electron Spectrosc. Rel. Phenomena*, 52, 797, 1990.

75. Wandelt, K., *J. Vac. Sci. Technol. A*, 2, 802, 1984.

76. (a) Colbert, J., Zangwill, A., Strongin, M., and Krummacher, S., *Phys. Rev. B*, 27, 1378, 1983. (b) Caprile, C., Franciosi, A., Wielickza, D., and Olson, C. G., *J. Vac. Sci. Technol. A*, 4, 1526, 1986.

77. (a) Kuppers, J., Nitschké, F., Wandelt, K., and Ertl, G., *Surface Sci.*, 88, 1, 1979. (b) Kaindl, G., Chiang, T.-C., Eastman, D. E., and Himpsel, F. J., *Phys. Rev. Lett.*, 45, 1808, 1980.

78. Jacobi, K. and Rotermund, H. H., *Surface Sci.*, 116, 435, 1982.

79. Daiser, S. and Wandelt, K., *Surface Sci.*, 128, L213, 1983.

80. Miranda, R., Daiser, S., Wandelt, K., and Ertl, G., *Surface Sci.*, 131, 61, 1983.

81. Wandelt, K., Hulse, J., and Kuppers, J., *Surface Sci.*, 104, 212, 1981.

82. Kuppers, J., Wandelt, K., and Ertl, G., *Phys. Rev. Lett.*, 43, 928, 1979.

83. Dolle, P., Markert, K., Heichler, W., Armstrong, N. R., Wandelt, K., Kim, K. S., and Fiato, R. A., *J. Vac. Sci. Technol. A*, 4, 1465, 1986.

84. Barr, T. L., *J. Vac. Sci. Technol. A*, 7, 1677, 1989.

85. Mason, M. G., *Phys. Rev. B*, 27, 748, 1983.

86. Citrin, P. H. and Wertheim, G. K., *Phys. Rev. B*, 27, 3176, 1983.

87. Parmigiani, F., Kay, E., Bagus, P. S., and Nelin, C. J., *J. Electron Spectrosc. Rel. Phenomena*, 36, 257, 1985.

88. Fraissard, J. and Ito, T., *J. Chem. Phys.*, 76, 5225, 1982.

89. Rotermund, H. H. and Jacobi, K., *Solid State Comm.*, 44, 493, 1982; Jacobi, K., Hsu, Y.-P., and Rotermund, H. H., *Surface Sci.*, 114, 683, 1982.

90. Rotermund, H. H. and Jacobi, K., *Surface Sci.*, 126, 32, 1983.

91. Jacobi, K., *Surface Sci.*, 192, 499, 1987; Astaldi, C. and Jacobi, K., *Surface Sci.*, 200, 15, 1988; Jacobi, K., *Phys. Rev. B*, 38, 5869, 1988; Jacobi, K., *Phys. Rev. B*, 38, 6291, 1988.

92. Wandelt, K., in *Chemistry and Physics of Solid Surfaces*, Vol. 8, Vanselow, R. and Howe, R., Eds., Springer-Verlag, Berlin, 1990, 289.

93. Wertheim, G. and Creselius, G., *Phys. Rev. Lett.*, 40, 813, 1978; Creselius, G., Wertheim, G. K., and Buchanan, D. N. E., *Phys. Rev. B*, 18, 6519, 1978.

94. Citrin, P. H., Wertheim, G. K., and Baer, Y., *Phys. Rev. Lett.*, 41, 1425, 1978.

95. Duc, T. H., Guillot, C., Lasailly, Y., Lecante, J., Jugnet, Y., and Vedrine, J. C., *Phys. Rev. Lett.*, 43, 789, 1979.

96. **Van der Veen, J. F., Himpsel, F. J., and Eastman, D. E.,** *Phys. Rev. Lett.,* 44, 189, 1980; **Heinmann, P., Van der Veen, J. F., and Eastman, D. E.,** *Solid State Commun.,* 38, 595, 1981.

97. **Alvarado, S. F., Campagna, M., and Gudat, W.,** *J. Electron Spectroc. Rel. Phenomena,* 18, 43, 1980; **Kammerer, R., Barth, J., Flodstrom, A., and Johansson, L. I.,** *Solid State Commun.,* 41, 435, 1982.

98. **Johansson, B. and Martensson, N.,** *Phys. Rev. B,* 21, 4427, 1980; **Rosengren, A. and Johansson, B.,** *Phys. Rev. B,* 23, 3852, 1981; **Rosengren, A. and Johansson, B.,** *Phys. Rev. B,* 22, 3706, 1980; **Johansson, B. and Martensson, N.,** *Phys. Rev. B,* 24, 4484, 1981.

99. **Fuggle, J. C., Campagna, M., Zolnierek, Z., Lasser, R., and Platon, A.,** *Phys. Rev. Lett.,* 45, 1597, 1980.

100. **Mason, M. G. and Baetzold, R. C.,** *J. Chem. Phys.,* 64, 271, 1976.

101. **Nilsson, A.,** Ph.D. thesis, No. 182, Faculty of Science, University of Uppsala, 1989.

102. **Citrin, P. H. and Wertheim, G. K.,** *Phys. Rev. B,* 27, 3176, 1983.

103. (a) **Houston, J. E., Park, R. L., and Laramore, G. E.,** *Phys. Rev. Lett.,* 30, 846, 1973. (b) **Eastman, D. E.,** *Phys. Rev. B,* 2, 1, 1970.

104. **Citrin, P. H., Wertheim, G. K., and Baer, Y.,** *Phys. Rev. B,* 27, 3160, 1983.

105. **Johansson, B. and Martensson, N.,** *Helv. Phys. Acta,* 56, 405, 1983.

106. **Arlinghaus, F. J., Gay, J. G., and Smith, J. R.,** *Phys. Rev. B,* 20, 1332, 1980; **Eastman, D. E., Chiang, T. C., Heinmann, P., and Himpsel, F. J.,** *Phys. Rev. Lett.,* 45, 656, 1980.

107. **Liang, K. S., Salanek, W. R., and Aksay, I. A.,** *Solid State Commun.,* 19, 329, 1976.

108. **Hufner, S., Wertheim, G. K., and Wernick, J. H.,** *Phys. Rev. B,* 8, 4511, 1973.

109. **Siegbahn, K.,** UUIP-1136, University of Uppsala, Uppsala, Sweden, lecture given at the Royal Society, 1985.

110. **Jolly, W. and Hendrickson, D. N.,** *J. Am. Chem. Soc.,* 92, 1863, 1970.

111. **Miedema, A. R.,** *Z. F. Metallkunde,* 69, 455, 1978; **Miedema, A. R. and Dorleijn, J. W. F.,** *Surface Sci.,* 95, 447, 1980; **Gerkemo, J. and Miedema, A. R.,** *Surface Sci.,* 124, 351, 1983.

112. **Stenborg, A.,** Ph.D. thesis, No. 231, Faculty of Science, University of Uppsala, 1989.

113. **Martensson, N., Stenborg, A., Bjorneholm, O., Nilsson, A., and Andersen, J. N.,** *Phys. Rev. Lett.,* 60, 1731, 1988.

114. **Nilsson, A., Martensson, N., Hedman, J., Ericksson, B., Bergman, R., and Gelius, U.,** *Surface Sci.,* 162, 51, 1985.

115. **Nilsson, A., Ericksson, B., Martensson, N., Andersen, J. N., and Onsgaard, J.,** *Phys. Rev. B,* 38, 10357, 1988.

116. **Stenborg, A., Bjorneholm, O., Nilsson, A., Martensson, N., Andersen, J. N., and Wigren, C.,** *Surface Sci.,* 211/212, 470, 1989.

117. **Andersen, J. N., Bjorneholm, O., Stenborg, A., Nilsson, A., Wigren, C., and Martensson, N.,** UUIP No. 1193, 1989, to be published.

118. **Nilsson, A., Morris, M. A., and Chadwick, D.,** *Surface Sci.,* 152/153, 247, 1985.

119. **Williams, F. L. and Nason, D.,** *Surface Sci.,* 45, 337, 1974; **Wynblatt, P. and Ku, R. C.,** *Proc. of ASM Materials Science Seminar on Interfacial Segregation,* Johnson, W. C. and Blakely, J. M., Eds., ASM, Metals Park, 1979, 115; **Seah, M. P.,** *Practical Surface Analysis,* 2nd ed., Briggs, D. and Seah, M. P., Eds., John Wiley & Sons, Chichester, England, 1990, chap. 7.

120. **Martensson, N., Nilsson, A., Stenborg, A., and Parashkevov, D.,** *Auger Spectroscopy and Electronic Structure,* Springer-Verlag, Berlin, to be published.

121. **Ericksson, B.,** Ph.D. thesis, No. 216, Faculty of Science, University of Uppsala, 1989.

122. **Nilsson, A., Ericksson, B., Martensson, N., Andersen, J. N., and Onsgaard, J.,** *Phys. Rev. B,* 36, 9308, 1987.
123. **Andersen, J. N., Onsgaard, J., Nilsson, A., Ericksson, B., and Martensson, N.,** *Surface Sci.,* 202, 183, 1988.
124. **Watson, P. R.,** *J. Phys. Chem. Ref. Data,* 16, 953, 1987.
125. **Estrup, J.,** *Chemistry and Physics of Solid Surfaces V,* Vanselow, R. and Howe, R., Eds., Springer-Verlag, Berlin, 1984.
126. **Fäldt, A. and Myers, H. P.,** *Solid State Commun.,* 48, 253, 1983.
127. **Stenborg, A., Andersen, J. N., Bjorneholm, O., Nilsson, A., and Martensson, N.,** *Phys. Rev. Lett.,* 63, 187, 1989.
128. **Stenborg, A., Bjorneholm, O., Nilsson, A., Martensson, N., Andersen, J. N., and Wigren, C.,** *Phys. Rev. B,* to be published.
129. **Himpsel, F. J., Heinmann, P., Chaing, T.-C., and Eastman, D. E.,** *Phys. Rev. Lett.,* 45, 1112, 1980.
130. **Brennan, S., Stöhr, J., Jaeger, R., and Rowe, J.,** *Phys. Rev. Lett.,* 45, 1414, 1980.
131. **Haneman, D.,** *Phys. Rev.,* 121, 1093, 1961.
132. **Pendry, J. C.,** *Phys. Rev. Lett.,* 47, 1913, 1981.
133. **Watson, R. E., Davenport, J. W., Perlman, M. L., and Sham, T. K.,** *Phys. Rev. B,* 24, 1791, 1981.
134. **Siegel, R. W. and Eastman, J. A.,** *Mater. Res. Soc. Proc. Issue,* 132, 3, 1989.
135. **Henry, C. R. and Chapon, C.,** *Surface Sci.,* 156, 952, 1985.
136. **Metois, J. J., Gauch, M., Masson, A., and Kern, R.,** *Thin Solid Films,* 46, 205, 1977; **Henry, C. R., Chapon, C., and Mutafotshiev, B.,** *Thin Solid Films,* 46, 157, 1977.
137. **Mason, M. G., Gerensen, L. J., and Lee, S. T.,** *Phys. Rev. Lett.,* 39, 288, 1977.
138. **Legare, P., Sakisaka, Y., Brucker, C. F., and Rhodin, T. N.,** *Surface Sci.,* 139, 316, 1984.
139. **Wertheim, G. K., Di Cenzo, S. B., and Youngquist, S. E.,** *Phys. Rev. Lett.,* 51, 2310, 1983.
140. **Egelhoff, W. F. and Tibbetts, G. G.,** *Solid State Commun.,* 29, 53, 1979; **Egelhoff, W. F. and Tibbetts, G. G.,** *Phys. Rev. B,* 19, 5028, 1979.
141. **Abbrati, L., Braicovich, L., Bertoni, C. M., Calandra, C., and Manglic, F.,** *Phys. Rev. Lett.,* 40, 469, 1978.
142. **Apai, G., Lee, S.-T., and Mason, M. G.,** *Solid State Commun.,* 37, 213, 1981; **Lee, S.-T., Apai, G., Mason, M. G., Benbow, R., and Hurych, Z.,** *Phys. Rev. B,* 23, 505, 1981; **Mason, M. G.,** *Phys. Rev. B,* 27, 748, 1983.
143. **Roulet, H., Mariot, J.-M., Dufour, G., and Hague, C. F.,** *J. Phys. F,* 10, 1025, 1981; **Oberti, L., Monot, R., Mathieu, H. J., Landolt, D., and Buttlet, J.,** *Surface Sci.,* 106, 301, 1981.
144. **Parmigiani, F., Kay, E., Bagus, P. S., and Nelin, C. J.,** *J. Electron Spectrosc. Rel. Phenomena,* 36, 257, 1985; **Bagus, P. S., Nelin, C. J., and Bauschlicher, C. W., Jr.,** *Surface Sci.,* 156, 615, 1985.
145. **Wertheim, G. K., DiCenzo, S. B., Buchanan, D. N. E., and Bennett, P. A.,** *Solid State Commun.,* 53, 377, 1985; **Wertheim, G. K., DiCenzo, S. B., and Buchanan, D. N. E.,** *Phys. Rev. B,* 33, 5384, 1986.
146. (a) **Bagus, P. S., Nelin, C. J., Kay, E., and Parmigiani, F.,** *J. Electron Spectrosc. Rel. Phenomena,* 43, C13, 1987. (b) **DiCenzo, S. B. and Wertheim, G. K.,** *J. Electron Spectrosc. Rel. Phenomena,* 43, C7, 1987.
147. **Lewis, R. T. and Kelly, M. A.,** *J. Electron Spectrosc. Rel. Phenomena,* 20, 105, 1980.
148. **Barr, T. L.,** *Crit. Rev. Anal. Chem.,* 22, 229, 1991.
149. **Barr, T. L.,** *Practical Surface Analysis,* 2nd ed., Briggs, D. and Seah, M. P., Eds., John Wiley & Sons, Chichester, England, 1990, chap. 8.
150. **Schlier, R. E. and Farnsworth, H. E.,** *J. Chem. Phys.,* 30, 917, 1959.

334

151. **Barr, T. L.,** to be published.
152. **Barr, T. L.,** *Surface Interface Anal.,* 4, 185, 1982.
153. **Barr, T. L.,** presentation NACE Conference, *Corrosion '90,* Las Vegas, May 1990, NACE Preprints, Houston, 1990, No. 296.
154. **Barr, T. L.,** *Zeolites,* 10, 760, 1990.
155. **Briggs, D.,** *Practical Surface Analysis,* 2nd ed., Briggs, D. and Seah, M. P., Eds., John Wiley & Sons, Chichester, England, 1990, chap. 9.
156. **Barr, T. L. and Yin, M. P.,** *J. Vac. Sci. Technol.* A10, 2788, 1992.
157. **Sparrow, G.,** private communication.
158. **Gibb, T. R. P., Jr.,** *Progress in Inorganic Chemistry,* Vol. 3, Cotton, F. A., Ed., Wiley-Interscience, New York, 1962, 315.
159. **Luder, W. F. and Zuffanti, S.,** *The Electronic Theory of Acids and Bases,* John Wiley & Sons, New York, 1946.
160. **Cotton, F. A. and Wilkinson, G.,** *Advanced Inorganic Chemistry,* 5th ed., John Wiley & Sons, New York, 1988, chap. 3.
161. **Roberts, M. W. and McKee, C. S.,** *Chemistry of Metal-Gas Interface,* Clarendon Press, Oxford, 1978, chap. 10.
162. **Khoobiar, S.,** *J. Phys. Chem.,* 68, 411, 1964; **Boudart, M.,** *Adv. Catal. Related Subjects,* 20, 153, 1969.
163. **Davis, L. C.,** *Phys. Rev. B,* 25, 2912, 1982; **Davis, L. C.,** *J. Appl. Phys.,* 59, R25, 1986; **Aono, M., Chiang, T. C., Himpsel, F. J., and Eastman, D. E.,** *Solid State Comm.,* 37, 471, 1981; **Hecht, M. H. and Lindau, I.,** *Phys. Rev. Lett.,* 47, 821, 1981.
164. **Lad, R. J. and Henrich, V. E.,** *J. Vac. Sci. Technol. A,* 7, 1893, 1989.
165. **Lad, R. J. and Henrich, V. E.,** *Phys. Rev. B,* 38, 10860, 1988.
166. **Shen, Z.-X., Lindberg, P. A. P., Spicer, W. E., and Lindau, I.,** *High T_c Superconducting Thin Films, Devices and Applications,* Margaritando, G., Joynt, R., and Onellion, M., Eds., (AVS Series 6, AIP Conf. Proc. No. 182), American Institute of Physics, New York, 1989, 330.
167. **Siegbahn, H. and Siegbahn, K.,** *J. Electron Spectrosc. Rel. Phenomena,* 2, 319, 1973.
168. **Siegbahn, H., Asplund, L., Kelfve, P., Hamrin, K., Karlsson, L., and Siegbahn, K.,** *J. Electron Spectrosc. Rel. Phenomena,* 5, 1059, 1974.
169. **Fellner-Feldegg, H., Siegbahn, H., Asplund, L., Kelfve, P., and Siegbahn, K.,** *J. Electron Spectrosc. Rel. Phenomena,* 7, 421, 1975.
170. **Siegbahn, H., Svensson, S., and Lundholm, M.,** *J. Electron Spectrosc. Rel. Phenomena,* 24, 205, 1981.
171. (a) **Siegbahn, H.,** *J. Phys. Chem.,* 89, 897, 1985. (b) **Siegbahn, K.,** UUIP-1136 invited lecture, Royal Society, 1985; **Siegbahn, K.,** *J. Electron Spectrosc. Rel. Phenomena,* 51, 1, 1990.
172. **Siegbahn, H. and Lundholm, M.,** *J. Electron Spectrosc. Rel. Phenomena,* 28, 135, 1982.
173. **Baron, B., Chartier, P., Delahay, P., and Lugo, R.,** *J. Chem. Phys.,* 51, 2562, 1969; **Baron, B., Delahay, P., and Lugo, R.,** *J. Chem. Phys.,* 53, 1399, 1970; **Nemec, L., Baron, B., and Delahay, P.,** *Chem. Phys. Lett.,* 16, 278, 1973.
174. **Aulich, H., Delahay, P., and Nemec, L.,** *J. Chem. Phys.,* 59, 2354, 1973.
175. **Stoletow, M. A.,** *Compt. Rend.,* 106, 1593, 1888.
176. **Delahay, P.,** *Electron Spectroscopy: Theory, Techniques and Application,* Vol. 5, Brundle, C. R. and Baker, A. D., Eds., Academic Press, New York, 1984, chap. 2.
177. **Watanabe, I., Flanagan, J. B., and Delahay, P.,** *J. Chem. Phys.,* 73, 2057, 1980.
178. **Nemec, L., Chia, L., and Delahay, P.,** *J. Phys. Chem.,* 79, 2935, 1975.
179. **Nemec, L., Gaehrs, H. J., Chia, L., and Delahay, P.,** *J. Chem. Phys.,* 66, 4450, 1977.

180. **Siegbahn, H., Lundholm, M., Arbman, M., and Holmberg, S.,** *Phys. Scr.,* 30, 305, 1984.
181. **Siegbahn, H., Lundholm, M., Arbman, M., and Holmberg, S.,** *Phys. Scr.,* 27, 241, 1983.
182. **Agren, H. and Siegbahn, H.,** *J. Chem. Phys.,* 81, 488, 1984.
183. **Holmberg, S., Moberg, R., Yuan, Z. C., and Siegbahn, H.,** *J. Electron Spectrosc. Rel. Phenomena,* 41, 337, 1986.

General References

A selected list of related texts presented in a chronological order.

A. TEXTS THAT HELPED ESTABLISH THE FOUNDATION— THE FIRST 15 YEARS (1967–1982)

1. (i) **Siegbahn, K. et al.,** *ESCA*, Almquist, Uppsala, 1967; (ii) **Siegbahn, K. et al.,** *ESCA Applied to Free Molecules*, North Holland, Amsterdam, 1969. The "founding" books by the Nobel Prize winning creator of ESCA and his group. Heavily devoted to gas phase ESCA.
2. **Baker, A. D. and Bitteridge, D.,** *Photoelectron Spectroscopy*, Pergamon, 1972. Introductory.
3. **Eland, H. D.,** *Photoelectron Spectroscopy*, Butterworths, London, 1974. Introductory.
4. **Carlson, T. A.,** *Photoelectron and Auger Spectroscopy*, Plenum, New York, 1976. Deals extensively with Auger and not applied.
5. **Briggs, D., Ed.,** *Handbook of X-ray and Ultraviolet Photoelectron Spectroscopy*, Heyden, London, 1977. Good general discussion, but not strong in key applied areas.
6. **Ibach, H.,** *Electron Spectroscopy for Surface Analysis*, Springer-Verlag, Berlin, 1977. Many other surface analysis methods besides ESCA. No practical applications.
7. **Cardona, M. and Ley, L.,** *Photoemission in Solids*, Springer-Verlag, Berlin, 1979. Good theory. Applications only to chemisorption and microelectronics.
8. **Wagner, C. D., Riggs, W. M., Davis, L. E., Moulder, D. F., and Muilenberg, G. E.,** *Handbook of XPS*, PHI, Perkin Elmer, Eden Prairie, MN, 1979. A useful book of spectra and numbers. Almost no discussion.
9. **Kane, P. F. and Larrabee, G. B., Eds.,** *Characterization of Solid Surfaces*, Plenum Press, New York, 1974. Introductory.
10. **Czanderna, A. W.,** *Methods of Surface Analysis*, Elsevier, New York, 1975. Good comparison of different methods.
11. **McIntyre, W. S., Ed.,** *Quantitative Surface Analysis of Materials*, ASTM STP 643, Philadelphia, 1978. Early consideration of major problem areas.
12. **Barr, T. L., and Davis, I. E., Eds.,** *Applied Surface Analysis*, ASTM, STP 699, Philadelphia, 1980. Early use in applications.
13. **Windawi, H. and Ho, F., Eds.,** *Applied Electron Spectroscopy for Chemical Analysis*, John Wiley, New York, 1982. Nice introductory treatment of applications.
14. **Brundle, C. R. and Baker, A. D., Eds.,** *Electron Spectroscopy Theory, Techniques and Applications*, Academic Press, New York, 1976–1982. A good, detailed source up to 1980.
15. **Siegbahn, H. and Karlsson, L.,** *Handbook of Physics,* Vol. 31, Springer-Verlag, Berlin, 1982, 215. A fine update of ESCA accomplishments at Uppsala and elsewhere.

B. SPECIALIZED TEXTS START TO APPEAR IN THE EARLY 1980s

1. **Thomas, J. M. and Lambert, R. M., Eds.,** *Characterization of Catalysis*, Wiley-Interscience, New York, 1980.

2. **Dwight, D. W., Fabesh, T. J., and Thomas, H. R., Eds.,** *Photon, Electron and Ion Probes of Polymer Structure and Properties*, ACS Symp. Series, 162, 1981.
3. **Powell, C. J. and Casper, L., Eds.,** *Industrial Applications of Surface Analysis*, ACS Symp. Series, 99, Washington, D.C., 1982.

C. THE NEW AGE OF PHOTOELECTRON SPECTROSCOPY: 1983–PRESENT

1. **Briggs, D. and Seah, M. P., Eds.,** *Practical Surface Analysis*, John Wiley, 1st ed., Chichester, England, 1983. The standard text of the 1980s. Both ESCA and Auger. Extensive discussion of applications, but just by one method or the other.
2. **Gosh, P. K.,** *Introduction to Photoelectron Spectroscopy*, ACS Monograph Series, Vol. 67, John Wiley & Sons, New York, 1983. A good general discussion of ESCA, with extensive UPS and little consideration of application.
3. **Briggs, D. and Seah, M. P., Eds.,** *Practical Surface Analysis*, 2nd ed., John Wiley, Chichester, England, 1990. An update of first edition for the 1990s, with both ESCA and Auger. Extensive, but selective, discussion of applications. Omits theory and many advances in methodology.

D. TEXTS DEALING WITH SPECIAL APPLICATIONS OF ESCA AND RELATED FIELDS

1. **Mittal, K., Ed.,** *Physicochemical Aspects of Polymer Surfaces*, Plenum Press, New York, 1983.
2. **Auciello, O. and Kelly, R., Eds.,** *Ion Bombardment Modification of Surfaces*, Elsevier, Amsterdam, 1984.
3. **Oechsner, H., Ed.,** *Thin Films and Depth Profile Analysis*, Topics in Current Physics, Springer-Verlag, Berlin, 1984.
4. **Andrade, J. D., Ed.,** *Surface and Interfacial Aspects of Biomedical Polymers*, Plenum Press, New York, 1985.
5. **Woodruff, D. P. and Delchar T. A.,** *Modern Techniques of Surface Analysis*, Cambridge University Press, Cambridge, 1986.
6. **Powell, C., Ed.,** Workshop on Quantitative Surface Analysis, NBS, Gaithersberg, MD, 1986.
7. **King, D. A. and Woodruff, D. P., Eds.,** *The Chemical Physics of Solid Surfaces and Heterogeneous Catalysis*, Vols. 1–5, Elsevier, Amsterdam, 1981–1988.
8. **Minachev, K. M. and Shapiro., E. S.,** *Catalyst Surface Physical Methods of Studying* (Mir Publishers, Moscow), CRC Press, Boca Raton, FL, 1990.

E. GENERAL SURFACE SCIENCE TEXTS

1. **Roberts, M. W. and McKee, C. S.,** *Chemistry of the Metal-Gas Interface*, Clarendon Press, Oxford, 1978.
2. **Somorjai, G. A.,** *Chemistry in Two Dimensions*, Cornell University Press, Ithaca, NY, 1981.
3. **Adamson, A. W.,** *Physical Chemistry of Surfaces*, John Wiley & Sons, New York, 4th ed., 1982.
4. **Zangwill, A.,** *Surface Physics*, Cambridge University Press, Cambridge, 1988.
5. **Christmann, K.,** *Introduction to Surface Physical Chemistry*, Springer-Verlag, New York, 1991.

Table of binding energies for select metals and their key oxides

Element	Core level	Binding energy 0.2 eV for elemental metal (M^o)	Binding energy ±0.2 eV for key metal oxide M_xO_y (1) EC(1s) = 284.9 eV	Binding energy ±0.2 eV O(1s) for M_xO_y using EC(1s) = 284.9 eV	Type of oxide SC ≡ semicovalent I ≡ ionic VI ≡ very ionic (see text)
I	II	III	IV	V	VI
Mg	2p	49.5	50.2	531.2	SC
Al	2p	72.8	74.1	531.5	SC
Si	2p	99.4	102.9	532.9	SC
Ca	$2p_3$	—	347.5	529.8	I
Ti	$2p_3$	453.3	458.5	529.7	I
Fe	$2p_3$	706.6	710.8	529.9	I
Co	$2p_3$	778.2	780.9	529.9	I
Ni	$2p_3$	852.7	854.6	530.0	I
Cu	$2p_3$	932.5	933.7	529.8	I
Zn	$2p_3$	1021.7	1022.0	530.7	SC
Ga	$3d_5$	—	20.2	530.9	SC
Ge	$3d_5$	30.0	32.7	531.2	SC
Sr	$3d_5$	—	134.6	529.0	VI
Y	$3d_5$	155.8	156.7	529.3	VI
Zr	$3d_5$	178.9	182.2	529.9	I[a]
Mo	$3d_5$	227.9	232.4	530.4	I
Rh	$3d_5$	307.2	308.4	530.0	I
Pd	$3d_5$	335.4	336.9	530.1	I
Ag	$3d_5$	368.2	367.8	529.4	VI
Cd	$3d_5$	404.5	404.2	528.9	VI
In	$3d_5$	443.9	444.3	530.1	I
Sn	$3d_5$	484.9	486.3	530.1	I
Ba	$3d_5$	—	778.7	528.5	VI
La	$3d_5$	—	834.0	528.8	VI
Ce	$3d_5$	—	881.5	529.1	VI
W	$4f_7$	31.4	35.5	530.4	I
Pt	$4f_7$	71.0	74.4	530.0	I

Note: In general, results for maximum valent (common) oxides are given. Where this is not practical the most common oxide is chosen.

[a] ZrO_2 is nearly a VI oxide.

340

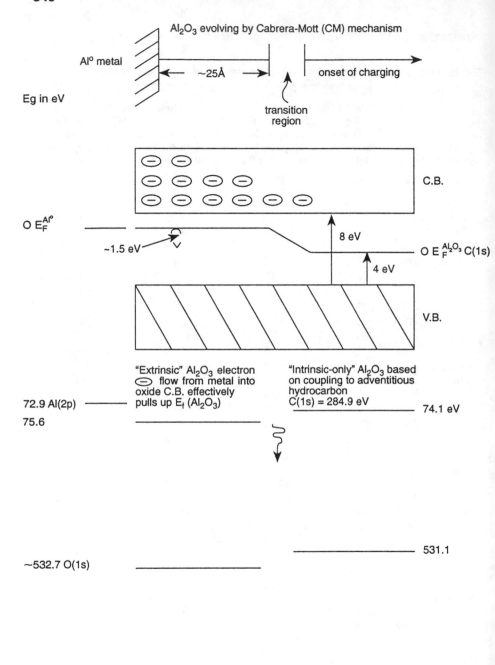

Figure A-1 Idealized Sc oxide on elemental metal (Al_2O_3 on $Al°$).

EXPLANATION OF THESE BINDING ENERGIES

The tabulations represent examples where the binding energies of the oxide compounds are determined by one of the two most common procedures presently employed.

Column IV, thus, presupposes that the compounds in question are, if thick enough, nonconductors, and, therefore, the binding energies of these species are adjusted to a common value of the C(1s) peak for the (supposedly common) adventitious hydrocarbon species (in this case 284.9 eV was employed) adsorbed onto the surface of these (before measurement) air exposed compounds.

The binding energies for the oxides in question may also be determined for a thin film of that compound on top of its elemental metal. In this case the compound film is thin enough so that the key ESCA peaks of the elemental species may still be detected through the film. Evidence suggests that in this case the two species (elemental and compound) are often Fermi edge coupled by a short range (Schottky) flow of electrons and despite the fact that the compounds are often nonconductors, they are thin enough so that no charging shift is *yet* produced. Thus binding energies determined by this method represent an example of a detailed use of the classic, Siegbahn, Fermi edge coupling method for binding energy determination.

It is perhaps not surprising that there are some differences in the resulting values for these two methods. What is surprising is that for many compounds these differences are fairly small (generally <0.4 eV), whereas for a select few the differences are much larger, e.g., for Al_2O_3, Fermi edge-coupled thin film Al(2p) = 75.6 eV, column IV Al(2p) = 74.1 eV.

Reflection dictates that only one of the latter can be the "correct" binding energy. Therefore a statement as to the reason for this apparent difference is in order. It should be apparent to all that the Fermi edge coupling procedure is per force the more *precise* method of measurement (based on the use of the word "precision" by analytical chemists). It is also true, however, that one must seriously doubt that the majority of the hydrocarbon or graphite adsorbed during the formation of the thin film oxide on certain select elemental metals and metalloids retains the same C-H and C-C bonds, and yet somehow is subject to a substantial alteration seemingly producing hydrocarbons with unprecedented increases in binding energies of as much as 1.5 to 2 eV, e.g., in the case of Al_2O_3 thin film the adventitious carbon is found at 286.5 eV.

It should be noted that the onset of low temperature oxide growth on these elements is known generally to extract electrons out of the elemental metal into the newly formed oxide film. This fact (establishing the Cabrera-Mott field) may result in an effective shift of the Fermi edge of the evolving oxide. This appears to be the case for the oxides referred to as semicovalent. These generally have wide, unobstructed band gaps between the O(2p) and their conduction bands. Most of the common "ionic" oxides, however, have band gaps partially filled with relatively discrete d and f bands that apparently mitigate this effect. Thus, as demonstrated in the enclosed figure, the Fermi edge-coupled binding energy result for an SC oxide may, in fact, not be that for the "intrinsic-only" mid-gap oxide, and a shift of the oxide Fermi-edge back to that mid gap point (as approximately dictated by choosing C(1s) for the adventitious hydrocarbon at 284.9 eV) may be in order. The reader must decide whether to go with values of greater precision or those of more conventional definition (column IV), but care must be employed not to mix the two.

It is also instructive to notice evidence in the table of what has been labeled as the "Brundle shift" (after C. R. Brundle, IBM), i.e., the fact that apparently all "ionic" oxides produce O(1s) binding energies of 530.0 ± 0.4 eV.

A more complete discussion of the types of oxides (column VI) and their revealing valence bands is contained in Barr and Brundle, *Phys. Rev. B* 46, 9199, 1992.

INDEX